Laser Physics

by A. MAITLAND and M. H. DUNN

Lecturers in Physics, University of St. Andrews, Scotland

1969

NORTH-HOLLAND PUBLISHING COMPANY
AMSTERDAM - LONDON

Library of Congress Catalog Card Number 76–97205
ISBN 7204 0153 4

Publishers:

NORTH-HOLLAND PUBLISHING COMPANY – AMSTERDAM

NORTH-HOLLAND PUBLISHING COMPANY, LTD. – LONDON

Sole distributors for the Western Hemisphere:

WILEY INTERSCIENCE DIVISION
JOHN WILEY & SONS, INC. – NEW YORK

Printed in The Netherlands

Laser Physics

PREFACE

This book introduces the fundamentals of laser physics so that a reader with a background corresponding to an honours degree in physics or electronic engineering will develop sufficient understanding to enable him to tackle most of the laser literature which adopts the semi-classical approach. Since it is an introductory text, we have tried to keep the book "self-contained" and to help this we have included a number of appendices covering those topics which are essential background to the main text and which are not usually found in one book. The aspects we consider are biased towards gas discharge lasers with emphasis on general principles rather than detailed information about the performance of the various types of laser available. One consequence of this approach is that credits and references are only made to those sources which we have consulted directly. The vast amount of literature forming the basis of the papers to which we refer is unacknowledged, and for this we apologise. With very few exceptions, we present the concepts and developments of laser physics from "first principles"; all steps involving physical principles are given; only algebra is omitted. We have departed from this policy only when the physical principles or results are important but the detailed quantitative treatment of their development is extensive and specialized (e.g. high order Lamb theory).

The problem of units is a hardy perennial. The various disciplines combined in laser physics have developed their own convenient systems, with the result that a familiarity with these is almost a prerequisit. The units we have used vary according to general practice in the topics treated.

Authors are generally indebted to a large number of people to whom adequate words of thanks cannot be found; we are no exception. This book could not have been written without the encouragement of our respective families. We warmly appreciate the patience, and helpful guidance of North-Holland Publishing Company. We are particularly indebted to Professor J. F. Allen for making available the facilities of the Physics Department of

St. Andrews University. Mr. J. Spark and Mr. T. McQueen have helped us cheerfully and very substantially in producing diagrams and copies of the typed manuscript. Finally, our greatest thanks are due to Mrs. Mary Maitland for undertaking the tedium of all the typing.

CONTENTS

Appendices

CHAPTER 1

THE LASER

1. Introduction

In this first chapter our intention is to develop certain physical ideas about the operation of the laser, the character of its radiation, and those features that set it apart from other sources of radiation in the optical region of the spectrum. In so doing we shall provide references to other chapters in the book where these various aspects are discussed more rigorously and in much greater detail. One of our aims at this point, therefore, is to provide a broad, if sketchy picture, that places other sections of the book in perspective and provides the necessary links between them.

Our approach throughout this book is in terms of what is known as "semi-classical theory". In this approach the optical radiation field is treated as a classical entity obeying Maxwell's equations, whereas the atoms of the active medium are described by the laws of quantum mechanics. The semi-classical theory may be approached in varying degrees of sophistication, particularly with regard to the manner in which the coupling between the radiation field and the atoms of the active medium is described. We shall consider the various approaches more fully in ch. 3, but for the present we embark on developing the simplest possible picture consistent with the ideas of semi-classical theory. Firstly, though, we shall summarize some of the properties of conventional line sources in the visible region of the spectrum, so that from these we can develop the concepts underlying the operation of the laser.

2. Conventional sources of line spectra

2.1 *Factors influencing line profile*

We consider the atoms of the source to be in the gaseous (or vapour) state, so that, approximately, the atoms are free from interacting neighbours and

References on page 24

we can associate characteristic energy levels with individual atoms. In order to excite the atoms to energy levels high above the ground state, energy must be fed into the system, let us say by passing an electric current through the gas. The actual energy levels excited by this process depend in a complex way on the parameters associated with the discharge (pressure and temperature of the gas, dimensions of the containing vessel, electric current and field strength, etc.) and on the parameters associated with the atomic states (cross-sections for excitation to various states, either by electrons, ions or atoms, relaxation rates of the states under consideration, etc.). Broadly speaking, in order to describe the system with regard to the populations in the various excited levels, we try to determine the distribution of kinetic energy amongst the electrons, atoms and ions in the discharge, and its dependence on the discharge parameters, and then use this information together with the atomic parameters (cross-sections for excitation as a function of kinetic energy of particles involved, relaxation rates by radiative decay, etc.) to determine the populations of the atoms associated with the various energy levels. Of course, with particular energy levels special effects (e.g. resonance effects) may occur which alter their populations significantly from what would be expected in this simple approach.

When an atom in an excited level decays spontaneously to a lower level, radiation is emitted, the frequency of which is related to the energy gap between the levels thus,

$$E_{mn} = (E_m - E_n) = h\nu_{mn}. \tag{1.1}$$

The intensity of the radiation emitted at this frequency is linearly dependent on the spontaneous transition probability, A_{mn}, which is the probability, per unit time, that an atom in the upper excited state (m) decays spontaneously to the lower state (n). The transition probability for a particular pair of states is derived by quantum mechanical arguments from the wave functions describing the two states. In so far as a particular excited state may decay by spontaneous emission to several lower states, its total radiative decay rate is obtained by summing over the spontaneous transition probabilities associated with these different transitions,

$$\gamma_m = \sum_p A_{mp}. \tag{1.2}$$

With this decay rate we can associate a "lifetime" for the excited state to decay by spontaneous emission,

$$\tau_m = \gamma_m^{-1}. \tag{1.3}$$

When we are concerned with the relative intensity of components of the radiation that are associated with transitions involving different upper levels, then the populations of these different states need to be considered. The intensity of radiation emitted by a particular transition is, therefore, given by

$$I_{mn} = K \cdot N_m \cdot A_{mn} \cdot h\nu_{mn}, \tag{1.4}$$

where K is some geometric factor.

From our discussion so far, we see that the radiation emitted by the source consists of discrete frequency components whose frequencies are related to the energy levels of the atoms concerned in the emission by (1.1) and whose intensities depend both on atomic parameters (through A_{mn}) and on discharge parameters (through N_m), being given by (1.4).

In so far as the probability of the atom being in the upper state associated with a particular transition decreases with time, as a consequence of all the radiative decay processes from that state, the wavetrain associated with the transition is damped, and its amplitude may be regarded as being of the form

$$\exp(-\gamma t/2) \cdot \sin(2\pi\nu_{mn} t). \tag{1.6}$$

In order to determine the frequency spectrum associated with this wavetrain, we take the Fourier transform of (1.6), and so obtain the following (normalized) Lorentzian function (appendix J) describing the intensity distribution across the line profile,

$$g(\nu, \nu_{mn}) = \gamma \cdot \{4\pi^2(\nu-\nu_{mn})^2+(\gamma/2)^2\}^{-1}. \tag{1.7}$$

In the case of spontaneous emission, different atoms emit independently of each other, so the intensity distribution associated with all the atoms in the source, that are in the relevant excited state, is obtained by summing over the intensity distributions (1.7) associated with individual atoms. Clearly, (1.7) describes the intensity distribution seen by a spectrometer, say, viewing the source as a whole.

The question now arises as to the meaning of the decay constant in (1.7). It might at first be thought that this decay constant is just that associated with the upper state of the transition, and hence given by (1.2). However, it may be shown by a detailed quantum mechanical treatment (ch. 3, § 3.4) that the decay rate for the lower level of the transition is involved as well, and that in fact

$$\gamma = (\gamma_m+\gamma_n). \tag{1.8}$$

A short lived lower state can therefore broaden the linewidth associated

References on page 24

with the transition. The type of broadening that we have discussed above is referred to as homogeneous broadening, in that each and every atom in the relevant upper state emits across the whole of the linewidth, and it is not possible to associate particular frequency components of the linewidth with particular atoms in the source. We have so far regarded homogeneous broadening as being determined by the radiative decay rates of the states involved in the transition (natural broadening). Under certain circumstances, depending on the environment of the atom, these states may be destroyed by collisional de-excitation at a faster rate than that due to radiative processes. In these cases, the decay rates in (1.8) must be modified to take this into account. The homogeneous linewidth of a transition may, therefore, become a function of the discharge parameters, as opposed to being solely dependent on atomic parameters as in the case of radiative decay considered above.

The homogeneous linewidth $\Delta\nu_H$ (defined as the full width at the half intensity points) may readily be derived from (1.7) as

$$\Delta\nu_H = \gamma/2\pi. \tag{1.9}$$

For a transition where the lifetimes of both the levels are of the order of 10^{-8} sec, the homogeneous linewidth is of the order of 30 MHz. It is of interest to note that the expression for the linewidth (1.9) may be obtained by an argument involving the Uncertainty Principle. In so far as the oscillator exists only for a time $\tau(=\gamma^{-1})$, on account of its radiative decay, the Uncertainty Principle states that the energy of the oscillator must be uncertain by an amount

$$\Delta E \sim \hbar\gamma.$$

The observation of the emitted radiation constitutes a measurement of the oscillator energy, in so far as its frequency is determined by (1.1). Hence the frequency of the emitted radiation is uncertain to an extent given by (1.9).

In the gas discharge considered, the radiating atoms are not stationary as assumed above, but have kinetic energy. In many circumstances, the distribution of kinetic energy amongst the atoms corresponds to the thermal equilibrium case, and is hence, described by the Maxwell-Boltzmann distribution (appendix J). The radiation emitted by a moving atom as viewed in the laboratory frame of reference is shifted in frequency by the Doppler effect, such that if v is the component of velocity of the atom towards the observer, the frequency of the radiation corresponding to an unshifted component of frequency ν_0 is given by

$$\nu = \nu_0(1+v/c) \tag{1.10}$$

(where we have assumed that $v \ll c$). As a consequence of the Doppler effect, therefore, the width of the spectral line (as viewed in the laboratory frame of reference) is increased as the emitting atoms have a distribution of velocities corresponding to their temperature.

In order to obtain the form of the spectral line therefore, we divide the emitting atoms (those that are in the upper state of the transition) into groups, or populations, depending on their velocity component (v) towards the observer. Suppose the population of atoms with their velocity component lying in the range v to $(v+\delta v)$ is $\delta N_m(v)$, then this group of atoms emits over the homogeneous linewidth given by (1.7) but the centre frequency is now shifted from ν_{mn} (the resonance frequency associated with a stationary atom) to a new value $\nu'_{mn}(v)$ given by (1.10). The intensity of the radiation associated with this group of atoms is therefore given by

$$\delta I(\nu, \nu'_{mn}) = \text{const.}\ \delta N_m(v) \cdot A_{mn} \cdot h\nu_{mn} \cdot g(\nu, \nu'_{mn}). \tag{1.11}$$

In order to derive the line profile, (1.11) must be summed over all the populations $\delta N_m(v)$. From the Maxwell Boltzmann distribution, we can readily determine the distribution of the upper state atoms amongst these populations, and further from (1.10) we can readily label the populations according to their apparent resonance frequencies (ν'_{mn}) as opposed to their velocities. We therefore have

$$\delta N_m(\nu'_{mn}) = N_m \cdot \mathscr{D}(\nu'_{mn}, \nu_{mn}) \cdot \delta \nu'_{mn}, \tag{1.12}$$

for the population of atoms whose apparent resonance frequency lies in the range ν'_{mn} to $(\nu'_{mn}+\delta\nu'_{mn})$. The function $\mathscr{D}(\nu'_{mn}, \nu_{mn})$ is derived in appendix J.

On substitution of (1.12) into (1.11) we obtain on integration the following expression that describes the line profile of the radiation emitted by a collection of moving atoms.

$$I(\nu, \nu_{mn}) = \text{const.}\ N_m \cdot A_{mn} \cdot h\nu_{mn} \cdot \int_{-\infty}^{+\infty} \mathscr{D}(\nu'_{mn}, \nu_{mn}) \cdot g(\nu, \nu'_{mn}) \cdot \mathrm{d}\nu'_{mn}. \tag{1.13}$$

In many cases the broadening of the line due to the function $\mathscr{D}(\nu'_{mn}, \nu_{mn})$ exceeds that due to the function $g(\nu, \nu'_{mn})$. Under these circumstances, the line is referred to as being inhomogeneously broadened. The important distinction between homogeneous and inhomogeneous broadening is that in the former case, as we have seen above, each and every atom (in the

References on page 24

appropriate state, of course) emits across the whole of the line profile, and it is not possible to identify certain frequency components with certain atoms, whereas in the latter case different atoms emit into different parts of the line profile, and it is possible to identify a particular population of atoms with a particular range of frequencies in the line profile. The width of the function $\mathscr{D}(\nu'_{mn}, \nu_{mn})$ may readily be determined in terms of the kinetic temperature of the emitting atoms, their mass etc. (appendix J),

$$\Delta\nu_I = 2\nu_{mn} \cdot \sqrt{\frac{2kT \ln 2}{mc^2}}. \tag{1.14}$$

A typical value for $\Delta\nu_I$ in the optical region of the spectrum corresponding to a gas of atomic weight 40, with a kinetic temperature of 800° C would be 1 GHz.

2.2 *Factors influencing coherence*

So far in our description of line spectra, we have determined the frequency distribution of the radiation energy. We now briefly turn our attention to a consideration of the coherence of this spontaneously emitted radiation. In so far as the different atoms in the source act independently, the wavetrains emitted by the different atoms are uncorrelated both in phase and in amplitude.

We shall now examine the radiation field in a volume of space, Δv, which is at a distance, R, from a source of area ΔA_s. In particular let us choose two points in this volume and examine the phase relation at any one time between the radiation field at the two points. It may be shown by simple physical arguments (ch. 10 § 2) that provided the volume in which the two points are chosen is less than a certain size, ΔV_c, known as the "volume of coherence", then it is possible to define an average phase relation, which is maintained over time, between the radiation sampled at the two points, and such a phase relation may be shown up by interference effects. However, if we move our points so that they are separated by a volume greater than the coherence volume, the phase relation between the radiation at the two points begins to fluctuate randomly in time, and it is no longer possible to produce interference effects. Of course, the idea of a sharp boundary is only an approximate one because, in reality, there is a gradual transition from one behaviour to the other. The volume of coherence may be related to the dimensions of the source and its associated bandwidth ($\Delta\nu$) by (10.16)

$$\Delta V_c \sim \frac{\lambda^2 R^2 c}{\Delta A_s \cdot \Delta\nu}.$$

This expression tells us that as we move further from the source, the volume of coherence increases (being proportional to R^2). However, in so far as the energy density of the radiation field varies as R^{-2}, the energy of radiation in the volume of coherence remains constant; at larger distances from the source it is smeared out over a larger volume of space.

So far, in our discussion of the volume of coherence associated with a normal spectral source, we have implied that any arbitrary shape may be associated with the volume. However, as will be demonstrated later (ch. 10 § 2) for the geometry we have considered above, the volume of coherence has a dimension, s_c, known as the "coherence length", along the direction of propagation of a magnitude $(c/\Delta\nu)$; and an area, known as the "coherence area" perpendicular to this direction of a magnitude $(\lambda^2 R^2/\Delta A_s)$. If we choose our two points so as to lie in a plane perpendicular to the direction of propagation, then we are sampling the radiation field at two points on a common wavefront, and the coherence we observe is referred to as "spatial coherence". Spatial coherence is therefore described by the coherence area.

If we choose our two points to lie along the direction of propagation we use the coherence length to describe the coherence behaviour. In this latter circumstance we may look at the coherence in another way. Suppose we consider the case where our two points merge in space, and now examine the phase relation between the radiation field at this single spatial point, but at two times separated by an interval, τ (say t and $t+\tau$). We would find that provided τ was approximately less than a certain value, τ_c, known as the "coherence time" of the radiation, a definite phase relation would be maintained over time. If the time interval were greater than this value, the phase relation would fluctuate randomly in time. In the former case we would be able to observe interference effects, but not in the latter. Now, since the radiation is propagating parallel to the axis along which the coherence length is defined, coherence length and coherence time are simply related by the velocity of propagation of the radiation ($s_c = c\tau_c$). Sampling the radiation at two spatial points along the direction of propagation at the same time is equivalent to sampling the radiation at one spatial point but at two different times, provided that the time interval and spatial separation are related by the velocity of propagation of the radiation. Coherence of this kind is referred to as "temporal coherence".

References on page 23

3. The optical cavity

The optical cavity associated with the laser oscillator differs from more familiar forms of cavity (e.g. the microwave cavity) in having dimensions that are large compared to the wavelength of the radiation involved. If we consider an enclosed optical cavity (fig. 1.1a), length d and cross-sectional

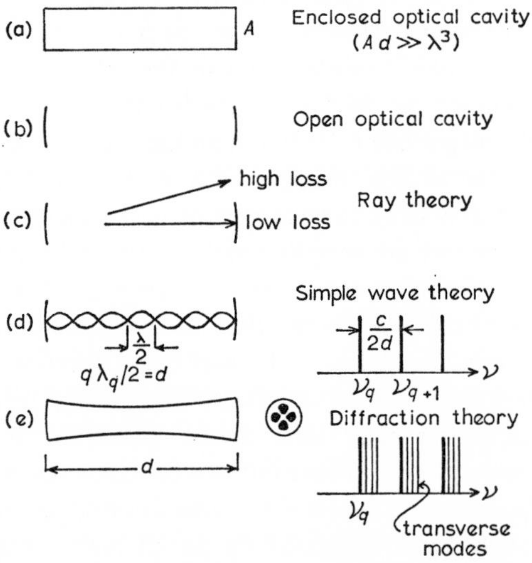

Fig. 1.1. The passive optical cavity.

area A, then the number of cavity modes lying in the frequency interval ν to $(\nu + \delta\nu)$ is (from 2.20):

$$8\pi\nu^2 \cdot V \cdot \delta\nu / c^3,$$

where $V = A \cdot d$. If we take typical values for V ($d = 100$ cm, $A = 1$ cm^2) and ν(10^{15} Hz) appropriate to the optical cavity, then the number of modes lying across a typical inhomogeneous line profile ($\delta\nu \sim 1000$ MHz) is very large indeed, being of the order of 10^{11}.

The special feature associated with the optical cavity of the laser oscillator is that it is an open structure, consisting of two, widely spaced, high reflectivity mirrors, and of the form illustrated in (fig. 1.1b). This has the consequence that by far the greater majority of the modes given by (2.20) have a very high loss, while comparatively few of them are low loss modes. In the case of the laser oscillator, the energy source for the radiation field is an "active" medium placed inside the laser cavity. We discuss the nature of the active medium in more detail shortly, but for the present all that we need to know is that the active medium is excited in such a way as to be capable of supplying energy to those cavity modes whose frequencies lie under the line profile of the laser transition. The process whereby this energy is supplied is that of stimulated emission, and the rate at which stimulated emission supplies energy to a mode is proportional to the intensity of the radiation field of the mode. It is therefore apparent that as the excitation of the active medium is increased, a point is reached at which the rate of stimulated emission into a low loss mode exceeds the losses associated with that mode. At this point the radiation field of the mode builds up in intensity, and so the rate of stimulated emission into the mode increases (i.e. the mode oscillates), until some point is reached at which the intensity of the radiation field of the mode saturates (ch. 8). The modes of lowest loss reach oscillation threshold first and thus gain energy from the active medium in preference to the high loss modes.

When the optical cavity contains an active medium it is referred to as an active cavity. In the remainder of this section we will, in the main, be concerned with the properties of the passive cavity only (i.e. one in which there are no energy sources present), with the aim of understanding the manner in which high and low loss modes arise.

A convenient way in which to do this is to regard the cavity as a filter. We suppose that at some arbitrary time (t_0) there is an arbitrary radiation field within the cavity, which is generated by some unspecified means. After the time t_0 we consider all energy sources of this radiation field to be removed, and we then investigate the nature of the radiation field remaining in the cavity at a much later time. Obviously, as time progresses, the low loss modes of the cavity will predominate more and more and we can regard the cavity as a filter, in the sense that it filters from an arbitrary radiation field those components corresponding to the low loss modes. If we consider the cavity from the point of view of geometrical optics (in so far as $A \gg \lambda^2$ this is a valid approach), then we immediately see that the low loss modes are associated with radiation propagating along the axis of the cavity (fig. 1.1(c)), as opposed to radiation propagating transverse to the axis which has a high loss. The fraction of those modes associated with the closed cavity (given by 2.20) that have a low loss in the case of the open cavity is, from this point of view, given by the solid angle subtended by the mirror surfaces at the centre of the cavity, i.e.

$$(A/\pi d^2) \sim 10^{-4}.$$

References on page 24

From this geometrical analysis we can also reach other conclusions about the cavity. In general the cavity mirrors are curved (for reasons to be discussed shortly), and one condition that obviously must be fulfilled, if there are to be any low loss modes associated with an open cavity, is that there must be certain directions of propagation of radiation such that, on repeated reflection backwards and forwards between the mirrors, the radiation does not "walk out" of the cavity. Obviously such a requirement places restrictions on the radii of curvature of the cavity mirrors in relation to their separation. Such restrictions are referred to as "stability requirements", and a cavity is said to be stable when walk-out does not occur on repeated reflection, and unstable in the converse case. We derive these stability requirements in ch. 5 § 5.1 where we consider the geometric properties of cavities. (Walk-out after several passes might be acceptable in high gain systems, ch. 5 § 7.)

The approach to the laser cavity in terms of geometric optics demonstrates the possibility of high and low loss modes existing for an open cavity, provided certain restrictions are placed on the geometry of the cavity. So far, however, we have said nothing about the field configurations or frequencies associated with the important low loss modes, nor have we made any estimate of the losses of these modes. We have further neglected diffraction effects which lead to losses from the radiation field on each transit of the open cavity. The importance of diffraction effects is assessed by consideration of the Fresnel number (N) of the open cavity

$$N = (A/\lambda d)$$

which specifies the number of Fresnel zones which can be seen over the surface of one of the mirrors from the centre of the other.

We have therefore, to demonstrate that, for the open cavity, stable low loss modes exist in the presence of diffraction losses, then to determine the field configurations and losses that characterize these modes. There are several approaches to this problem which are discussed fully in later chapters, and which we summarize below:

(a) Using the scalar formulation of Huygens' Principle, an expression relating the radiation field configuration produced over the surface of one of the mirrors through illumination by a particular field configuration on the surface of the other mirror is derived. Starting with an arbitrary field configuration, this process is repeated for hundreds of transits of the radiation between the mirrors to determine whether the resulting field configuration eventually becomes self-reproducing after a finite number of transits.

By "self reproducing", we mean that the amplitude and phase pattern of the field remains unaltered on propagating from one mirror to the other and back again, apart from a constant factor describing an overall loss and phase shift per transit. Such a self reproducing field configuration represents a transverse mode of the open cavity. This is essentially the approach of FOX and LI [1961].

(b) An alternative approach to (a), which is still based on Huygens' Principle, is to derive an equation relating the field configuration on one of the mirrors, to the field configuration that it gives rise to after propagation to the other mirror and back again. By looking for solutions where these two field configurations are the same, apart from the loss and phase shift factor mentioned above, the self reproducing field configurations are derived. This is the approach of BOYD and GORDON [1961], which we discuss in detail in ch. 6.

(c) A third approach is directly through Maxwell's equations. Solutions to these equations are derived, which describe narrow, propagating beams. The phase fronts of such beams are then fitted to the curved mirror surfaces by adjusting certain beam parameters. When such a condition is fulfilled, the beam is reflected directly back along itself, and hence describes a self-reproducing field configuration of the cavity. This is the approach we adopt in our description of the Gaussian beam in ch. 7.

The round trip phase shift must be an integral multiple of 2π and so to each transverse mode of the cavity, there corresponds a series of longitudinal modes. Since the transverse modes of an open cavity approximate to plane waves, we can readily determine the approximate frequencies corresponding to different longitudinal modes from this condition, namely (fig. 1.1(d))

$$(\lambda_q/2)q = d;$$

$$\nu_q = \frac{qc}{2d}. \qquad (1.15)$$

The different longitudinal modes are designated by the number of nodes between the cavity mirrors (i.e. by q).

The longitudinal modes corresponding to the same transverse mode, are uniformly spaced in frequency space, their frequency separation being inversely proportional to the separation of the cavity mirrors. For a typical laser cavity ($d \sim 1$ m), the longitudinal modes are separated by 150 MHz.

The different transverse modes represent different field configurations (self-reproducing) across the cavity mirrors. We consider the form of these patterns and the manner in which they are designated in ch. 4 § 6 and in more detail in ch. 6. The different transverse modes corresponding to a particular longitudinal mode have different frequencies and different round-trip losses. The former depend on the radii of curvature of the cavity mirrors and on the mirror separation, while the latter depend on the Fresnel number of the cavity as well (ch. 4 §§ 3, 5.2, ch. 6 § 1). In cavities where the mirror separation is small compared to their radii of curvature, the frequency separation of the different transverse modes associated with a particular longitudinal mode is small compared to the frequency separation of the longitudinal modes; i.e. the transverse modes bunch together at the frequencies determined by (1.15). This is the condition illustrated in (fig. 1.1(e)). With other cavity configurations the separation of the transverse modes may become large, and in the case of the confocal cavity (where the mirror separation is equal to the radius of curvature of the mirrors) they become degenerate with the longitudinal modes, in the sense that the frequency separation between adjacent transverse modes is equal to half the frequency separation of adjacent longitudinal modes.

We have mentioned that different transverse modes have different diffraction losses associated with them. This is the case because the different transverse modes describe different distributions of energy across the cavity mirrors. The diffraction losses for a mode having a large proportion of its energy concentrated towards the centre of the cavity mirrors, are smaller than for a mode having a larger concentration of energy towards the edge of the mirrors. (See, however, ch. 8 § 5.1 for active medium effects.)

The diffraction losses are a function of the radii of curvature of the mirrors as well as of their area and separation. This property is of considerable importance to the laser cavity, for by using the focussing properties of a concave mirror surface, diffraction losses can be significantly reduced below what would be expected from a simple consideration of the Fresnel number. The focussing effect produced by the concave mirror helps to off set the spreading of the field due to diffraction effects. As an example, if we consider two cavities (which both have a Fresnel number of unity), one consisting of circular plane mirrors and the other of confocal spherical mirrors, in the former case the power loss per transit for a particular transverse mode due to diffraction is 20%, while in the latter case it is reduced to 0.6 % (ch. 6).

A detailed diffraction treatment of the open optical cavity shows, therefore, that provided certain stability requirements are fulfilled, a series of low loss transverse modes can be associated with the structure. To each transverse mode there corresponds a series of longitudinal modes, (described by their number of nodes along the cavity axis). Different transverse modes, in general, have different frequencies and different diffraction losses, and these are both determined by the cavity parameters. In the case of a passive cavity, these low loss modes would be "filtered out" after a certain time from an arbitrary field configuration by the processes of propagation and diffraction.

If we return to the case of an active cavity, then as the excitation of the active medium of the laser is increased, it is apparent that, other things being equal, the low loss transverse modes will commence oscillation first. In general the particular longitudinal mode (of all the longitudinal modes associated with a given transverse mode) that reaches oscillation threshold first depends on the variation of the gain of the active medium with frequency. We return to this point in § 5. By suitable adjustment of the parameters associated with the cavity and the active medium, it is possible to obtain oscillation on one cavity mode alone.

In discussing the low loss modes of an open cavity, we have assumed that only geometric and diffraction losses are significant. In practice, of course, there are also absorption and transmission losses associated with the cavity mirrors; the transmission loss being the method by which useful energy is coupled out of the cavity. In so far as these mirror losses affect all transverse modes equally, they do not alter the field configurations or frequencies associated with the modes as derived by a consideration of diffraction effects alone. However, to the diffraction loss associated with each mode, there must now be added an additional loss associated with the mirrors. If the total fractional loss in energy per transit of the radiation through the cavity is f, then the decay of the intensity of a passive cavity mode is given by

$$\frac{1}{I}\frac{\mathrm{d}I}{\mathrm{d}t} = -\frac{fc}{d}, \tag{1.16}$$

so that

$$I = I_0 \exp(-fct/d). \tag{1.17}$$

By analogy with the argument we used in deriving the homogeneous linewidth of a transition (1.7), we can associate a linewidth with the damped mode of a passive cavity, which is given by

$$\Delta\nu_c = (fc/2\pi d) \tag{1.18}$$

and also a quality factor (Q) (ch. 4 § 5)

$$Q = 2\pi \cdot d \cdot \nu/fc. \tag{1.19}$$

References on page 24

Since the amplitude and phase across the wavefront of a transverse mode are completely defined, the whole wavefront is coherent (i.e. the coherence area of § 2.2 is just the area of the wavefront of the mode). In ch. 10 § 9, we demonstrate, more generally, that the processes of propagation and diffraction of radiation (such as occur in an open cavity) lead to spatially coherent radiation being filtered out from an initially incoherent radiation field. The temporal coherence of the radiation from a passive cavity is determined by the linewidth of the cavity modes. The coherence time is just the inverse of the cavity linewidth (i.e. the time constant of the radiation in the cavity),

$$\tau_c \sim d/fc. \tag{1.20}$$

In § 6 we consider how the temporal coherence of the modes of an active cavity is modified through energy being supplied to the radiation field by spontaneous and stimulated emission.

If, in the case of the passive cavity, the losses are too large, the linewidth of a damped mode might be greater than the separation of adjacent modes, so that the modes begin to lose their identity. This is equivalent, in the Fox and Li approach, to the statement that the radiation field does not undergo sufficient transits, before it is damped away, for the processes of propagation and diffraction to filter out the modes.

4. Stimulated emission

In 1917, Einstein demonstrated that the state of thermodynamic equilibrium between radiation and matter, where the distribution of radiation energy with frequency is described by Planck's Law and the distribution of atoms amongst their various excited states by the Boltzmann distribution, could be explained by postulating the following coupling processes between the two:

(i) An atom can undergo a transition from an upper level to a lower level through the process of spontaneous emission. In this case the probability per unit time of an atom in the upper level undergoing a transition to the lower level is independent of the intensity of the radiation field, and depends only on the details of the atomic states involved in the transition. As such it may be described by a coefficient, A_{mn}, such that the rate at which atoms in the upper level (m) decay to the lower level (n) is

$$N_m \cdot A_{mn},$$

where N_m is the population of atoms in the upper level. In the previous section we have discussed the nature of the radiation emitted during spontaneous emission.

(ii) An atom in the upper level may also emit radiation by stimulated emission. In this case the probability per unit time of the transition occurring is proportional to the energy density (per unit volume per unit frequency interval) in the radiation field at a frequency equal to the resonance frequency associated with the two atomic states of the transition. The rate of stimulated emission is therefore

$$N_m \cdot B_{mn} \cdot \rho_{\mathrm{T}}(\nu_{mn}),$$

where the subscript on the radiation energy density indicates that it describes the case of thermal equilibrium.

(iii) An atom in the lower level may absorb radiation energy, undergoing a transition to the upper level. This process is analogous to (ii) above, so that the absorption rate may be written

$$N_n \cdot B_{nm} \cdot \rho_{\mathrm{T}}(\nu_{mn}).$$

In so far as thermal equilibrium is a steady state condition, detailed balancing must exist between the processes tending to populate or depopulate the various energy states of the atoms, etc., and so we can write

$$N_m \cdot A_{mn} + N_m \cdot B_{mn} \cdot \rho_{\mathrm{T}}(\nu_{mn}) = N_n \cdot B_{nm} \cdot \rho_{\mathrm{T}}(\nu_{mn}).$$

By making use of the Boltzmann distribution, which determines the ratio (N_m/N_n) and Planck's radiation formula which describes $\rho_{\mathrm{T}}(\nu)$, it may be shown that the above coupling processes are sufficient to explain the observed distributions at thermal equilibrium, and, further, relationships between the various coefficients may be obtained. These are summarized below

$$\begin{aligned} B_{mn}/B_{nm} &= (g_n/g_m), \\ B_{mn}/A_{mn} &= c^3(8\pi h\nu_{mn}^3)^{-1}, \end{aligned} \tag{1.21}$$

where g_m and g_n are the statistical weights of the two states, (ch. 2. §§ 4.6, 4.7).

In the case of the laser, as we have seen from our discussion on optical cavities (§ 3), we are concerned with a radiation field which is highly monochromatic, and which is propagating in a well-defined direction (along the axis of the cavity). The Einstein coefficients have been derived under very different circumstances, namely those of thermal equilibrium, where we are dealing with a radiation field which is isotropic, and where the

References on page **24**

intensity of radiation changes very slowly with frequency. In this latter case, the intensity of the radiation field is constant over that range of frequencies for which it can interact with the atom to induce a transition between the two levels (i.e. the radiation density is constant over the "response function" of the atom). In the former case, the intensity of the radiation field approximates to a delta function in changing rapidly compared to the width of the "response function" of the atom. If we assume that the transition rates for stimulated emission associated with the different frequency components of the black body field are additive, then the transition rate associated with monochromatic radiation of frequency, ν, and intensity I_ν may be written

$$B_{mn}(I_\nu/c) \cdot g(\nu, \nu_{mn}), \tag{1.22}$$

where $g(\nu, \nu_{mn})$ is some normalized response function associated with the atom (of resonance frequency ν_{mn}). The conditions under which the above assumption leading to (1.22) is valid, and a rigorous derivation of (1.22) by semi-classical theory, are both given in ch. 3. The normalized response function of the atom to monochromatic radiation is, in fact, given by (1.7), which we derived in connection with homogeneous line broadening in spontaneous emission. Recourse to a classical consideration of the interaction of radiation with matter, in terms of the Lorentzian theory of the electron (ch. 2 § 4.5), also leads to a response function of the form of (1.7). The process of stimulated emission retains the phase, wavelength, plane of polarization and every other attribute associated with the radiation field producing the stimulated emission, its effect being to increase the energy associated with this field.

5. Gain profile and threshold condition for homogeneously broadened transition

We are now in a position to derive the gain profile for a homogeneously broadened transition. If the population, per unit volume, of atoms in the upper level is N_m and in the lower level N_n, then the net increase in intensity of monochromatic radiation in propagating a distance δz through the medium is

$$\delta I_\nu = (B_{mn} N_m - B_{nm} N_n) \cdot g(\nu, \nu_{mn}) \cdot (I_\nu/c) \cdot h\nu \cdot \delta z. \tag{1.23}$$

Using the relationships between the Einstein coefficients (1.21), (1.23) may be written

$$\frac{1}{I_\nu} \cdot \frac{\mathrm{d}I_\nu}{\mathrm{d}z} = \frac{A_{mn} \cdot c^2}{8\pi\nu^2} \cdot \left[N_m - \frac{g_m}{g_n} N_n \right] \cdot g(\nu, \nu_{mn}) = \alpha_H(\nu), \tag{1.24}$$

so that

$$I = I_0 \exp \{\alpha_H(\nu) \cdot z\}.$$

In order for the medium to exhibit gain (i.e. for stimulated emission to dominate over absorption), then

$$N_m > \frac{g_m}{g_n} N_n. \tag{1.25}$$

If this condition is fulfilled, a "population inversion" is said to exist between the two levels associated with the transition. For a medium in thermal equilibrium, examination of the Boltzmann distribution shows that

$$N_m < \frac{g_m}{g_n} N_n; \tag{1.26}$$

in other words, absorption always exceeds stimulated emission. This is to be expected as there is the additional process of spontaneous emission, and under conditions of thermal equilibrium the total loss rate from the upper level to the lower level, due to stimulated and spontaneous emission combined, must balance the excitation rate from the lower level to the upper level through the absorption of radiation. In the active medium of a laser, therefore, the populations of the levels of the laser transition must be removed from the state of thermal equilibrium.

The "gain profile" of the active medium of a homogeneously broadened laser is described by the Lorentzian linewidth function (1.7). From our previous discussion on the optical cavity, we have seen that there exist certain low loss field configurations (the low loss cavity modes) determined by the cavity boundary conditions. When the gain per transit of the active medium, for such a low loss mode, exceeds the loss per transit (due to transmission through the cavity mirrors, or to diffraction losses), then the energy in the cavity mode builds up until some "gain saturation" effect limits its value (we discuss this in ch. 8). In other words the mode begins to oscillate. The threshold condition for oscillation is readily derived from (1.24), which describes the gain of the active medium, and (1.16) describing the cavity loss, as:

$$\frac{A_{mn} c^2}{8\pi \nu^2} \left[N_m - \frac{g_m}{g_n} \cdot N_n \right] \cdot g(\nu_q, \nu_{mn}) \geqslant f/d, \tag{1.27}$$

which may be written in terms of the cavity Q (1.19) as:

References on page 24

$$\frac{A_{mn}c^3}{8\pi\nu^2}\left[N_m - \frac{g_m}{g_n}N_n\right]\cdot g(\nu_q, \nu_{mn}) \geqslant \frac{2\pi\nu_q}{Q}\,.^{\dagger} \tag{1.28}$$

As the population inversion on the transition is increased, by increasing the excitation of the active medium, the lowest loss transverse mode of the longitudinal cavity mode closest to the resonance frequency of the atomic transition reaches threshold first and begins to oscillate. A further increase in the excitation may then lead to other transverse and longitudinal modes reaching threshold, but this depends on the cavity parameters (the separation of the modes) and the atomic parameters (the width of the atomic resonance). In so far as the different cavity modes are gaining energy from the same atoms in the case of a homogeneously broadened transition, coupling effects are to be expected between the modes – so that they cannot be regarded as totally independent oscillators.

Since the probability of stimulated emission into a cavity mode is proportional to the intensity of radiation in that mode, energy supplied to the active medium to create the population inversion, is preferentially "funnelled" into the oscillating modes. This state of affairs in the laser oscillator is to be contrasted with an arrangement in which spontaneous emission, from a line source say, is filtered using a passive cavity. In this latter case a narrow range of frequencies (given by $\Delta\nu_c$) may be filtered out by the cavity, but the remaining energy in the radiation field is rejected and hence lost. In the case of the laser oscillator, where the active medium is inside the cavity, the process of stimulated emission leads to energy being preferentially fed into the low loss modes of the cavity. Also, as we shall see shortly, in the case of the laser oscillator the spectral width of the radiation is (theoretically) many orders of magnitude down on the spectral width of the radiation that could be filtered out from spontaneously emitted light by use of a conventional passive cavity.

The interplay between the optical cavity and the active medium may therefore be summarized as follows:

(i) Through the processes of propagation and diffraction a passive cavity imparts spatial coherence to the radiation through what is essentially a filtering process.

(ii) By repeated reflections from mirror to mirror the cavity retains the radiation energy associated with the low loss modes within the volume of the

† In the above expressions, (1.27) and (1.28), we have confined our attention to the behaviour of the longitudinal modes associated with a particular, unspecified, transverse mode.

active medium, so enabling the process of stimulated emission, which depends on the radiation field intensity, to become dominant.

(iii) The process of stimulated emission, which exceeds the associated process of absorption in the active medium once a population inversion has been achieved on the transition, channels energy from the active medium into the cavity modes, i.e. it feeds radiant energy into the cavity which is correlated in amplitude and phase etc., with the radiation field producing the stimulated emission.

(iv) The population inversion must be great enough so that the gain associated with the active medium of the mode under consideration is sufficient to overcome the cavity losses. Under these circumstances the radiation field can build up in the cavity mode. The condition when the gain is just equal to the cavity loss is referred to as the threshold condition.

6. Spectral width of the laser oscillator

In deriving the threshold condition for oscillation in a particular mode, we equated the cavity loss to the gain of the active medium. On this basis the quality factor associated with the oscillation, which includes both negative losses (gain) as well as positive losses, is infinite, and this implies an infinitely narrow spectral width (i.e. a delta function) for the active mode. If we return to the procedure we adopted for deriving the quality factor (or spectral width) of a passive cavity (1.19), we can regard the gain of the active medium as maintaining the amplitude of the harmonic output against decay due to damping, without thereby perturbing its phase. This is, of course, equivalent to a delta function describing the frequency spectrum.

What we have so far disregarded, is the influence of spontaneous emission which is also feeding energy into the cavity mode. Since the oscillation does not increase without limit, the gain (due to stimulated emission) must therefore be less than the cavity loss, namely

$$[\text{Gain due to active medium}] + [\text{Spontaneous emission}] = [\text{Cavity loss}];$$

$$\{[\text{Gain}] - [\text{loss}]\} \propto Q^{-1}. \tag{1.29}$$

The Q of the oscillation, although it may still be very large, is now no longer infinite, and hence the spectral width of the oscillation has a finite value. Retaining the electrical analogy we can say that the laser oscillator is a regenerative (positive feedback) amplifier of spontaneous emission noise.

We can also regard the influence of spontaneous emission from the point

References on page 24

of view of classical coherence theory by noting that since it introduces energy into the radiation field that is uncorrelated with the field, it thereby introduces random amplitude and phase fluctuations which manifest themselves in a finite spectral width.

In ch. 4 § 5.1, the spectral width of a laser oscillator is derived by simple arguments involving the Q of an active cavity, as

$$\delta\nu = \frac{8\pi(\Delta\nu_c)^2 \cdot h\nu}{P}, \tag{1.30}$$

where P is the power of the coherent radiation leaving the cavity. This expression was derived by SCHAWLOW and TOWNES [1958] by analogy with an earlier expression of GORDON et al. [1955] relating to the linewidth of the maser. We consider this expression, and the influence of gain saturation (ch. 8) on amplitude and phase fluctuations associated with the laser oscillator in more detail in ch. 10 §§ 11–13.

From (1.30) it can be seen that the bandwidth of the laser radiation is dependent on the coherent power extracted from the cavity – the more power extracted, for given cavity parameters, the narrower the bandwidth. This is to be expected from our discussion above since the greater the power extracted, the greater the intensity of radiation inside the cavity, and hence the more stimulated emission dominates over spontaneous emission as the mechanism supplying energy to the cavity mode.

7. Mode pulling effects

An active medium exhibits a frequency dependent gain profile and it may be shown, by recourse to the "dispersion relations", that it must also exhibit a frequency dependent refractive index. In ch. 2 § 4.5 where we consider the Lorentzian theory of the electron, we derive an expression (2.62) for the refractive index of a medium consisting of classical electron oscillators, as

$$\mu = 1 + \frac{Ne^2\pi}{m\nu_0} \cdot \frac{(\nu_0 - \nu)}{[4\pi^2(\nu - \nu_0)^2 + (\gamma/2)^2]}. \tag{1.31}$$

We can write (1.31) in terms of the absorption coefficient (2.69) associated with such a medium. This relationship between refractive index and absorption coefficient has a more general applicability than the classical theory of the Lorentzian electron (ch. 2 § 4.10.1), and therefore we can replace the absorption coefficient in the resulting expression by the gain coefficient for the active medium, $\alpha(\nu_q)$, (1.24), to obtain the following

$$\mu(\nu_q) = \left[1 - \frac{\alpha(\nu_q)\cdot(\nu_{mn}-\nu_q)\cdot c}{\nu_q\cdot\gamma}\right], \tag{1.32}$$

where the minus sign appears because we have negative absorption, i.e. gain.

At threshold we can equate the gain coefficient to the cavity loss, and so write the refractive index in terms of the cavity parameters using (1.9), (1.18), (1.24) and (1.27) to get

$$\mu(\nu_q) = \left[1 - \frac{\Delta\nu_c}{\nu_q\cdot\Delta\nu_H}(\nu_{mn}-\nu_q)\right]. \tag{1.33}$$

In an active cavity the mode frequencies are given by (1.15), but with the refractive index of the active medium included, namely:

$$_{\mathrm{L}}\nu_q = \frac{qc}{2d\mu(\nu_q)} = \frac{\nu_q}{\mu(\nu_q)}, \tag{1.34}$$

where $_{\mathrm{L}}\nu_q$ refers to the oscillation condition. Combining (1.33) and (1.34) and re-arranging slightly, we obtain the following expression, which describes linear mode-pulling in an active cavity at threshold

$$\frac{(_{\mathrm{L}}\nu_q-\nu_q)}{(\nu_{mn}-\nu_q)} = \frac{\Delta\nu_c}{\Delta\nu_H} = \sigma. \tag{1.35}$$

The pulling of the passive mode is always towards the line centre (ν_{mn}) and is linear in the sense that the pulling is proportional to the distance of the passive cavity mode (ν_q) from the line centre. The coefficient σ in the above, which is the ratio of the cavity linewidth to the homogeneous linewidth, is referred to as the "stabilization ratio" and typically has values between $\frac{1}{10}$ and $\frac{1}{100}$. The situation is analogous to two coupled simple harmonic oscillators of different resonance frequency (ν_q, ν_{mn}) where one oscillator is more heavily damped than the other. The frequency of oscillation is essentially determined by the oscillator with the smaller damping, but is pulled towards the frequency of the other oscillator.

8. Inhomogeneously broadened laser transition

We have so far restricted our attention to a laser transition which is homogeneously broadened (i.e. where all the atoms exhibit the same resonance frequency, ν_{mn}). In most real circumstances, the atoms exhibit a

References on page 24

distribution of resonance frequencies. For example, in the case of the gas discharge considered in § 2.1, the atoms exhibit a distribution of apparent resonance frequencies on account of their distribution of thermal velocities.

In order to derive the threshold condition for laser oscillation under these circumstances, we divide the atoms in the upper and lower laser levels into groups or populations according to their apparent resonance frequencies (ν'_{mn}). By analogy with (1.12) the population of atoms in the upper laser level with an apparent resonance frequency in the range ν'_{mn} to $(\nu'_{mn}+\delta\nu'_{mn})$ is

$$\delta N_m(\nu'_{mn}) = N_m \mathscr{D}(\nu'_{mn}, \nu_{mn})\delta\nu'_{mn} \tag{1.36}$$

and for the lower laser level

$$\delta N_n(\nu'_{mn}) = N_n \mathscr{D}(\nu'_{mn}, \nu_{mn})\delta\nu'_{mn}. \tag{1.37}$$

We have assumed that the same thermal distribution characterizes the atoms in both the upper and lower laser levels.

The net gain in intensity of a monochromatic wave of frequency ν, due to these populations is given by analogy with (1.23) as

$$\delta I_\nu(\nu'_{mn}) = [B_{mn}\delta N_m(\nu'_{mn}) - B_{nm}\delta N_n(\nu'_{mn})] \times$$
$$\times g(\nu, \nu'_{mn}) \cdot (I_\nu/c) \cdot h\nu \cdot \delta z. \tag{1.38}$$

In order to derive the total net gain due to all the atoms in the upper and lower laser levels, (1.38) is summed over the inhomogeneous profile, to obtain, finally

$$\frac{1}{I_\nu}\frac{\mathrm{d}I_\nu}{\mathrm{d}z} = \frac{A_{mn}c^2}{8\pi\nu^2}\left[N_m - \frac{g_m}{g_n}N_n\right] \cdot \int_{-\infty}^{\infty} g(\nu, \nu'_{mn})\mathscr{D}(\nu'_{mn}, \nu_{mn})\mathrm{d}\nu'_{mn} =$$
$$= \alpha_I(\nu). \tag{1.39}$$

In the case when the inhomogeneous linewidth ($\Delta\nu_I$) is very much greater than the homogeneous linewidth ($\Delta\nu_H$), $\mathscr{D}(\nu'_{mn}, \nu_{mn})$ is essentially constant over the range of values of ν'_{mn} contributing to the integral in (1.39), so that

$$\alpha_I(\nu) = \frac{A_{mn}c^2}{8\pi\nu^2}\left[N_m - \frac{g_m}{g_n}N_n\right] \mathscr{D}(\nu, \nu_{mn}). \tag{1.40}$$

This expression for the gain of an inhomogeneously broadened transition should be contrasted with that for a homogeneously broadened transition (1.24). If, for simplicity, we consider the case when the frequency of the monochromatic radiation corresponds to the line centre (ν_{mn}), then we have

$$\frac{\alpha_I(\nu_{mn})}{\alpha_H(\nu_{mn})} \propto \frac{\Delta\nu_H}{\Delta\nu_I}\,. \tag{1.41}$$

This result is capable of a simple physical interpretation. In the case of an inhomogeneously broadened transition the radiation couples only to those atoms whose apparent resonance frequencies lie no further from the frequency of the radiation than the homogeneous linewidth. In so far as the fraction of the atoms fulfilling this requirement is given by the ratio of the homogeneous to the inhomogeneous linewidth, the gain of an inhomogeneously broadened transition compared to a purely homogeneously broadened transition is reduced in this ratio.

The threshold condition for an inhomogeneously broadened transition is given by analogy with (1.28) as

$$\frac{A_{mn}}{8\pi\nu^2}\,c^3\left[N_m - \frac{g_m}{g_n}\,N_n\right]\mathscr{D}(\nu_\nu\,,\,\nu_{mn}) \geqslant \frac{2\pi\nu_q}{Q}\,. \tag{1.42}$$

The linear mode pulling associated with an inhomogeneously broadened transition is

$$\frac{\nu_q - \nu_q}{\nu_{mn} - \nu_q} = \frac{\Delta\nu_c}{\Delta\nu_I}\,. \tag{1.43}$$

This expression, should be contrasted with the linear mode pulling expression for a homogeneously broadened transition (1.35). In ch. 9 § 6, we derive the linear mode pulling expression at threshold for an inhomogeneously broadened transition, and show that it is only the first term in a series expansion that describes non-linear effects as well.

9. Gain saturation

So far we have restricted our attention to the threshold conditions for laser oscillation, which we obtained by equating the gain per transit of the active medium to the cavity loss. The question now arises as to what happens when the population inversion is increased still further, so that the gain of the active medium exceeds the cavity loss for a particular mode. The gain expressions that we have derived predict that the intensity of the radiation field under these circumstances builds up without limit. Clearly this cannot happen in practice, and when a certain field intensity is reached the gain must decrease, or saturate, so as to equal the cavity loss.

As in all oscillators, the gain re-adjusts itself through effects non-linear

References on page **24**

in the radiation intensity. So far we have developed relations that indicate that the gain is a linear function of the field intensity in the sense that

$$\delta I_\nu = \alpha(\nu) I_\nu \delta z.$$

This is a small signal approximation, and a more general description involves terms non-linear in the field intensity

$$\delta I_\nu = [\alpha(\nu) I_\nu - \beta(\nu) I_\nu^2] \delta z.$$

The physical mechanism that leads to the appearance of the non-linear terms in the gain expression is the following. The active medium feeds energy into the radiation field to maintain it against the cavity losses. This energy is available because transitions from the upper laser level to the lower laser level predominate over transitions in the reverse direction. This additional loss mechanism from the upper laser level to the lower laser level decreases the population inversion, which now depends on the intensity of the radiation field. In other words, the gain becomes non-linear in the field intensity, because the population inversion is a function of the field intensity. By equating the saturated gain to the cavity loss, the field intensity at saturation (steady state oscillation) can be derived.

In order to derive relations for the saturated gain, it is necessary to write down rate equations describing the population inversion as a function of field intensity, and then to substitute the population inversion expression so derived into the gain expressions.

In ch. 8 we consider gain saturation for both homogeneously and inhomogeneously broadened laser transitions, and derive explicit expressions for the power output of the laser oscillator as a function of cavity, atomic and discharge parameters. Gain saturation is also dealt with at length in ch. 9 in the formalism of Lamb theory.

References

Boyd, G. D. and J. P. Gordon, 1961, Bell Syst. Tech. J. **40**, 489.

Fox, A. G. and T. Li, 1961, Bell Syst. Tech. J. **40**, 453.

Gordon, J. P., H. J. Zeiger and C. H. Townes, 1955, Phys. Rev. **99**, 1264.

Schawlow, A. L. and C. H. Townes, 1958, Phys. Rev. **112**, 1940.

CHAPTER 2

REVIEW OF RADIATION THEORY

1. Introduction

The study of electromagnetic radiation has been one of the most fascinating and fruitful fields of physics and has played a significant role in the development of such momentous concepts as relativity and quantum theory, which, in turn, led EINSTEIN in 1917 to postulate the process of stimulated emission, thus laying the foundation for the invention and development of lasers. Laser operation involves the interaction of atoms with a radiation field and itself leads to the creation of radiation of such intensity that many new and strange effects appear. A survey of relevant radiation theory is presented in the following sections.

2. Radiation in vacuum

2.1 *Intensity of radiation*

The intensity in a specified direction (also called specific intensity) of radiation is the energy flowing through unit area normal to the direction of flow per unit solid angle per unit frequency interval in unit time. Referring to fig. 2.1 the intensity at point P is given by

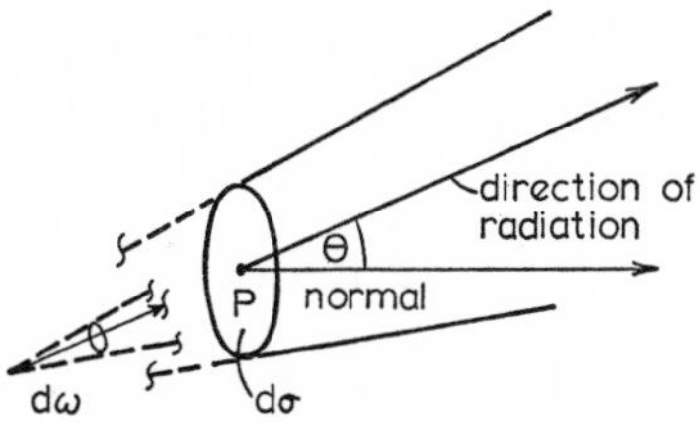

Fig. 2.1. The specific intensity of radiation is the energy flowing through unit area normal to the direction of flow per unit solid angle per unit frequency interval in unit time.

References on page 54

$$I_\nu = \mathrm{d}E_\nu/\mathrm{d}\sigma \cos\theta \mathrm{d}\omega \mathrm{d}\nu \mathrm{d}t, \tag{2.1}$$

where $\mathrm{d}E_\nu$ is the amount of radiant energy passing through area $\mathrm{d}\sigma$ in directions within solid angle $\mathrm{d}\omega$ in time $\mathrm{d}t$ in the frequency interval $(\nu, \nu+\mathrm{d}\nu)$. The radiation within $\mathrm{d}\omega$ described in fig. 2.1 forms a pencil of radiation.

The monochromatic intensity integrated over frequency gives the integrated intensity.

For polar coordinates with z-axis in the direction of the outward normal to $\mathrm{d}\sigma$, we have,

$$\mathrm{d}\omega = \sin\theta \mathrm{d}\theta \mathrm{d}\phi. \tag{2.2}$$

The mean intensity averaged over all angles is

$$J_\nu = \frac{1}{4\pi}\int_{\theta=0}^{\pi}\int_{\phi=0}^{2\pi} I_\nu(\theta, \phi) \sin\theta \mathrm{d}\theta \mathrm{d}\phi. \tag{2.3}$$

For an axially symmetric radiation field such as that of a laser

$$J_\nu = \frac{1}{2}\int_0^\pi I_\nu \sin\theta \mathrm{d}\theta. \tag{2.4}$$

A radiation field is homogeneous if I is the same at any point; it is isotropic if I is independent of direction at any point. If I is the same at all points and in all directions, the radiation field is homogeneous and isotropic.

Equation (2.1) may be used to compare radiation from lasers with that from conventional sources. However, such comparisons should be made always with a clear view as to their purpose in relation to the physics of the given problem because (2.1), due to the extremely narrow bandwidth of laser radiation, may give an answer which indicates the use of a laser in a situation where the narrow bandwidth is unnecessary or cannot be utilized.

Further, it should be noted that (2.1) says nothing about the polarization which is an important characteristic of a laser radiation field.

2.2 *Flux of radiation*

The net rate of flow of energy through $\mathrm{d}\sigma$ about P, fig. 2.1 per unit area and unit frequency interval is the flux and is given by

$$F_\nu = \int \frac{\mathrm{d}E_\nu}{\mathrm{d}\sigma \mathrm{d}\nu \mathrm{d}t}, \tag{2.5}$$

where the integral is over all solid angles. From (2.1), this gives

$$F_\nu = \int I_\nu \cos\theta \mathrm{d}\omega. \tag{2.6}$$

Using polar coordinates with the z-axis in the direction of the outward normal to $\mathrm{d}\sigma$, we get

$$F_\nu = \int_0^\pi \int_0^{2\pi} I(\theta, \phi) \cos\theta \sin\theta \,\mathrm{d}\theta \mathrm{d}\phi = \pi \mathscr{F}_\nu. \tag{2.7}$$

(For a black body at temperature T, the total flux $\pi\mathscr{F}_T$ is given by $\pi\mathscr{F}_T = $ $= \text{constant} \times T^4$.) The factor π is introduced for convenience rather like a similar introduction in electrostatics. This monochromatic flux can be integrated over the frequency range to give the integrated flux.

Consider a cartesian reference frame in the radiation field and consider the respective radiation fluxes across surface elements normal to the axes. If the direction cosines of the radiation direction are l_1, l_2, l_3 and F_x denotes the flux per unit time flowing across unit area normal to the x-axis, and similarly for the y and z directions, we have (dropping suffix ν for convenience),

$$F_x = \int I l_1 \mathrm{d}\omega, \quad F_y = \int I l_2 \mathrm{d}\omega, \quad F_z = \int I l_3 \mathrm{d}\omega. \tag{2.8}$$

The flux across a surface element orientated at angle θ with the radiation direction and with direction cosines m_1, m_2, m_3 with respect to the chosen cartesian frame is given by

$$\begin{aligned} F &= \int I \cos\theta \,\mathrm{d}\omega = \int I(l_1 m_1 + l_2 m_2 + l_3 m_3)\mathrm{d}\omega = \\ &= m_1 \int I l_1 \mathrm{d}\omega + m_2 \int I l_2 \mathrm{d}\omega + m_3 \int I l_3 \mathrm{d}\omega = \\ &= m_1 F_x + m_2 F_y + m_3 F_z = \\ &= \boldsymbol{F} \cdot \boldsymbol{n}, \end{aligned} \tag{2.9}$$

where $\boldsymbol{n}$ is the unit vector normal to the surface element.

Laser radiation generally has an axis of symmetry so

$$\pi\mathscr{F}_\nu = 2\pi \int_0^\pi I_\nu(\theta) \sin\theta \cos\theta \,\mathrm{d}\theta. \tag{2.10}$$

Flux and intensity are terms which are often confused. If there is no absorption, the intensity of a beam remains constant along its path even though the beam may be divergent. It is the flux which decreases according to the inverse square law.

References on page 54

2.3 *Energy density of radiation*

A region with electromagnetic radiation passing through it will, at any instant, contain some radiation energy in transit. The energy density, u, of radiation is the radiative energy contained in unit volume and is given by

$$u = (E^2 + H^2)/8\pi, \tag{2.11}$$

where E and H are the electric and magnetic fields respectively.
Its relation to intensity is shown below.

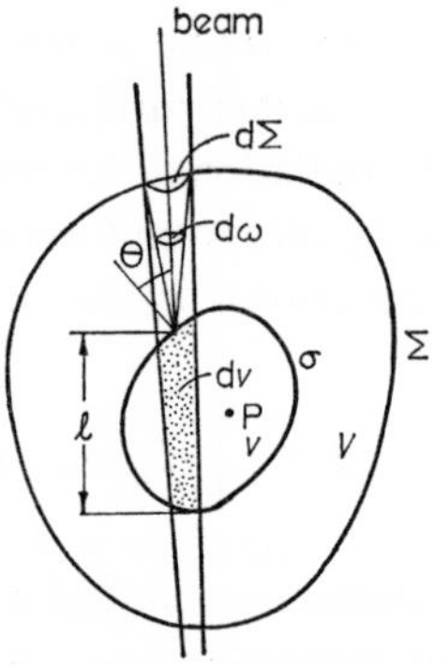

Fig. 2.2. Energy density of radiation in the region surrounding P is related to the specific intensity of the radiation.

Let P, fig. 2.2 be the point in the radiation field at which the energy density of radiation is to be determined and consider a convex surface σ of small linear dimensions surrounding P. Let this small volume v be in the central region of a volume V bounded by another convex surface Σ of much greater dimensions but so close to v that I_ν may be regarded as constant throughout V. Let a beam pass through the system, entering volumes V at element $\mathrm{d}\Sigma$ and v at $\mathrm{d}\sigma$, and fill element of volume $\mathrm{d}v$ of v. The energy passing through $\mathrm{d}\sigma$ in time $\mathrm{d}t$ within solid angle $\mathrm{d}\omega$ and frequency interval $(\nu, \nu+\mathrm{d}\nu)$ is, from (2.1)

$$\mathrm{d}E_\nu = I_\nu \mathrm{d}\sigma \cos\theta \,\mathrm{d}t \mathrm{d}\omega \mathrm{d}\nu. \tag{2.12}$$

If $\mathrm{d}t = l/c$ is the time taken by the radiation to cross volume v the energy present within $\mathrm{d}v$ at any instant is

$$\mathrm{d}E_\nu = I_\nu \mathrm{d}\sigma \cos\theta \,\mathrm{d}\omega \mathrm{d}\nu l/c,$$

but $l\mathrm{d}\sigma \cos\theta = \mathrm{d}v$, so

$$\mathrm{d}E_\nu = I_\nu \mathrm{d}\omega \mathrm{d}v \mathrm{d}\nu/c.$$

The radiation from all directions present in v at any instant is obtained by integrating this equation over all solid angles and the whole volume v;

$$\frac{\mathrm{d}\nu}{c}\int \mathrm{d}v \int \mathrm{d}\omega I_\nu = \frac{v}{c}\,\mathrm{d}\nu \int I_\nu \mathrm{d}\omega.$$

Dividing by v and by $\mathrm{d}\nu$ we get the required energy density per unit frequency interval

$$u_\nu = (1/c)\int I_\nu \mathrm{d}\omega. \tag{2.13}$$

Equation (2.13), gives the density resulting from radiation in all directions, whereas lasers generally give a radiation field consisting of a narrow beam within solid angle $\mathrm{d}\omega$; in this case the energy density of radiation is

$$u_\nu = I_\nu \mathrm{d}\omega/c. \tag{2.14}$$

For isotropic radiation

$$u_\nu = 4\pi I_\nu/c. \tag{2.15}$$

We may express u in terms of frequency or wavelength to describe a given region of the spectrum but, of course, $u_\nu \mathrm{d}\nu = u_\lambda \mathrm{d}\lambda$ provided that $\mathrm{d}\nu/\nu = \mathrm{d}\lambda/\lambda$ neglecting the minus sign which is not relevant in this context. Combining these equations we get

$$u_\nu \nu = u_\lambda \lambda \tag{2.16}$$

where ν and λ are in arbitrary units but their product is the velocity of light.

It is sometimes useful to convert frequency to wavelength using $|\mathrm{d}\nu| = c|\mathrm{d}\lambda|/\lambda^2$.

2.4 *Radiation pressure*

Radiation powers of $10^7 - 10^8$ watt from solid-state lasers are commonplace. Power levels of $10^9 - 10^{10}$ watt have been achieved. At such power levels it becomes necessary to consider the role of radiation pressure in the interaction of radiation and matter, particularly when considering the motion of particles at the focus of a giant pulse laser beam.

Classical theory of Maxwell shows that electromagnetic radiation has linear momentum in the direction in which the waves are travelling. If the energy density of the radiation is u, the momentum is u/c and is in the same direction as the pencil of radiation. The same result can be obtained by applying the mass-energy relationship of relativity. The energy u has a mass of u/c^2 and hence a momentum of uc/c^2, which is u/c as before. Considered

References on page 54

from a quantum point of view, an atom absorbing $h\nu$ from a beam, acquires momentum $h\nu/c$ in the direction of the beam. The atom which has absorbed the photon is excited and may emit it spontaneously after a certain time and so acquire a recoil momentum $h\nu/c$ in the direction opposite to that of the photon. Spontaneous emission from an excited atom is in no way related to the direction of any other photons present in the system and so occurs in any direction. However, a photon $h\nu$ may interact with an excited atom and so stimulate it to emit an identical photon $h\nu$. Stimulated emission is always in the same direction as the stimulating photon and so the atom acquires a recoil momentum which is always in the opposite direction. In general, whenever an atom absorbs or emits a photon it gains or loses momentum.

In so far as radiation impinging on a system can cause changes in momentum it may be regarded as exerting a force on the system. A force is exerted when the momentum of the radiation field is altered. Consider a group of atoms absorbing radiation from a beam and emitting spontaneously. Since a group of atoms emitting spontaneously do so in random directions, the average momentum change in the system is zero but the photons are from a beam and so cause momentum changes in the beam direction which constitute a force on the system.

A perfect absorber normal to a beam of electromagnetic waves of energy density u absorbs energy uc per second per unit area. Since the energy per second crossing unit area normal to the direction of propagation of electromagnetic radiation is the Poynting vector $\boldsymbol{P}$, we have $uc = \boldsymbol{P}$.

If electromagnetic waves of energy density u are incident at angle θ on a perfectly absorbing surface of area $\mathrm{d}\sigma$, the momentum absorbed by the surface per second (i.e. rate of transfer of momentum to the surface) is $(uc\,\mathrm{d}\sigma \cos\theta)/c = u\mathrm{d}\sigma \cos\theta$, and so p, the radiation pressure at the surface is given by

$$p = u \cos\theta = (P/c)\cos\theta.$$

For normal incidence on a perfect absorber, the radiation pressure is equal to the energy density. The radiation pressure is one atmosphere if the incident beam has 10^9 watt cm^{-2}, a value easily attained in focused laser beams.

The force exerted by radiation on a surface is $u \cos\theta \mathrm{d}\sigma$ in the direction of the beam. The force normal to the surface is thus $u\mathrm{d}\sigma \cos^2\theta$, and that tangential to the surface is $u\mathrm{d}\sigma \cos\theta \sin\theta$. When the radiation is isotropic, the average force normal to the surface is $u\mathrm{d}\sigma/3$ and that tangential is zero, since the average value of $\cos^2 \theta$ over a sphere is $\frac{1}{3}$ and of $\cos \theta \sin \theta$ is zero.

Consider radiation of energy density u travelling towards a perfect reflector and incident normally upon it; the pressure on the reflector due to loss of momentum of the incident beam is u and due to the recoil momentum of the reflected beam is also u giving a total pressure of $2u$ which is just the total radiation density at the reflector. For a general surface, some energy will be absorbed, some reflected, and some transmitted. Then, we have $u = u_a + u_r + u_t$. Both u_r and u_t cause recoil pressures. The total pressure on the front surface is $u_T = u + u_r$. The total pressure on the back surface is u_t.

In terms of intensity I_ν, the energy crossing $d\sigma$ is $I_\nu \cos \theta d\omega d\nu d\sigma dt$. This energy has momentum in the direction I_ν of amount $I_\nu \cos \theta d\omega d\nu d\sigma dt/c$. The total momentum crossing $d\sigma$ is $d\sigma dt \int_{4\pi} I_\nu \cos^2 \theta d\omega d\nu/c$.

Expressing $d\omega$ in polar coordinates, the pressure p_ν per unit frequency interval due to radiation in the frequency interval $d\nu$ is

$$p_\nu d\nu = \int_0^{2\pi} \int_0^{\pi} I_\nu d\nu \cos^2 \theta \sin \theta d\theta d\phi/c.$$

For isotropic radiation

$$p_\nu = 2\pi I_\nu \int_0^{\pi} \cos^2 \theta \sin \theta d\theta/c = 4\pi I_\nu/3c.$$

From (2.15) we get

$$p_\nu = u_\nu/3.$$

Electromagnetic radiation may also carry angular momentum ranging from zero for plane polarization to a maximum for circular polarization given by $I/\omega c$ where I is the intensity, and ω is the angular frequency.

2.5 *Radiative stress tensor*

The transfer of momentum across a surface is that which causes pressure in a system, whether the momentum is transferred by radiation as in the case considered above, or by particle motion as in a gas. When there is non-isotropic radiation or an unsymmetrical molecular velocity distribution, the appropriate pressure in the system loses its "hydrostatic" character, and varies according to the location and direction considered.

The energy densities of laser radiation can be so high that radiation pressure becomes significant in considering the interaction of the radiation with matter. Also, the radiation fields to be considered are anisotropic with energy

References on page 54

transport in the direction of the laser radiation. Thus, the laser radiation will exert forces in directions appropriate to its direction as described in section 2.4. Let the intensity of radiation I at a point P have direction cosines l_1, l_2, l_3 with respect to a cartesian co-ordinate system and let the normal to a perfectly absorbing surface element $\mathrm{d}\sigma$ around P have direction cosines m_1, m_2, m_3, and be at angle θ to the direction of I where

$$\cos\theta = l_1 m_1 + l_2 m_2 + l_3 m_3 .$$

Energy crossing $\mathrm{d}\sigma$ per second in solid angle $\mathrm{d}\omega$ is $I\cos\theta\,\mathrm{d}\omega\,\mathrm{d}\sigma$. The momentum imparted per second to the surface is $I\cos\theta\,\mathrm{d}\omega\,\mathrm{d}\sigma/c$ and is directed along I. The resulting force per unit area has components

$$l_1\int I\cos\theta\,\mathrm{d}\omega/c,\quad l_2\int I\cos\theta\,\mathrm{d}\omega/c,\quad l_3\int I\cos\theta\,\mathrm{d}\omega/c$$

along the x-, y-, z-axes, respectively. If the surface is taken to be normal to the x-axis, then $\cos\theta = l_1$ and the force acting on unit area has the three components,

$$\int I l_1 l_1 \mathrm{d}\omega/c \equiv p_{xx},\quad \int I l_1 l_2 \mathrm{d}\omega/c \equiv p_{xy},\quad \int I l_1 l_3 \mathrm{d}\omega/c \equiv p_{xz},$$

where the first suffix gives the direction of the normal to the surface considered, and the other gives the direction in which the component of the force is taken.

Similarly, we have

$$\int I l_2 l_1 \mathrm{d}\omega/c \equiv p_{yx},\quad \int I l_2 l_2 \mathrm{d}\omega/c \equiv p_{yy},\quad \int I l_2 l_3 \mathrm{d}\omega/c \equiv p_{yz},$$

and

$$\int I l_3 l_1 \mathrm{d}\omega/c \equiv p_{zx},\quad \int I l_3 l_2 \mathrm{d}\omega/c \equiv p_{zy},\quad \int I l_3 l_3 \mathrm{d}\omega/c \equiv p_{zz}.$$

Thus, the radiative stress tensor is

$$\mathbf{P} \equiv \begin{bmatrix} p_{xx} & p_{xy} & p_{xz} \\ p_{yx} & p_{yy} & p_{yz} \\ p_{zx} & p_{zy} & p_{zz} \end{bmatrix},$$

with

$$p_{ij} = \int I l_i l_j \mathrm{d}\omega/c.$$

3. Radiation in a cavity

3.1 *Modes of oscillation in an enclosed cavity*

A plane wave propagating in a given direction can be represented by

$$A = A_0 \cos(\omega t - \theta - \boldsymbol{k} \cdot \boldsymbol{r}). \tag{2.17}$$

In this equation, A_0 is the amplitude; θ is the phase angle which is independent of position and time, and depends upon the chosen location of the origin of coordinates; $\boldsymbol{k}$, the wave vector, specifies the direction of propagation (we discuss its magnitude below); and $\boldsymbol{r}$ is the vector locating an arbitrary point in the wave surface relative to the chosen origin. Let the distance from the origin along the direction of propagation be s. For points in this direction $\boldsymbol{k} \cdot \boldsymbol{r} = ks$. The wavelength λ is the distance between successive maxima $A_m = A_0$. Maxima are located at s_m and $s_m + \lambda$, so

$$\cos(\omega t - \theta - ks_m) = 1$$

and

$$\cos\big(\omega t - \theta - k(s_m + \lambda)\big) = 1.$$

The same result could be obtained by adding 2π, thus

$$\cos\big(\omega t - \theta - (ks_m + 2\pi)\big) = 1.$$

Hence, we have

$$ks_m + k\lambda = ks_m + 2\pi,$$

which gives

$$k = 2\pi/\lambda. \tag{2.18}$$

In free space with no boundaries, electromagnetic waves may propagate with any wavelength. In an enclosure, with perfectly conducting walls (for simplicity) the waves are reflected at the walls and so interfere to build up standing waves. The possible wavelengths of these standing waves are restricted by the effect of the boundary conditions of zero tangential electric intensity and zero normal magnetic intensity at the walls. Each standing wave system is a mode of oscillation of the cavity. To calculate the number of modes we shall consider a "rectangular" cavity of sides a, b, c (c is also used for the velocity of light, but this should not lead to any confusion.)

The possible wavelengths in the a-, b-, c-directions are given respectively by

$$\lambda/2 = a/l, \quad \lambda/2 = b/m, \quad \lambda/2 = c/n,$$

where l, m, n are positive integers. Therefore, the possible magnitudes of the

References on page 54

components of a propagation vector $\boldsymbol{k}$ in these directions are

$$k_{l,m,n} = 2\pi l/2a, \quad 2\pi m/2b, \quad 2\pi n/2c \tag{2.19}$$

and each value of l, m, n refers to a separate mode. To complete the information, there are two mutually perpendicular polarizations $i = 1, 2$. To study the modes we now use a coordinate system defining a k-space in which the magnitudes $k_{l,m,n}$ are recorded. In this space, the distance between modes is

$$\Delta k_l = k_{l+1,m,n} - k_{l,m,n} = 2\pi/2a.$$

Similarly, we have

$$\Delta k_m = 2\pi/2b, \quad \text{and} \quad \Delta k_n = 2\pi/2c.$$

Thus, each mode has a volume $(2\pi)^3/8abc$ in k-space. The mode density is $8abc/(2\pi)^3$. In general the mode density for any volume is $8 \times \text{volume}/(2\pi)^3$. Let the frequencies in the cavity be in the range $0 \to \nu$. These have wave vectors in the range $0 \to 2\pi/\lambda = 2\pi\nu/c$. This means that, in k-space, these modes of oscillation are within the volume of the octant of a sphere of radius $2\pi\nu/c$, since only the magnitudes of k are recorded in the k-space. The total number of modes in this octant is

$$N = 2 \times (\text{volume of sphere}/8) \times \text{mode density}.$$

The factor 2 occurs because of the two polarizations. We then have

$$N = 2 \times \frac{1}{8}\,\frac{4\pi}{3}\left(\frac{2\pi\nu}{c}\right)^3 \times \frac{8V}{(2\pi)^3},$$

$$N = \frac{8\pi}{3}\,\frac{\nu^3}{c^3}\,V.$$

The number of modes, ρ, per unit real volume (mode density) and per unit frequency interval is

$$\rho = (1/V)(\mathrm{d}N/\mathrm{d}\nu) = 8\pi\nu^2/c^3. \tag{2.20}$$

The number of modes per unit real volume and per unit frequency interval with solid angle $\delta\Omega$ is, therefore

$$\frac{\rho}{4\pi}\,\delta\Omega = \frac{2\nu^2\delta\Omega}{c^3}.$$

The black-body radiation density u per unit frequency interval is the product

ρE_{av}, where E_{av} is the average energy per mode obtained from Planck's law,

$$E_{av} = \frac{\Sigma_0^\infty nh\nu \exp(-nh\nu/kT)}{\Sigma_0^\infty \exp(-nh\nu/kT)} = \frac{h\nu}{\exp(h\nu/kT)-1}. \qquad (2.21)$$

Hence, we have

$$u = \frac{8\pi h\nu^3}{c^3} \frac{1}{\exp(h\nu/kT)-1}. \qquad (2.22)$$

In terms of wavelength this is

$$u = \frac{8\pi hc}{\lambda^5} \frac{1}{\exp(hc/kT\lambda)-1}. \qquad (2.23)$$

The ratio of the Einstein coefficients (section 4.7) is

$$A/B = 8\pi h\nu^3/c^3. \qquad (2.24)$$

Hence, we have

$$A = h\nu\rho B. \qquad (2.25)$$

If $A = uB$ i.e. equal rates of spontaneous and stimulated emission, then we have

$$u = h\nu\rho \qquad (2.26)$$

i.e., for this condition, the energy density is that resulting from one photon per mode.

3.2 *Blackbody radiation into a single transverse mode*

In order to compare laser radiation with that from other sources we need to describe the radiation in terms of modes of oscillation. In § 3.1 we have described blackbody radiation in terms of the modes of oscillation in an enclosed cavity; we now consider the blackbody radiation into a single transverse mode of an optical resonator. Such radiation may originate spontaneously in the active medium.

The number of photons radiated into the mode designated by m, n, q is

$$\delta_{mnq} = 2/[\exp(h\nu/kT)-1], \qquad (2.27)$$

where the factor 2 comes from the two planes of polarization, and m, n refer to the transverse modes and q to the longitudinal modes (ch. 4 § 6). Each transverse mode consists of the sum of all the longitudinal modes. The total energy in such a transverse mode is given by summing (2.27) over q and multiplying by $h\nu$. In a resonator of length L, the longitudinal mode

References on page 54

separation is $\Delta\nu = c/2L$. The number of modes separated by $\Delta\nu$ in unit frequency interval is $1/\Delta\nu = 2L/c$. This is $\mathrm{d}q/\mathrm{d}\nu$ which is then given by

$$\mathrm{d}q/\mathrm{d}\nu = 2L/c.$$

Thus, the total energy in a single transverse mode is

$$E_{mn} = \int_0^\infty g_{mnq} h\nu 2L \mathrm{d}\nu/c.$$

Hence, the energy in a single transverse mode, mn, of a resonator is

$$E_{mn} = \int_0^\infty \frac{4L}{c} \frac{h\nu \mathrm{d}\nu}{\exp(h\nu/kT)-1}.$$

Letting $h\nu/kT = x$, $\mathrm{d}\nu = (kT/h)\mathrm{d}x$, we get

$$E_{mn} = \frac{4L}{hc}(kT)^2 \int_0^\infty \frac{x\mathrm{d}x}{\mathrm{e}^x-1}.$$

This integral is the Riemann Zeta function of argument 2 and has the value $\pi^2/6$.

Therefore, the energy per unit length of the resonator is

$$E_{mn}/L = \rho_{mn} = 2\pi^2(kT)^2/3hc. \tag{2.28}$$

The energy which passes per second through a cross section of the resonator travelling in a given direction along the cavity axis is

$$W = \rho_{mn}c/2 = \pi^2(kT)^2/3h. \tag{2.29}$$

We can put this in more familiar form by expressing the mean energy kT in terms of a mean photon energy; by definition, we have $kT = h\bar{\nu}$. Therefore, we have $\bar{\lambda} = hc/kT$, which can be written

$$(kT\bar{\lambda}/hc)^2 = 1. \tag{2.30}$$

Multiplying (2.29) by (2.30) then gives, finally

$$W = \sigma T^4(5\bar{\lambda}^2/2\pi^3), \tag{2.31}$$

where σ is Stephen's constant.

Equation (2.31) gives the blackbody power which can be radiated into a single transverse mode of an optical resonator and was given by REMPEL [1963] in comparing lasers and conventional radiators.

3.3 *Number of photons emitted by a blackbody into a quantum state*

Photons in a given quantum state occupy the same cell in phase space. The number of photons in a given state is the degeneracy parameter (δ). There is no limit to the number of photons which can occupy the same cell in phase space, just as there is no limit to field intensity in classical theory. The number of photons in a given state is given by the number of quanta which crosses the coherence area A_c within the coherence time τ_c (ch. 10 § 2). From (10.1), it follows that

$$\tau_c = 2\pi/\Delta\omega. \tag{2.32}$$

From Planck's law, ch. 2 § 3.1, we can obtain the number of photons of frequency in the range (ω, $\omega+\Delta\omega$) emitted by a blackbody source of area ΔA_s per second in all directions; it is (from (2.22)),

$$\frac{\Delta A_s}{\hbar\omega} \cdot \frac{\hbar\omega^3 \Delta\omega}{\pi^2 c^2} \cdot \frac{1}{\exp(\hbar\omega/kT)-1}.$$

Of these, the fraction which passes through the coherence area A_c at a distance R is $A_c/4\pi R^2$. Thus, the number (δ) of photons crossing A_c in time τ_c is given by

$$\delta = \frac{A_c \Delta A_s}{2R^2} \cdot \frac{\omega^2}{\pi^2 c^2} \cdot \frac{1}{\exp(\hbar\omega/kT)-1}. \tag{2.33}$$

From (10.16), (10.15) and (10.1) it can be shown that $A_c = \lambda^2 R^2/\Delta A_s$, and when this is substituted into (2.33) we obtain (2.27); i.e. the degeneracy parameter and the number of photons per mode are the same. The volume of coherence, a cell in phase space, and a mode are thus equivalent descriptions. The number of photons emitted by a laser into a given quantum state is simply $P\tau_c/\hbar\omega$, where P is the power emitted into a single mode.

4. Radiation and matter

4.1 *Mass absorption coefficient*

The interaction between radiation and matter is usually described in terms of an emission and an absorption coefficient. The various processes which lead to broadening of emission lines also operate during the absorption process, so both emission and absorption coefficients are functions of frequency.

A beam of radiation of intensity I_ν incident normally to the surface of an absorber of thickness $\mathrm{d}h$ is reduced in intensity by $\mathrm{d}I_\nu$ on passing through it.

References on page 54

We do not distinguish between the various types of absorption and we include all loss processes in the term since we are, for present purposes, only concerned with calculating the reduction in intensity. The decrease is proportional to I_ν and to $\mathrm{d}h$. The constant of proportionality is the absorption coefficient. The density ρ of the material may be introduced into this coefficient, so arriving at the mass absorption coefficient, $k_{m\nu}$. Thus, considering unit cross section, the energy $\mathrm{d}I_\nu \mathrm{d}\nu$ absorbed per second in $\mathrm{d}h$ is given by

$$\mathrm{d}I_\nu \mathrm{d}\nu = -I_\nu \mathrm{d}\nu\, k_{m\nu} \rho\, \mathrm{d}h. \tag{2.34}$$

The coefficient $k_{m\nu}$ may be calculated by quantum mechanics.

If the absorber is of thickness h, and $I_{\nu 0}$ is the initial intensity, we may integrate this equation to determine the intensity of the beam after traversing distance h in the medium

$$I_{\nu h} = I_{\nu 0} \exp\left[-\int_0^h k_{m\nu} \rho\, \mathrm{d}h\right]. \tag{2.34a}$$

This equation can be written in the form

$$I_{\nu h} = I_{\nu 0} \exp\left[-\tau_\nu\right], \tag{2.35}$$

where τ_ν is the optical thickness.

4.2 *Atomic absorption coefficient*

Let m be the mass of the atom, and let there be n atoms in unit volume. Using the relationship $\rho = mn$ we have

$$\mathrm{d}I_\nu \mathrm{d}\nu = -I_\nu \mathrm{d}\nu\, k_{m\nu} mn\, \mathrm{d}h.$$

The atomic absorption coefficient may be defined by the equation

$$k_{a\nu} = k_{m\nu} m. \tag{2.36}$$

Then, we have

$$\mathrm{d}I_\nu = -I_\nu k_{a\nu} n\, \mathrm{d}h. \tag{2.37}$$

The dimensions of $k_{a\nu}$ are (length)^2.

4.3 *Number of absorption processes per second*

Consider an elementary cylinder of absorbing material of density ρ with base $\mathrm{d}A$ and height $\mathrm{d}h$. Let radiation travel along the axis of the cylinder within the solid angle $\mathrm{d}\omega$ and frequency interval $\mathrm{d}\nu$ about ν. The energy absorbed in time $\mathrm{d}t$ by volume $\mathrm{d}V(= \mathrm{d}h\,\mathrm{d}A)$ is

$$\mathrm{d}V\Delta E = -I_\nu k_{m\nu}\rho\,\mathrm{d}h\,\mathrm{d}A\,\mathrm{d}\nu\,\mathrm{d}\omega\,\mathrm{d}t.$$

But $\rho\,\mathrm{d}h\mathrm{d}A$ is the mass of material in the elementary cylinder, so we have

$$\Delta E = -I_\nu k_{a\nu} n_n \mathrm{d}\nu\,\mathrm{d}\omega\,\mathrm{d}t,$$

where n_n is the number of absorbing atoms in unit volume. The number of absorption processes per second per unit volume in which energy $h\nu$ is absorbed to cause a transition from level n to level m is

$$\Delta E/h\nu\mathrm{d}t = -I_\nu k_{a\nu} n_n \mathrm{d}\omega\,\mathrm{d}\nu/h\nu, \tag{2.38}$$

$$n_{n\to m}\mathrm{d}\nu = n_n k_{a\nu}\mathrm{d}\nu \int I_\nu \mathrm{d}\omega/h\nu. \tag{2.39}$$

In terms of radiation density we get

$$n_{n\to m}\mathrm{d}\nu = n_n k_{a\nu}\mathrm{d}\nu\, c u_\nu/h\nu. \tag{2.40}$$

4.4 *Classical theory of emission*

An oscillating dipole consisting of an electron and a nucleus (a Lorentz oscillator) radiates energy at the rate given by

$$\frac{\mathrm{d}\varepsilon}{\mathrm{d}t} = -\frac{2e^2(\dot{v})^2}{3c^3} \tag{2.41}$$

where e is the electron charge (esu) and $\dot{v}$ is its acceleration. The loss of energy may be equated to the work done per second by a radiative reaction force, thus

$$\boldsymbol{F}\cdot\boldsymbol{v} = -2e^2(\dot{v})^2/3c^3. \tag{2.42}$$

The average force for a cycle may be obtained as follows. Integrating by parts we get

$$\int_{t_1}^{t_2}\left(\boldsymbol{F}-\frac{2}{3}\frac{e^2}{c^3}\ddot{\boldsymbol{v}}\right)\cdot\boldsymbol{v}\mathrm{d}t = -\left[\frac{2}{3}\frac{e^2}{c^3}\boldsymbol{v}\cdot\dot{\boldsymbol{v}}\right]_{t_1}^{t_2}.$$

We can choose t_2-t_1 to be a cycle so that the contents of the square bracket are the same value at t_2 and t_1 and thus vanish giving

$$\boldsymbol{F}_{\mathrm{av}} = \frac{2}{3}\frac{e^2}{c^3}\ddot{\boldsymbol{v}}. \tag{2.43}$$

Let x be the displacement of the electron of a dipole. The velocity $v = \dot{x}$ and $\ddot{v} = \dddot{x}$. The equation of motion of the electron (mass, m) is

References on page 54

$$m\ddot{x} = -Kx + \frac{2}{3}\frac{e^2}{c^3}\dddot{x}, \tag{2.44}$$

where $-Kx$ is the restoring force (coulombic) on the dipole oscillator. The force due to the radiative reaction is much smaller than the restoring force and so the displacement x is given approximately by $x \approx x_0 \exp(-i\omega_0 t)$ which gives $\dddot{x} = -\omega_0^2 \dot{x}$ and finally, since $\omega_0 = (K/m)^{\frac{1}{2}}$ we get

$$\ddot{x} + \gamma\dot{x} + \omega_0^2 x = 0 \tag{2.45}$$

where γ is the classical radiation damping constant given by

$$\gamma = 2e^2\omega^2/3c^3 m. \tag{2.46}$$

Since γ is small, the solution of (2.45) is given by

$$x = x_0 \exp(-\gamma t/2) \exp(-i\omega_0 t).$$

The rate of energy loss averaged over a cycle may be obtained from (2.41) which gives

$$\overline{\frac{d\varepsilon}{dt}} = -\frac{e^2 x_0^2 \omega_0^4}{3c^3} \exp(-\gamma t). \tag{2.47}$$

This shows that the mean rate of radiation from a dipole is proportional to the square of the dipole moment. It is clear that $\tau = 1/\gamma$ can define a damping time, or a radiative lifetime of the oscillator. This expression gives lifetimes $\sim 10^{-8}$ sec for visible wavelengths.

4.5 *Classical theory of atoms interacting with electromagnetic radiation*

The study of classical oscillators was well established before quantum theory was developed and many of the ideas and descriptive terms used in classical theory found application in the quantum theory by way of the correspondence principle. We shall introduce some of this background by developing the classical theory of absorption and shall then show how the classical theory is related to quantum theory by introducing the concept of oscillator strength which is so useful in spectroscopy.

In the classical theory of Lorentz, atoms are regarded as harmonic oscillators in which, in the simplest case, the electrons oscillate with frequency ω_0. Consider a monochromatic electromagnetic wave of wavelength much greater than the dimensions of the atom so that the field may be regarded as uniform over the atomic volume. Let this wave be incident upon a Lorentz atom with a single electron. We shall neglect the effects of the magnetic field since the magnitude of the force it exerts on an electron is only the fraction v/c of the force due to the electric field on the electron. Let the electric field be given by

$$\boldsymbol{E} = \boldsymbol{E}_0 \exp(-\mathrm{i}\omega t). \tag{2.48}$$

The application of an electric field results in a displacement of the electron from its equilibrium position in the direction of the vector $\boldsymbol{E}$. The equation of motion of the electron of mass m and charge $-e$ is given by

$$m\ddot{\boldsymbol{x}} = -e\boldsymbol{E}_0 \exp(-\mathrm{i}\omega t) - K\boldsymbol{x} - g\dot{\boldsymbol{x}}, \tag{2.49}$$

where $K\boldsymbol{x}$ is the restoring force and $g\dot{\boldsymbol{x}}$ is a damping force which is assumed to be proportional to the velocity.

Let $g/m = \gamma$ and then, since $K/m = \omega_0^2$, we get

$$\ddot{\boldsymbol{x}} + \gamma\dot{\boldsymbol{x}} + \omega_0^2 \boldsymbol{x} = (-e\boldsymbol{E}_0/m) \exp(-\mathrm{i}\omega t). \tag{2.50}$$

We neglect solutions of this equation which represent transient effects and try the solution $\boldsymbol{x} = \boldsymbol{x}_0 \exp(-\mathrm{i}\omega t)$. (This solution is tried because it is likely that the electron displacement will follow the applied field.) Substituting this expression into (2.50) we obtain

$$\boldsymbol{x}_0 = \frac{-e\boldsymbol{E}_0/m}{(\omega_0^2 - \omega^2) - \mathrm{i}\gamma\omega}.$$

Therefore, we have

$$\boldsymbol{x} = -(e\boldsymbol{E}/m)/[(\omega_0^2 - \omega^2) - \mathrm{i}\gamma\omega]. \tag{2.51}$$

The complex form of $\boldsymbol{x}_0$ arises because of the damping term and means that the displacement of the electron is not in phase with the applied field.

From elementary theory we know that the dipole moment $\boldsymbol{p}$ of an atom induced by an applied field is given by

$$\boldsymbol{p} = \alpha\boldsymbol{E} \tag{2.52}$$

where the constant α is the polarizability of the atom. Therefore, we get

$$-e\boldsymbol{x} = \boldsymbol{p} = \alpha\boldsymbol{E} = (e^2\boldsymbol{E}/m)/[(\omega_0^2 - \omega^2) - i\gamma\omega],$$
$$\alpha = +(e^2/m)/[(\omega_0^2 - \omega^2) - i\gamma\omega]. \tag{2.53}$$

By quantum theory, we obtain the following for the polarizability (in the absence of damping),

$$\beta = f_{nm}(e^2/m)/[\omega_{nm}^2 - \omega^2], \tag{2.54}$$

References on page 54

showing that f_{nm} should be introduced in the classical expression to make it identical with the quantum equation. This factor is the effective number of electrons in the classical system and is called the *oscillator strength.*

If the number of dipoles per unit volume is N, the dipole moment per unit volume ($\boldsymbol{P}$) of gases is obtained simply by summing all the dipole moments (this is only true for gases), thus

$$\boldsymbol{P} = N\boldsymbol{p} = -Ne\boldsymbol{x}. \tag{2.55}$$

Using the equations $\varepsilon = 1+4\pi N\alpha$ and $\mu_c^2 = \varepsilon$ where ε is the dielectric constant and μ_c is the refractive index of the medium, we obtain

$$\varepsilon = \mu_c^2 = 1+\frac{4\pi Ne^2/m}{(\omega_0^2-\omega^2)-\mathrm{i}\gamma\omega}. \tag{2.56}$$

This shows that the dielectric constant (and refractive index) is complex and means that the medium will absorb the radiation. Let the refractive index be separated into its real and imaginary parts thus

$$\mu_c = \mu+\mathrm{i}\kappa, \tag{2.57}$$

$$\mu+\mathrm{i}\kappa = \left[1+\frac{4\pi Ne^2/m}{(\omega_0^2-\omega^2)-\mathrm{i}\gamma\omega}\right]^{\frac{1}{2}}. \tag{2.58}$$

The second term in the square brackets for gases is small compared with unity and so by expanding and neglecting terms beyond the second we get

$$\mu+\mathrm{i}\kappa = 1+\frac{2\pi Ne^2/m}{(\omega_0^2-\omega^2)-\mathrm{i}\gamma\omega}. \tag{2.59}$$

Separating the real and imaginary parts we get

$$\mu = 1+\frac{(2\pi Ne^2/m)(\omega_0^2-\omega^2)}{(\omega_0^2-\omega^2)+\gamma^2\omega^2}; \tag{2.60}$$

$$\kappa = \frac{(2\pi Ne^2/m)\gamma\omega}{(\omega_0^2-\omega^2)+\gamma^2\omega^2}. \tag{2.61}$$

Near resonance, $\omega \approx \omega_0$ and we may replace $(\omega_0^2-\omega^2)$ by $2\omega(\omega_0-\omega)$. Hence, we have

$$\mu = 1+\frac{(\pi Ne^2)(\omega_0-\omega)/\omega m}{(\omega_0-\omega)^2+\gamma^2/4}; \tag{2.62}$$

$$\kappa = \frac{\pi Ne^2\gamma/2\omega m}{(\omega_0-\omega)^2+\gamma^2/4}. \tag{2.63}$$

We shall now derive the relationship between κ and the mass absorption coefficient $k_{m\nu}$ (ch. 2 § 4.1) for a medium of density ρ. Consider a plane electromagnetic wave propagating in the x direction through a medium and let the initial intensity be I_0 and the intensity at point x be I corresponding to electric field intensities of E_0 and E. The wave equation for E is $\partial^2 E/\partial x^2 = (1/v^2)\partial^2 E/\partial t^2$ where v is the phase velocity. A solution of this equation is

$$E = E_0 \exp\left[-\mathrm{i}\omega\left(t - \frac{x}{v}\right)\right]. \tag{2.64}$$

The velocity v is given by $v = c/\mu_c$, hence, the solution becomes

$$E = E_0 \exp\left(-\frac{\omega\kappa x}{c}\right) \exp\left[-\mathrm{i}\omega\left(t - \frac{\mu}{c}x\right)\right]. \tag{2.65}$$

The factor $\exp(-\omega\kappa x/c)$ shows that the amplitude decreases as the wave propagates through the medium. Since intensity is proportional to the square of the amplitude, we have

$$I = I_0 \exp(-2\omega\kappa x/c). \tag{2.66}$$

Comparing this equation with (2.34a) which we rewrite here as

$$I = I_0 \exp(-k_m \rho x) \tag{2.67}$$

we see that

$$k_m \rho = 2\omega\kappa/c. \tag{2.68}$$

Therefore, we have

$$k_m \rho = \frac{\pi N e^2 \gamma/mc}{(\omega_0 - \omega)^2 + \gamma^2/4}. \tag{2.69}$$

Since $k_m \rho$ is the absorption coefficient per unit volume and N is the number of bound electrons per unit volume with a natural frequency ω_0, we have, per oscillator, the classical atomic absorption coefficient (classical atomic absorption cross section), k_a, given by

$$k_a = \frac{\pi e^2}{mc}\left[\frac{\gamma}{(\omega - \omega_0)^2 + \gamma^2/4}\right]. \tag{2.70}$$

At the frequency given by $(\omega_0 - \omega) = \gamma/2$, the absorption coefficient is half its value at ω_0. Hence, γ is the total half-width of the absorption coefficient. The line profile described by (2.70) is Lorentzian.

The damping described by the factor γ may be due to radiation or collisions. That this is so for radiation damping is shown in ch. 2 § 4.4. The analysis which shows that collisional damping of classical oscillators also may be described by γ is too long to present here. The interested reader may refer to STONE [1963 p. 264] who shows that

$$\gamma = \gamma_{\text{nat}} + \gamma_{\text{coll}}, \tag{2.71}$$

where γ_{nat} is due to radiation damping (natural) and γ_{coll} is due to collision damping.

The basis of the analysis is that the collisions cause arbitrary changes in phase and amplitude of the vibrations of the Lorentz atoms. In this case $\gamma_{\text{coll}} = 2/\tau_{\text{coll}}$ where τ_{coll} is the mean free time between collisions (the factor 2 arises because each collision is assumed to terminate the life of 2 oscillators, assuming only excited species present). The mean free time is given by l/v where l is the mean free path and v is the mean velocity. For a cross section $\pi\sigma^2$ and pressure p the mean free path is $l = kT/4\sqrt{2}\pi\sigma^2 p$; the mean velocity is $v = (8kT/\pi M)^{\frac{1}{2}}$. Therefore γ_{coll} can be estimated from the equation

$$\gamma_{\text{coll}} = \frac{32\sqrt{\pi}\sigma^2 p}{(MkT)^{\frac{1}{2}}}. \tag{2.72}$$

The duration of a collision may be taken to be the time for the colliding particles to travel a few diameters. This is $\sim 10^{-8}/10^5$ sec $= 10^{-13}$ sec which involves many periods of optical oscillation.

4.6 *Differential Einstein coefficients*

The Einstein coefficients are concerned with the probabilities of transitions between energy levels of atomic and molecular systems. We shall distinguish between *energy states* and *energy levels.* An energy level is made up of a set of $(2J+1)$ energy states whether or not the degeneracy has been removed. A *line* is the radiation associated with all possible transitions between the states belonging to two levels. A line is made up of components resulting from transitions between pairs of states. In conventional light sources the populations of the states within a level are equal because the excitation and de-excitation processes are quite random and isotropic; this is natural excitation (CONDON and SHORTLEY [1935]). In the case of a laser, the intense, polarized, unidirectional radiation field produces anisotropic de-excitation (or selective depopulation) by stimulated emission and so may produce large departures from the population distribution of natural excitation. The differential Einstein coefficients are essential for the analysis of this type of problem. They allow the influence of the direction of propagation and polarization of the radiation to be considered; they apply only to the case of broadband radiation, i.e. where the energy density of the radiation is essentially constant with respect to frequency over the absorption profile of the atom (2.70).

Consider transitions (in an atom, say) occurring between levels m and n of energies such that $E_m > E_n$ and $E_m - E_n = \hbar\omega$ where ω is the frequency of the photon involved. Various processes cause a frequency spread of $d\omega$ about ω for the emitted photons. The photon emitted by the transition $m \to n$ travels in a definite direction into solid angle $d\Omega$ and has a definite polarization, α. Since any polarization can be represented by two polarizations superimposed at right angles to each other, we have $\alpha = 1, 2$. Let dp_s be the probability per second, per unit frequency interval of a spontaneous transition $m \to n$. The probability of emission into frequency $d\omega$ about ω and into $d\Omega$ and with polarization α is proportional to $d\Omega$ and $d\omega$ so we have,

$$dp_s d\omega = a_{mn}^{\alpha} d\Omega d\omega$$

which, we write

$$dp_s = a_{mn}^{\alpha} d\Omega. \tag{2.73}$$

A photon from the transition $m \to n$ passing near another atom may be absorbed by causing a transition $n \to m$, or, if the atom is already in level m, the photon may stimulate (induce) the transition $m \to n$. Let the respective probabilities per second per unit frequency interval, for absorption and stimulated emission be dp_a and dp_i. These probabilities are proportional to the number of photons, of energy $\hbar\omega$, present in unit volume (photon density) at any given instant. The energy density of radiation propagated into $d\Omega$ with polarization α and of frequency in the range ω, $(\omega + d\omega)$ is $u_{\omega}^{\alpha}(\Omega) d\Omega d\omega$ so the photon density is $u_{\omega}^{\alpha}(\Omega) d\Omega d\omega / \hbar\omega$. Therefore, again dropping $d\omega$ we have

$$dp_a = b_{nm}^{\alpha} u_{\omega}^{\alpha}(\Omega) d\Omega \tag{2.74}$$

and

$$dp_i = b_{mn}^{\alpha} u_{\omega}^{\alpha}(\Omega) d\Omega. \tag{2.75}$$

The constants a_{mn}^{α}, b_{mn}^{α}, b_{nm}^{α} are differential Einstein coefficients.

Energy levels are usually degenerate in that several different states within a given energy level may be occupied with equal probability. (Degeneracy may be removed by the application of electric or magnetic fields.) If the total number density of atoms in energy level m is N_m and the number density of atoms in one of the states of level m is n_m, then we have

$$g_m n_m = N_m, \tag{2.76}$$

where g_m is the degree of degeneracy, or the statistical weight of level m. If the number of atoms per unit volume in energy states m and n are n_m and

References on page 54

n_n, respectively, then the number of photons emitted per second in transitions $m \to n$ is $n_m(\mathrm{d}p_s + \mathrm{d}p_i)$ and the number absorbed per second in transitions $n \to m$ is $n_n \mathrm{d}p_a$. If the population of level m is to maintain a steady value, the absorption and emission rates must be equal. Therefore, from (2.73), (2.74) and (2.75) we have

$$n_n b^\alpha_{nm} u^\alpha_\omega(\Omega) = n_m [b^\alpha_{nm} u^\alpha_\omega(\Omega) + a^\alpha_{mn}]. \tag{2.77}$$

Black-body radiation is independent of the particular properties of matter with which it is in equilibrium so if we analyse a system in thermal equilibrium in terms of the differential Einstein coefficients we may obtain relationships between the coefficients which are generally valid.

For a system in thermal equilibrium

$$(N_m/N_n) = (g_m/g_n) \exp[-(E_m - E_n)/kT]. \tag{2.78}$$

Therefore, using (2.76) and substituting (2.77) we get finally (use of (2.78) in (2.77) involves the approximation that the absorption line is narrow)

$$u^\alpha_\omega(\Omega) = a^\alpha_{mn}/b^\alpha_{nm}[\exp(\hbar\omega/kT) - (b^\alpha_{mn}/b^\alpha_{nm})]. \tag{2.79}$$

Planck's law for radiation of polarization α in the direction $\mathrm{d}\Omega$ about Ω is

$$u^\alpha_\omega(\Omega) = \frac{\hbar\omega^3}{8\pi^3 c^3} \frac{1}{\exp(\hbar\omega/kT) - 1}. \tag{2.80}$$

The more usual form of Planck's law (2.22) may be obtained from (2.80) by integrating over $\Omega = 4\pi$ and summing over the two polarizations $\alpha = 1, 2$. Planck's law is

$$u_\omega = \frac{\hbar\omega^3}{\pi^2 c^3} \frac{1}{\exp(\hbar\omega/kT) - 1}.$$

To work in terms of frequency rather than angular frequency we remember $2\pi\nu = \omega$, therefore, $2\pi\,\mathrm{d}\nu = \mathrm{d}\omega$. The radiation energy density must be the same however we chose to express it so $u_\omega \mathrm{d}\omega = u_\nu \mathrm{d}\nu$. Hence, we get $2\pi u_\omega = u_\nu$. Planck's law in terms of frequency is

$$u_\nu = \frac{8\pi h\nu^3}{c^3} \frac{1}{\exp(h\nu/kT) - 1}. \tag{2.22}$$

Energy density of radiation $U = \int_0^\infty u_\omega \mathrm{d}\omega$.

Comparing (2.79) and (2.80), we find

$$b^\alpha_{nm} = b^\alpha_{mn}, \quad \text{and} \quad a^\alpha_{mn}/b^\alpha_{mn} = \hbar\omega^3/8\pi^3c^3. \tag{2.81}$$

These differential Einstein coefficients can be applied in discussing the individual members of a degenerate level separately. They refer to transitions between energy states and are valid whether or not the system is in equilibrium.

4.7 *Einstein coefficients*

The energy density of radiation of frequency $\omega = \omega_{mn}$ and of given polarization taking into account all directions relevant to the radiation field present is

$$u_\omega^\alpha = \int u_\omega^\alpha(\Omega)\mathrm{d}\Omega. \tag{2.82}$$

If the radiation field is such that its direction is within a small angle $\mathrm{d}\Omega$, as is the case in laser beams, the density $u_\omega^\alpha(\Omega)$ is nearly a δ-function of the angle Ω. Accordingly, by integration of (2.74) taking account of all relevant directions, the absorption probability per second is $p_a = b_{nm}^\alpha \int u_\omega^\alpha(\Omega)\mathrm{d}\Omega$,

$$p_a = b_{nm}^\alpha u_\omega^\alpha. \tag{2.83}$$

Equation (2.83) applies to beams with plane wave fronts or to divergent beams.

The probability per second for a transition from level n to m (non-degenerate) is given by (cf. ch. 3 § 3.3 using (3.88) and (C.2))

$$p_a = u_\omega^\alpha \frac{4\pi^2}{\hbar^2} |\boldsymbol{D}_{mn}|^2 \cos^2\theta_{mn}, \tag{2.84}$$

where $\boldsymbol{D}_{mn}$ is the electric moment matrix, θ_{mn} is the angle between the vector $\boldsymbol{D}_{mn}$ and the electric vector of the radiation (i.e. the direction of polarization). Equations (2.83) and (2.84) both give the probability per second of a transition between n and m, so we find

$$b_{nm}^\alpha = (4\pi^2/\hbar^2)|\boldsymbol{D}_{mn}|^2 \cos^2\theta_{mn}. \tag{2.85}$$

Consider a pencil of radiation travelling in direction $\mathrm{d}\Omega$ and the vector $\boldsymbol{D}_{mn}$ as shown in fig. 2.3. Let one of the chosen directions of polarization

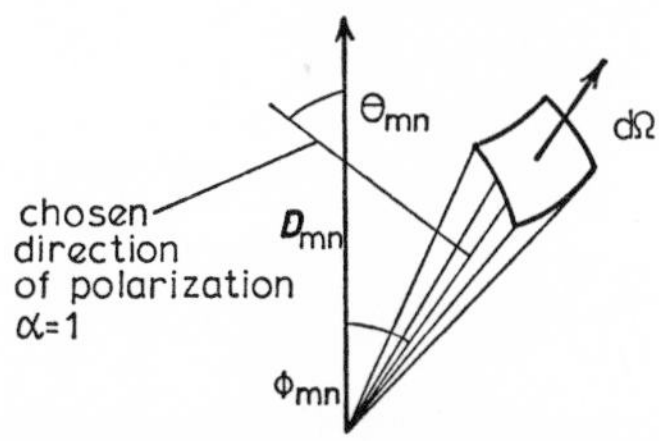

Fig. 2.3. Classical view of relative orientation of electric dipole and electric vector of radiation field.

References on page 54

($\alpha = 1$) be normal to the direction of the pencil and lie in the plane containing the central ray of the pencil and the vector $\boldsymbol{D}_{mn}$ and let the other direction, $\alpha = 2$, be normal to this plane. Since the angle between the direction of polarization $\alpha = 1$ and $\boldsymbol{D}_{mn}$ is θ_{mn}, and ϕ_{mn} is the angle between $\boldsymbol{D}_{mn}$ and the direction of propagation of the absorbed radiation, we have

$$\theta_{mn} = (\pi/2) - \phi_{mn}.$$

Considering direction $\alpha = 1$, (2.85) gives

$$b_{nm}^{(1)} = (4\pi^2/\hbar^2)|\boldsymbol{D}_{mn}|^2 \sin^2 \phi_{mn}. \tag{2.86}$$

For direction $\alpha = 2$, $\theta_{mn} = \pi/2$ and from (2.85) we have

$$b_{nm}^{(2)} = 0.$$

The probability of spontaneous emission of a photon of polarization α into $\mathrm{d}\Omega$ is obtained by combining (2.81), (2.73), and (2.86) which gives for radiation polarized in direction $\alpha = 1$,

$$\mathrm{d}p_s^{(1)} = (\omega_{mn}^3/2\pi c^3\hbar)|\boldsymbol{D}_{mn}|^2 \sin^2 \phi_{mn}\, \mathrm{d}\Omega \tag{2.87}$$

and

$$\mathrm{d}p_s^{(2)} = 0 \tag{2.88}$$

for radiation polarized in direction $\alpha = 2$.

If the levels m, n are degenerate, we must sum over all transitions $m \to n$. Spontaneous emission occurs in all directions so the total probability per second of spontaneous emission, obtained by integrating over all directions, is

$$A_{mn} = (4\omega_{mn}^3/3\hbar c^3)\Sigma|\boldsymbol{D}_{mn}|^2, \tag{2.89}$$

which is Einstein's coefficient for spontaneous emission of frequency ω_{mn}.

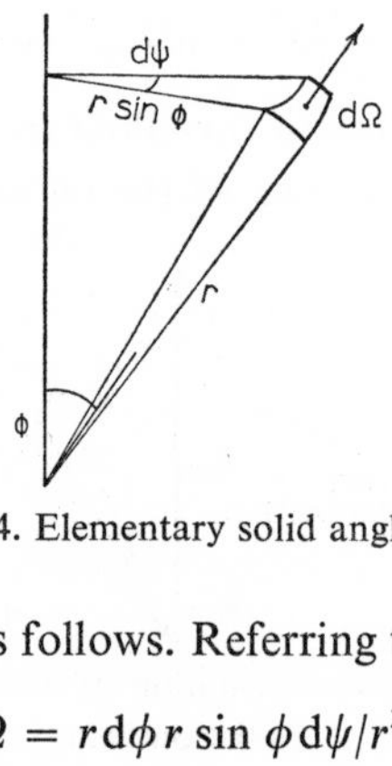

Fig. 2.4. Elementary solid angle $\mathrm{d}\Omega$.

Equation (2.89) is derived as follows. Referring to fig. 2.4 we have

$$\mathrm{d}\Omega = r\mathrm{d}\phi\, r \sin\phi\, \mathrm{d}\psi/r^2,$$

giving $d\Omega = d\phi \sin\phi \, d\psi$. Hence, we get

$$p_s^{(1)} = \frac{\omega_{mn}^3}{2\pi c^3 \hbar} |\boldsymbol{D}_{mn}|^2 \int_0^{2\pi} \int_0^{\pi} \sin^3 \phi_{mn} \, d\phi \, d\psi$$

from which we get (2.89).

The Einstein coefficient for absorption of isotropic unpolarized radiation of frequency ω_{mn} is

$$B_{nm} = \frac{1}{8\pi g_n} \Sigma \int_{4\pi} b_{nm}^{\alpha} \, d\Omega, \tag{2.90}$$

where the sum is to be taken over all transitions $n \to m$ and over the two polarizations $\alpha = 1, 2$. The Einstein coefficient for stimulated emission is

$$B_{mn} = \frac{1}{8\pi g_m} \Sigma \int_{4\pi} b_{mn}^{\alpha} \, d\Omega. \tag{2.91}$$

In practice there is generally a distribution of frequencies present in the radiation field and, therefore, the probability of stimulated emission or absorption, $p = Bu$, is given more accurately by $\int B(\nu)u(\nu)d\nu$, where $B(\nu)$ is an Einstein coefficient and $u(\nu)$ is the energy density expressed as a function of frequency. These may be slowly varying functions of frequency or they may vary rapidly depending upon the bandwidth of the frequency variations. If the radiation field has a broad frequency bandwidth compared with the width of the atomic levels to which $B(\nu)$ refers, we may regard $u(\nu)$ as a constant and take it outside the integral sign, thus, $p = u(\nu) \int B(\nu)d\nu$. If the radiation field is of a very narrow bandwidth, as is the laser radiation field, we may write, $p = B(\nu) \int u(\nu)d\nu$. These points are taken up more thoroughly in ch. 3. §§ 3.1–3.5.

4.8 *Relation between the absorption coefficient and the Einstein coefficient*

Let n_n and n_m be the numbers of atoms per unit volume in the lower and upper levels n and m of particles in a radiation field of density u_ν. The number of quanta absorbed per second in the frequency interval $d\nu$ is $n_n B_{\nu nm} u_\nu d\nu$. From (2.40) we have, therefore

$$\begin{aligned} B_{\nu mn} &= ck_\nu / h\nu, \\ \int k_\nu d\nu &= (h\nu/c) B_{mn}. \end{aligned} \tag{2.92}$$

According to these equations, the integral of the absorption coefficient is a constant and so it is independent of the line broadening process.

References on page 54

4.8.1. *Photoexcitation cross-section*

In considering trapping of radiation and many other problems connected with the interaction of radiation and the medium through which it is passing, it is useful to be able to estimate the order of magnitude of the cross-section for photoexcitation. This cross-section is related to the absorption coefficient by $\sigma_\nu = k_\nu/N$. The absorption coefficient, at line centre $\nu = \nu_0$ is given by

$$k_0 = \frac{1}{4\pi}\sqrt{\frac{\ln 2}{\pi}}\,\frac{c^2(g_m N_n - g_n N_m)}{\Delta\nu_D \nu_0^2 g_n \tau_s},$$

where n is the lower and m the upper states, $\Delta\nu_D$ is the full Doppler width of the line at half maximum, and τ_s is the natural radiative lifetime for the transition. At $\nu = \nu_0$, neglecting stimulated emission so that $N_n = N$, we have

$$\frac{k}{N} = \sigma = \frac{1}{4\pi}\sqrt{\frac{\ln 2}{\pi}}\,\frac{\lambda^2}{\Delta\nu_D \tau_s}\,\frac{g_m}{g_n}.$$

Typically $\tau_s = 10^{-8}$ sec, $\Delta\nu_D = 10^9$Hz, $\lambda = 10^{-4}$ cm, and if we assume $g_m = g_n = 1$ we then get $\sigma \approx 4\times 10^{-11}$ cm^2.

The value obtained for the cross-section should be compared with the atomic cross-section (hard sphere) which is generally $10^{-15}-10^{-16}$ cm^2.

4.9 *Lifetime of excited states*

Consider a system of excited atoms which do not interact and neglect effects due to photons present in the system. The excited atoms emit spontaneously. Let the number present in state E_m at time t be N_m. The average number of atoms which undergo the spontaneous transiton $m \to n$ in time $\mathrm{d}t$ is

$$\mathrm{d}N_m = -A_{mn} N_m \mathrm{d}t. \tag{2.93}$$

This equation gives

$$N_m = N_m^0 \exp(-t/\tau_{mn}),$$

where τ_{mn}, which is given by

$$\tau_{mn} = 1/A_{mn}, \tag{2.94}$$

is the mean life time of an atom in the excited state m. If there are two lower energy states n, p to which spontaneous transitions from m may be made, then the average number of atoms making these transitions in time $\mathrm{d}t$ is

$$\mathrm{d}N_m = -(A_{mn}+A_{mp})N_m \mathrm{d}t.$$

This equation gives

$$N_m = N_m^0 \exp\left[-t(A_{mn}+A_{mp})\right].$$

The mean life-time of an atom in state m is thus τ where

$$(1/\tau) = (1/\tau_{mn})+(1/\tau_{mp}). \tag{2.95}$$

4.10 *Refractive index and the dispersion relation*

From (2.57) and (2.59) we have

$$\mu_c = \mu+\mathrm{i}\kappa = 1+\frac{2\pi Ne^2/m}{(\omega_0^2-\omega^2)-\mathrm{i}\gamma\omega}. \tag{2.96}$$

This may be written

$$\mu_c-1 = \left(\frac{2\pi Ne^2}{m}\right)\frac{(\omega_0^2-\omega^2)+\mathrm{i}\gamma\omega}{(\omega_0^2-\omega^2)^2+\gamma^2\omega^2}, \tag{2.97}$$

which is of the form (cf. A.12)

$$\mu_c(\omega)-1 = \mathscr{R}(\omega)+\mathrm{i}\mathscr{I}(\omega), \tag{2.98}$$

where

$$\mathscr{R}(\omega) = \left(\frac{2\pi Nc^2}{m}\right)\frac{\omega_0^2-\omega^2}{(\omega_0^2-\omega^2)^2+\gamma^2\omega^2} \tag{2.99}$$

and

$$\mathscr{I}(\omega) = \left(\frac{2\pi Ne^2}{m}\right)\frac{\gamma\omega}{(\omega_0^2-\omega^2)^2+\gamma^2\omega^2}. \tag{2.100}$$

We see that $\mu_c(\omega)-1$ given by (2.96) is a function in the complex plane $\omega = \omega'+\mathrm{i}\omega''$ with poles in the lower half plane at the roots of the equation $(\omega_0^2-\omega^2)-\mathrm{i}\gamma\omega = 0$ and with no poles in the upper half-plane if γ is independent of ω †. The poles occur at

$$\omega = -\mathrm{i}(\gamma/2)\pm\{\omega_0^2-(\gamma/2)^2\}^{\frac{1}{2}}.$$

Also we note that as $|\omega| \to \infty$, $|\mu_c(\omega)-1| \to 0$.

Using the methods described in the section of appendix A which deals with dispersion relations we see that we require contours in the upper half-plane to avoid the poles in the lower half and that we need only consider that part of the contour which is the real axis. Inspection of (2.97) shows that

† Equation (2.46) shows that the poles lie at the roots of a cubic rather than a quadratic equation. For poles arising in the upper half-plane due to the dependence of γ on ω for a classical electron see TOLL [1952].

References on page 54

$\mu_c^*(\omega) = \mu_c(-\omega)$, thus obeying the crossing symmetry relation (appendix A) and enabling us to use (A.13b).

From (2.98), (2.99) and (A.13b) we obtain

$$\mathscr{R}[\mu_c(\omega_1)-1] = \frac{2}{\pi} P \int_0^\infty \frac{\omega \mathscr{I}[\mu_c(\omega)]\mathrm{d}\omega}{\omega^2-\omega_1^2}, \tag{2.101}$$

where $\mathscr{R}$ and $\mathscr{I}$ denote real and imaginary parts and

$$\mathscr{I}[\mu_c(\omega)-1] = \mathscr{I}[\mu_c(\omega)];$$

ω_1 is any frequency along the real axis. Equation (2.101) is one form of the Kramers-Kronig equation.

By (2.96), equation (2.101) can be written in the form

$$\mu(\omega_1) = \frac{2}{\pi} P \int_0^\infty \frac{\omega \kappa(\omega)\mathrm{d}\omega}{\omega^2-\omega_1^2}. \tag{2.102}$$

Since, by (2.68), the absorption coefficient k_m is related to κ, equation (2.102) enables the real part of the refractive index at a particular frequency to be calculated if the absorption coefficient is known for all frequencies. Alternatively, since the imaginary part of the refractive index is related to the total absorption cross-section by the equation

$$\kappa(\omega) = \frac{c}{2\omega} N\sigma_{\mathrm{tot}}(\omega),$$

where N is the number of atoms per unit volume, and $\sigma_{\mathrm{tot}}(\omega)$ is the total cross section of an atom for radiation of frequency ω we may write

$$\mu(\omega_1) = \frac{Nc}{\pi} P \int_0^\infty \frac{\sigma_{\mathrm{tot}}(\omega)\mathrm{d}\omega}{\omega^2-\omega_1^2}.$$

It will be noticed that, while we have considered the behaviour of a particular model of the atom interacting with radiation, the equation we have obtained does not involve any details of the interaction processes occurring. Dispersion relations were introduced by Kramers and Kronig in relating the refractive index and absorption coefficient but have proved to be of much wider applicability. The real importance of dispersion relations lies in the fact that it is now believed that a dispersion relation can be set up for any causal event. For a discussion of dispersion relations and causality see TOLL [1956]. Any quantity that can be analytically continued into the complex plane can be set up as a dispersion relation. We shall use this technique in the next section (§ 4.10.2) to derive an equation giving the phase shift per transit in a laser medium taking line shape into account. The equation has been shown by BENNETT [1962] to be important in understanding mode pulling.

4.10.1 *Phase shift and line shape*

Consider the radiation in the cavity of a laser operating at low gain and with linear amplification. We wish to calculate the phase shift per pass taking into account the line shape. Since line shape is concerned with amplitude (as a function of frequency) we are effectively looking for a relationship between amplitude and phase. The necessary connection is given by the dispersion relation. The problem will be formulated in terms which will enable us to use this relation.

As the radiation travels through the medium of the cavity from one mirror to the other making a single pass, it is amplified and undergoes an additional phase shift due to the medium. We follow BENNETT [1962] and represent the amplification and the additional phase shift resulting from a single pass by the complex fractional amplitude gain of the medium, $K(\omega)$, defined by

$$1+K(\omega) = [1+K_0(\omega)] \exp [\mathrm{i}\phi(\omega)], \tag{2.103}$$

where $K_0(\omega)$ is the fractional amplitude gain per pass and $\phi(\omega)$ is the additional single pass phase shift resulting from the amplifying medium. We shall consider the case in which K, $K_0 \ll 1$ i.e. amplification is small. Taking the logarithm of (2.103) and expanding, we get

$$K(\omega) \approx K_0(\omega)+\mathrm{i}\phi(\omega). \tag{2.104}$$

The fractional energy gain per pass is

$$g_0(\omega) \approx 2K_0(\omega). \tag{2.105}$$

Assuming that the process of amplification obeys causality, $K(\omega)$ is analytic in the upper half of the complex plane and we may use (A.14) to relate the real and imaginary parts of (2.104) and we get

$$\phi(\omega_0) = -\frac{P}{\pi}\int_{-\infty}^{\infty} \frac{K_0(\omega)\mathrm{d}\omega}{\omega-\omega_0}, \tag{2.106}$$

where ω_0 is not yet specified. For a Lorentzian line, inspection of (2.70) shows that we may assume that $K_0(\omega)$ is of form given by

$$K_0(\omega) \approx \frac{K_m(\Delta\omega)^2}{(\Delta\omega)^2+4(\omega_m-\omega)^2}, \tag{2.107}$$

References on page **54**

where subscript m refers to the line centre where the energy gain is maximum and $\Delta\omega$ is the full width of the line at half-maximum energy gain.

From (2.106) and (2.107) we get

$$\phi(\omega_0) = -\frac{P}{\pi}\int_{-\infty}^{\infty} \frac{K_m(\Delta\omega)^2 d\omega}{(\omega-\omega_0)[(\Delta\omega)^2+4(\omega_m-\omega)^2]}. \tag{2.108}$$

This has poles at ω_0, and $(\omega_m \pm i\Delta\omega/2)$.

We can evaluate the integral (2.108) by taking a counterclockwise contour along the real ω-axis, over the pole at ω_0, and in a semicircle of infinite radius in the upper half of the complex plane, thus excluding the pole at $(\omega_m - i\Delta\omega/2)$. The value of the integral is

$$2\pi i[\text{Residue at } (\omega_m - i\Delta\omega/2) + (\tfrac{1}{2}) \text{ Residue at } \omega_0].$$

Therefore, using (A.1) to calculate the residues, we get

$$\phi(\omega_0) = -\frac{2K_m(\Delta\omega)(\omega_m-\omega_0)}{[(\Delta\omega)^2+4(\omega_m-\omega_0)^2]};$$

$$\phi(\omega_0) = -\frac{2K_0(\omega_0)(\omega_m-\omega_0)}{\Delta\omega};$$

$$\phi(\omega_0) = -\frac{g_0(\omega_0)(\omega_m-\omega_0)}{\Delta\omega}. \tag{2.109}$$

References

BENNETT, Jr., W. R., 1962, Appl. Optics, Supplement on Optical Masers, pp. 24–61.

CONDON, E. U. and G. H. SHORTLEY, 1935, The Theory of Atomic Spectra (Cambridge Univ. Press, London) p. 97.

REMPEL, R. C., 1963, Optical Properties of Lasers as Compared to Conventional Radiators, Laser Technical Bulletin Number 1 (Spectra Physics, Inc., Mountain View, Calif., June, 1963) p. 9.

STONE, J. M., 1963, Radiation and Optics (McGraw-Hill, New York).

TOLL, J. S., 1956, Phys. Rev. **104**, 1760.

TOLL, J. S., 1952, Doctor of Philosophy Thesis, Princeton University.

CHAPTER 3

RADIATION AND ATOMIC SYSTEMS

1. Introduction

In this chapter we consider the interaction of radiation fields with atomic systems in the semi-classical approximation. In this approximation, the radiation field is treated classically, while the atomic system is treated quantum mechanically. The influence of the radiation field on the atomic system is described by an additional term (usually treated as a perturbation) in the Hamiltonian for the atomic system (§ 3.2), while the influence of the atomic system on the radiation field is described by a time-dependent polarization associated with the atomic system, which is included in Maxwell's equations (§ 3.4).

In the semi-classical approximation, and particularly when optical frequency fields are under consideration, it is usual to make a further approximation and to associate with the emission (or absorption) processes of radiation by the atomic systems, a time-independent transition probability, which is given by the "Fermi golden rule"

$$W_{ab} = \frac{2\pi}{\hbar^2} \cdot |\mathbf{D}_{ab}|^2 \cdot \rho(\omega), \tag{3.1}$$

where $\rho(\omega)$ is the spectral energy density of the radiation and $\mathbf{D}_{ab}$ is the relevant matrix element between the two levels a and b (§ 3.1). The probability of finding the atomic system in one or the other of the two states associated with the transition then obeys an exponential law, with a time constant given by the inverse of the transition probability defined by (3.1). We shall demonstrate in § 3.3 that this approximation is only valid when the coherence time of the radiation is short compared to this time constant. For laser radiation, the converse is usually true, and so the approximation of (3.1) cannot be made. In this case it is necessary to revert to the equations (3.61) and (3.62) which result from the introduction of the radiation field, as a perturbation, into the Schrödinger equation for the atomic system. In § 3.1

References on page 92

we derive such equations for a two level atomic system in a monochromatic radiation field.

The decay of a state by the process of spontaneous emission can be described by a time-independent transition probability which is derived from (3.1) by thermodynamic arguments (ch. 2 § 4.6). In order to take into account the finite lifetimes of the levels in our two level system due to their decay by spontaneous emission, we use a technique due to WEISSKOPF and WIGNER [1930] and introduce into equations (3.61) and (3.62) phenomenological damping constants (§ 3.1). In the absence of the monochromatic radiation field, these describe an exponential decay for the levels. The damping constants can be used to describe decay phenomena other than spontaneous emission; for example, they can be used to describe stimulated radiative decay processes to which the Fermi golden rule description applies, or collisional relaxation.

In § 3.5 we obtain solutions for equations (3.61) and (3.62) in the "Rabi strong signal" approximation, since this illustrates the behaviour of a two level system in a monochromatic radiation field for different values of the damping constants and for conditions on and off resonance.

In § 3.4 we shall derive forms of the equations (for the two level system) which are particularly convenient for the density matrix formalism found in Lamb's theory. The density matrix formalism is adopted in Lamb's theory since it provides a convenient method for averaging over all the atomic systems that make up an active medium and which are all interacting with the radiation field. In § 4.1 of this chapter we introduce the density matrix and discuss several of its properties, in particular those associated with "coherence" effects (§ 4.2).

We begin this chapter (§ 2) by reviewing those elements of quantum mechanics which will be particularly useful for an understanding of subsequent sections, and later chapters when we discuss various theoretical approaches to the laser. These early sections will also serve to define the formalism.

2. Basic postulates of quantum mechanics

In the following résumé of the quantum mechanical formalism to be employed, we introduce the basic postulates of quantum mechanics in an ad hoc fashion, and develop such expressions from them as are required in later sections. We adopt the Schrödinger representation and define the wavefunction in configurational space (3.3).

For those phenomena where a quantum mechanical, as opposed to a classical, description is required, it is no longer possible to determine well-defined values at each instant of time for all the variables associated with a system since the presence of the measuring device generally disturbs the dynamical state of the system in an unpredictable manner. It is only possible, in general, to associate a statistical distribution of values with a particular variable of the system. We accordingly introduce for a system with f-degrees of freedom, a probability density

$$W(q_1, q_2, \cdots, q_f, t)\mathrm{d}q_1 \cdots \mathrm{d}q_f = W(q, t)\mathrm{d}q, \tag{3.2}$$

which expresses the probability of finding the system at time, t, with its co-ordinates in the ranges q_1 to $(q_1+\mathrm{d}q_1)$ etc.

The probability density, by definition, must be a real, positive quantity, and we ensure this by expressing it as the square of the magnitude of a probability amplitude (hereafter referred to as a wavefunction), which in general may be complex; thus, we have

$$W(q, t) = |\Psi(q, t)|^2 = \Psi^*(q, t)\Psi(q, t). \tag{3.3}$$

(We shall use an asterisk generally to denote the complex conjugate of a function.) Since there must be unit chance of finding the system in some state, the probability density is normalized thus,

$$\int W(q, t)\mathrm{d}q = \int \Psi^*(q, t)\Psi(q, t)\mathrm{d}q = 1. \tag{3.4}$$

We have adopted here the configurational representation of the wave-function, where it is assumed to be a function only of the coordinates (q) of the system, together with time. An alternative approach is to define the wavefunction in terms of the momenta (p), and on this basis to develop an exactly parallel treatment to the one given below.

We postulate that the mean (or expectation) value of any observable quantity of the system, $F(q, p, t)$, which in general is a function both of the coordinates (q) and momenta (p), is given by

$$\langle F(t)\rangle = \int \Psi^*(q, t)\hat{F}\left(q, \frac{\hbar}{\mathrm{i}}\frac{\partial}{\partial q}, t\right)\Psi(q, t)\mathrm{d}q, \tag{3.5}$$

where $\hat{F}(q, \hbar/\mathrm{i}\ \partial/\partial q, t)$ is a linear operator which operates only on the coordinates (q). The operator is formed from the classical expression for the

References on page 92

observable by means of the following substitutions

$$q_r \to q_r; \qquad p_r \to \frac{\hbar}{\mathrm{i}} \frac{\partial}{\partial q_r} \qquad (r = 1, 2, \cdots, f). \tag{3.6}$$

However, in order to avoid ambiguities, the above substitutions must be carried out in a specific way. Firstly, the classical expression is written in terms of cartesian coordinates. In this form it will be found to be made up from terms quadratic in the p's (but independent of the q's) which are substituted as follows

$$p_r^2 \to -\hbar^2 \frac{\partial^2}{\partial q_r^2} \qquad (r = 1, 2, \cdots, f), \tag{3.7}$$

from terms involving only the q's, which remain unaltered, and possibly also from terms of the form

$$\sum_r [p_r f_r(q_1, \cdots, q_f)],$$

which are written in the symmetric form

$$\tfrac{1}{2} \sum_r [p_r f_r(q_1, \cdots, q_f) + f_r(q_1, \cdots, q_f) p_r] \tag{3.8}$$

before the substitutions (3.6) are made.

Operators corresponding to physical observables are linear and Hermitian. An operator, $\hat{F}$, is Hermitian if

$$\int f^*(q) \hat{F} g(q) \mathrm{d}q = \int g(q) [\hat{F} f(q)]^* \mathrm{d}q. \tag{3.9}$$

One of the reasons for requiring Hermitian operators, can be seen from (3.5), for if the mean value of the observable given by this equation is to be real, then

$$\langle F \rangle = \langle F \rangle^*$$

so that

$$\int \Psi(q, t) [\hat{F} \Psi(q, t)]^* \mathrm{d}q = \int \Psi^*(q, t) \hat{F} \Psi(q, t) \mathrm{d}q \tag{3.10}$$

which is the condition for $\hat{F}$ to be Hermitian.

The wavefunction specifies completely the dynamical state of the system, and all predictions about the system at a particular instant can be made from a knowledge of the wavefunction at that instant. In order to follow the change in the state of the system with time, therefore, we require an equation that describes the development of the wavefunction in time. This is provided by the time-dependent Schrödinger equation

$$\hat{H}(q,t)\Psi(q,t)+\frac{\hbar}{\mathrm{i}}\frac{\partial}{\partial t}\Psi(q,t)=0, \tag{3.11}$$

where $\hat{H}(q,t)$ is the Hamiltonian operator derived from the classical Hamiltonian for the system, according to the rules previously discussed.

Since this equation is linear and homogeneous, then if Ψ_1, and Ψ_2 are separate solutions of the equation, any linear combination of them

$$\Psi = a_1\Psi_1+a_2\Psi_2 \tag{3.12}$$

is also a solution. This is the property of superposition. It is possible, therefore, to treat a system in a state, Ψ, as being partly in the state Ψ_1, and partly in the state Ψ_2, and to use (3.12) to predict the relative probabilities of finding the system in either of the states on carrying out an appropriate experimental observation. Since the probability density depends on the product of the wavefunction with its complex conjugate, we have

$$\Psi^*\Psi = [|a_1|^2|\Psi_1|^2+|a_2|^2|\Psi_2|^2]+[a_1a_2^*\Psi_1\Psi_2^*+\text{c.c.}]. \tag{3.13}$$

Interference effects occur between the states Ψ_1, and Ψ_2, and it is not therefore generally possible to consider the expectation values corresponding to the state Ψ as the weighted mean of those corresponding to the states Ψ_1 and Ψ_2 (§ 4.2).

The equation is also of first order with respect to time, so it follows that once the wavefunction is specified at a given initial instant, the equation uniquely predicts the wavefunction at any later instant. This is, of course, in accordance with our requirement that the dynamical state of the system be completely specified by the wavefunction.

The classical Hamiltonian corresponds to the energy of the system, and, therefore, for a conservative system where the energy is a constant of the motion, the Hamiltonian is time-independent. Consequently, the corresponding Hamiltonian operator for a conservative system is not an explicit function of time. In this case the Schrödinger equation (3.11) can be integrated to give the following equation, which defines the operation to be carried out on the wavefunction $\Psi(q, t_0)$ at the initial time t_0, to obtain the wavefunction $\Psi(q, t)$ at a later time t

$$\Psi(q,t) = \exp[-\mathrm{i}(t-t_0)\hat{H}/\hbar]\Psi(q,t_0).* \tag{3.14}$$

* One way of dealing with (3.14) is to expand the exponential and evaluate each term.

References on page 92

Suppose we look for solutions to this equation of the form

$$\Psi = \psi(q) \exp(-iEt/\hbar), \tag{3.15}$$

where E is a real constant. Substitution of (3.15) into (3.14) (or (3.11)) leads to the following equation

$$\hat{H}\psi(q) = E\psi(q), \tag{3.16}$$

which is known as the time-independent Schrödinger equation. Since the function $\psi(q)$ must be single-valued, continuous and finite throughout the range of q, the values that the constant E can assume are limited. These values form a spectrum of eigenvalues that can be continuous, discrete, or partly both. We restrict our attention to the case when the spectrum of eigenvalues is entirely discrete, and designate the eigenvalues E_n ($n = 1, 2, \ldots$). We use (3.16) to find the relationship between an eigenvalue and its associated eigenfunction $u_n(q)$ where ($n = 1, 2, \ldots$):

$$\hat{H}u_n(q) = E_n u_n(q). \tag{3.17}$$

If more than one eigenfunction is found to correspond to a particular eigenvalue, such eigenfunctions are said to be degenerate. Eigenfunctions corresponding to different eigenvalues are necessarily orthogonal, and any set of degenerate eigenfunctions can be orthogonalized by taking suitable linear combinations. The eigenfunctions are also normalized, so that, in general, we have

$$\int u_m^*(q)u_n(q)\mathrm{d}q = \delta_{nm}, \tag{3.18}$$

where δ_{nm} is the Kronecker delta.

The eigenfunctions corresponding to any physical observable are assumed to form a complete, orthonormal set, so that we can, for example, expand the wavefunction at a given instant of time in terms of them

$$\Psi(q, t_0) = \sum_m a_m u_m(q), \tag{3.19}$$

where the a_m's are constants, which can be derived using the orthonormal properties of the eigenfunctions

$$a_m = \int u_m^*(q)\Psi(q, t_0)\mathrm{d}q. \tag{3.20}$$

If the expansion (3.19) is substituted into (3.14), we obtain the following expression for the wavefunction of a conservative system as a function of time,

$$\Psi(q, t) = \sum_m a_m u_m(q) \cdot \exp\left[-\mathrm{i}E_m(t-t_0)/\hbar\right]. \tag{3.21}$$

The expansion is similar in form to (3.19), since the Schrödinger equation is a linear equation.

The constants, a_m, can be derived from the wavefunction at a general time, t, by the use of an expression similar to (3.20), namely

$$a_m = \int u_m^*(q) \exp[\mathrm{i}E_m(t-t_0)/\hbar]\Psi(q, t)\mathrm{d}q. \tag{3.22}$$

So far the eigenvalues, E_m, have been introduced as constants that appear in the eigenfunction equation (3.17). We now demonstrate their physical significance by determining the expectation value of the energy of a system described by a wavefunction of the form

$$\Psi(q, t) = u_m(q) \exp(-\mathrm{i}E_m t/\hbar). \tag{3.23}$$

By substitution of (3.23) into (3.5), and using the Hamiltonian operator, we see that

$$\langle E \rangle = \int \Psi^*(q, t)\hat{H}\Psi(q, t)\mathrm{d}q = E_m. \tag{3.24}$$

The eigenvalues, E_m, therefore, specify the energy of the system, in that a system in a state described by the eigenfunction $u_m(q)$ has a mean energy given by the associated eigenvalue E_m. So far we have only determined the mean (or expectation) energy. Suppose we now make an estimate of the uncertainty in the energy of the system described by the wavefunction (3.23). This is characterized by the mean square deviation from the average given by

$$\langle(\Delta E)^2\rangle = \langle(E - \langle E \rangle)^2\rangle. \tag{3.25}$$

Using (3.5) to calculate this, then on substituting from (3.24) we obtain

$$\begin{aligned} \langle(\Delta E)^2\rangle &= \int u_m^*(q)\hat{H}^2 u_m(q)\mathrm{d}q - 2\langle E \rangle \int u_m^*(q)\hat{H}u_m(q)\mathrm{d}q + (\langle E \rangle)^2; \\ \langle(\Delta E)^2\rangle &= E_m^2 - 2E_m^2 + E_m^2 = 0. \end{aligned} \tag{3.26}$$

We have, therefore, demonstrated that for a system whose wavefunction is an eigenfunction of the Hamiltonian operator, the energy is precisely defined and is the eigenvalue associated with that eigenfunction. It can be seen from the wavefunction (3.23) for the system that the associated probability density is time-independent, because

$$W(q) = u_m^*(q)u_m(q), \tag{3.27}$$

and consequently the distributions of the coordinates of the system are independent of time. It can be shown that the same also applies to the momenta of the system, and it therefore follows that the result of a measurement of position or momentum carried out on the system, is time-independent. Such a system is referred to as being in a stationary state. A system in a stationary state has a precisely defined energy.

We have seen that, in general, the wavefunction of a conservative system can be described by the superposition of eigenfunctions of the Hamiltonian of the form given by (3.21). Suppose we consider the wavefunction of a system to be the superposition of only two eigenfunctions $u_a(q)$ and $u_b(q)$, then we have

$$\Psi(q, t) = au_a(q) \exp(-\mathrm{i}E_a t/\hbar) + bu_b(q) \exp(-\mathrm{i}E_b t/\hbar). \tag{3.28}$$

If we use (3.24) to calculate the mean value of the energy of the system, we find that

$$\langle E \rangle = aa^* E_a + bb^* E_b. \tag{3.29}$$

The mean energy of a conservative system is therefore time-independent, and is given by the weighted mean of the energies of the eigenstates. However, if we calculate the mean square deviation in the energy in a similar manner to (3.26), we obtain

$$\langle(\Delta E)^2\rangle = aa^* bb^* (E_a - E_b)^2, \tag{3.30}$$

showing that in this case the energy is no longer precisely defined. Also in this case the probability density for the system is no longer independent of time, as can readily be seen from the equation

$$W(q, t) = aa^* |u_a(q)|^2 + bb^* |u_b(q)|^2 + \\ + 2\,\mathscr{R}\,\{a^* b u_a^*(q) u_b(q) \exp[\mathrm{i}(E_a - E_b)t/\hbar]\}. \tag{3.31}$$

Examination of (3.31) shows that the probability density changes significantly over the time interval

$$\tau \sim \hbar/(E_a - E_b). \tag{3.32}$$

If measurements are carried out in a time interval short compared to this characteristic time, then the mean values of the measured quantity will be very nearly identical, whereas over time intervals comparable to the characteristic time, the mean values will differ significantly from one another. Using (3.30) and (3.32) we can relate the uncertainty in an energy measurement, $\langle(\Delta E)^2\rangle^{\frac{1}{2}}$ carried out on the system, to the time interval, τ, over which the properties of the system change considerably (the time-energy uncertainty relation).

Up to now we have considered the expansion of the wavefunction only in terms of the eigenfunctions of the Hamiltonian operator of the system (i.e. the energy eigenfunctions). Because of the unique role played by this operator in the equation of motion of the system (3.11), such an expansion is obviously important. However, the wavefunction can also be expanded in terms of the eigenfunctions associated with any observable of the system, since the eigenfunctions of an operator corresponding to any physical observable are assumed to form a complete set. We consider the case when the operator, $\hat{F}$, is not explicitly dependent on time, and suppose that its eigenfunctions form the complete set given by

$$\hat{F}(q)v_m(q) = F_m v_m(q), \qquad m = 1, 2, \ldots, \tag{3.33}$$

where F_m is the eigenvalue associated with the eigenfunction $v_m(q)$, and is such as to make this eigenfunction single-valued, finite, etc. The wavefunction can then be expressed in terms of these eigenfunctions as

$$\Psi(q, t) = \sum_m b_m(t)v_m(q). \tag{3.34}$$

Substitution of (3.34) into the expression for the mean value of an observable (3.5), then gives the mean value in terms of the eigenvalues of the observable

$$\langle F\rangle = \sum_m |b_m(t)|^2 F_m. \tag{3.35}$$

The coefficients, $b_m(t)$, involved in the expansion of the wavefunction can readily be calculated from a knowledge of the wavefunction by using the orthonormal properties of the eigenfunctions

$$b_m(t) = \int \Psi^*(q, t)v_m(q)\mathrm{d}q. \tag{3.36}$$

We have assumed in deriving (3.35) that the observable is not an explicit function of time, so that the eigenfunctions (and eigenvalues) defined by (3.33) are the same at all times. In this case, the expansion of the wavefunction defined by (3.34) also applies at all times, and involves coefficients, $b_m(t)$, which are functions only of time, and eigenfunctions, $v_m(q)$, which are functions only of position. If, on the other hand, the observable is an explicit

References on page 92

function of time, then the expansion (3.34) must be carried out at some prescribed time using the eigenfunctions as defined by (3.33) for that time. The expression for the mean value of the observable (3.35) only applies at the prescribed time, and its time variation is now more complicated in the sense that both the coefficients and the eigenvalues are functions of time.

Since the wavefunction can be expanded in terms of any complete set of orthonormal eigenfunctions, we now need to consider the manner in which we change from one of these sets to another. To do this, let us consider the expansion of a wavefunction in terms of two, complete sets of orthonormal eigenfunctions of the form

$$\Psi = \sum_n a_n u_n = \sum_p b_p v_p. \tag{3.37}$$

Since the eigenfunctions form complete sets, then any function of the coordinates can be represented in terms of them, and in particular we can expand the eigenfunctions of one set in terms of the eigenfunctions of the other set, thus

$$v_r = \sum_l u_l s_{lr}, \tag{3.38}$$

and therefore

$$u_l = \sum_r v_r s_{rl}^{-1}. \tag{3.39}$$

These equations define the transformation matrix for the two sets of eigenfunctions, in terms of its elements s_{lr}. Now from (3.38) and the definition (3.36), we have, using the orthonormal properties of the eigenfunctions

$$b_p = \sum_n a_n u_n v_p^* = \sum_n a_n \sum_r v_r s_{rl}^{-1} v_p^*, \tag{3.40}$$

and therefore, we have

$$b_p = \sum_n s_{rn}^{-1} a_n. \tag{3.41}$$

If we multiply throughout (3.37) by u_l^* and integrate over all q, then we get

$$s_{lr} = \int u_l^* v_r \mathrm{d}q, \tag{3.42}$$

and similarly if we multiply throughout the complex conjugate of (3.38) by v_r and integrate over all q, we obtain

$$(s_{rl}^{-1})^* = \int u_l^* v_r \mathrm{d}q \tag{3.43}$$

from which it follows that the transformation matrix has the following property

$$s_{lr}^* = s_{rl}^{-1}. \tag{3.44}$$

Such a matrix is said to be unitary, and the transformations effected by it are said to be unitary transformations.

This matrix can be used to transform the matrix elements of an operator in one representation

$$F_{nm}^{(1)} = \int u_n^* \hat{F} u_m \mathrm{d}q, \tag{3.45}$$

to those in another representation

$$F_{pq}^{(2)} = \int v_p^* \hat{F} v_q \mathrm{d}q, \tag{3.46}$$

since, we have

$$F_{pq}^{(2)} = \int \sum_l u_l^* s_{lp}^* \hat{F} \sum_r u_r s_{rq} \mathrm{d}q,$$

$$F_{pq}^{(2)} = \sum_{l,r} s_{pl}^{-1} F_{lr}^{(1)} s_{rq}, \tag{3.47}$$

which in matrix form can be written

$$\mathbf{F}^{(2)} = \mathbf{S}^{-1}\mathbf{F}^{(1)}\mathbf{S}. \tag{3.48}$$

The rate of change of the mean value of an observable of the system can be derived from (3.5) using the time-dependent Schrödinger equation (3.11). For the case when the observable is not an explicit function of time, then differentiation of (3.5) gives

$$\frac{\partial}{\partial t}\langle F\rangle = \int \left(\frac{\partial \Psi^*}{\partial t}\hat{F}\Psi + \Psi^* \hat{F}\frac{\partial \Psi}{\partial t}\right) \mathrm{d}q. \tag{3.49}$$

On substituting from (3.11) and using the properties of Hermitian operators, we obtain

$$\frac{\partial}{\partial t}\langle F\rangle = \frac{\mathrm{i}}{\hbar}\int \Psi^*(\hat{H}\hat{F} - \hat{F}\hat{H})\Psi \mathrm{d}q. \tag{3.50}$$

This can be written in the form

$$\frac{\partial}{\partial t}\langle F\rangle = \frac{\mathrm{i}}{\hbar}\int \Psi^*[\hat{H}, \hat{F}]\Psi \mathrm{d}q, \tag{3.51}$$

where the brackets [,] define the commutator of $\hat{H}$ and $\hat{F}$ thus,

$$[\hat{H}, \hat{F}] = \hat{H}\hat{F} - \hat{F}\hat{H}. \tag{3.52}$$

References on page 92

Expressions similar to (3.50) will be useful when we come to consider the density matrix representation later on (§ 4.1).

In our discussion so far we have shown how a quantum-mechanical system is completely described by a wavefunction whose time development is given by the time-dependent Schrödinger equation (3.11). We have further shown how the mean value, at a particular time, of some observable of the system can be deduced from a knowledge of the wavefunction at that time (3.5). We must now examine more closely the consequences of carrying out a measurement on some observable of the system, both as regards the value obtained in the measurement process itself, and the subsequent dynamical state of the system. Before doing this, however, we need to put forward the idea of an ideal experiment as being one in which it is possible to make allowance for all the calculable perturbations produced on the system by the measuring technique itself, so that in the limiting case the only perturbations left on the system are the uncontrollable (and causally unpredictable) quantum ones.

In the case of an ideal experiment being carried out on the system, such that the measurement is described completely by the observable, F, then the only values that such a measurement can yield are the eigenvalues associated with the corresponding operator $\hat{F}$, and which are given by (3.33). In any such measurement on a single system, the result will always be one of the eigenvalues, but if the measurement is carried out on a large number of identically prepared systems, then, although in each case the result will always be one or another of the eigenvalues, the mean value of all the results will correspond to that given by (3.5). From (3.35) which gives this mean value in terms of the eigenvalues, it is apparent that in a single measurement, the probability of the result being one particular eigenvalue is just the square of the modulus of the coefficient associated with the corresponding eigenfunction in the expansion of the wavefunction in terms of these eigenfunctions (3.34).

Immediately after the measurement has been carried out, the wavefunction of the system is the eigenfunction corresponding to the particular eigenvalue that the measurement has yielded. The subsequent development of the wavefunction is then described by the time-dependent Schrödinger equation as before. The measurement can therefore be looked upon as a "filtering" process in the sense that if the measurement gives the eigenvalue, F_m, the only part of the wavefuncion to "pass through" is that corresponding to the associated eigenfunction, $v_m(q)$.

From our discussion on the consequences of performing a measurement on a quantum mechanical system, it can be seen that if two observables share a common set of eigenfunctions, then they can both be measured simultaneously with absolute certainty. After the measurement of one of the observables, the system is in the state described by the eigenfunction corresponding to the eigenvalue obtained in the measurement. Since this is also an eigenfunction of the other observable, the outcome of its measurement is now predictable with certainty, being the eigenvalue of this observable corresponding to the eigenfunction. The state of the system is also unperturbed by the latter measurement in so far as it is still describable by the same eigenfunction, so that if the former measurement were repeated it would yield the same result as it did previously.

The above is only generally true in so far as the measurements described are all carried out at the same time, or one immediately after the other. This is because the evolution of the system during those periods when measurements are not being performed, is described by the time-dependent Schrödinger equation (3.11). Although immediately after a measurement, the system is describable by an eigenfunction of the observable, the wavefunction develops according to (3.11) i.e. as a mixture of eigenfunctions of the Hamiltonian operator. It is only when an observable has the same eigenfunctions as the Hamiltonian operator for the system, that the result of a measurement of that observable becomes independent of time.

It can be shown that a necessary and sufficient condition for two operators to have the same eigenfunctions is that they commute, i.e.,

$$\hat{A}(\hat{B}\Psi) = \hat{B}(\hat{A}\Psi). \tag{3.53}$$

When this is the case, the physical observables corresponding to the operators are simultaneously and precisely measurable. Examination of the expression for the rate of change of the mean value of an observable with time (3.50) shows that when the operator corresponding to the observable commutes with the Hamiltonian operator, the rate of change of the mean value is zero. This is, of course, exactly what we would expect from our discussion in the previous paragraph.

3. Interactions between atoms and radiation

3.1 *Two level atomic system*

We now wish to consider in detail the interaction of an atomic system with a radiation field. For the rest of this chapter we shall only be concerned

References on page 92

with the influence that the radiation field has on the atomic system, and shall leave until later a consideration of the way in which the radiation field is itself modified by the interaction. For simplicity we assume that the atomic system under investigation has only two energy levels, and that it is between these that the radiation field causes transitions.

If the two level system, in the absence of the radiation field, is a conservative system, then the energy eigenvalues and eigenfunctions are described by the time-independent Schrödinger equation

$$\hat{H}(q)u(q) = Eu(q),$$

where $\hat{H}(q)$ is the time-independent Hamiltonian operator for the isolated system. If the eigenvalues for the two energy levels are E_a and E_b with associated eigenfunctions $u_a(q)$ and $u_b(q)$ (we assume that the system is non-degenerate), then the wavefunction for the isolated system is given by analogy with (3.15) as

$$\Psi = au_a(q) \exp(-\mathrm{i}E_a t/\hbar) + bu_b(q) \exp(-\mathrm{i}E_b t/\hbar), \tag{3.54}$$

where the constants a and b are time-independent, being determined by the initial conditions imposed on the wavefunction (for example, by a measurement process). We have already discussed the behaviour of a system described by a wavefunction of the form of (3.54), and, in particular, have shown that the mean value of the energy of such a system is independent of time (3.29).

Although we are only considering a two level system we wish to take into consideration, in a phenomenological way, transitions that can take place from these two levels to other levels which we do not wish to consider explicitly. Such transitions might be purely radiative or brought about by collisions, but in both cases their effects are to set finite lifetimes to the two levels considered explicitly. Obviously, when we come to apply our conclusions to a treatment of the laser, such effects are of significance both for the upper laser level, where they compete with stimulated emission as a relaxation mechanism, and for the lower laser level where they help to maintain an inverted population in the presence of the populating action of the radiation field. Also they play an important role in determining the homogeneous linewidth of the transition.

Following a technique due to WEISSKOPF and WIGNER [1930], the decay mechanisms are allowed for by the introduction of damping constants, γ_a and γ_b, into the wavefunction. For a two level system this is now assumed to take the form

$$\Psi = au_a(q) \exp [(-\mathrm{i}E_a t/\hbar)-(\gamma_a t/2)]+ \\ +bu_b(q) \exp [(-\mathrm{i}E_b t/\hbar)-(\gamma_b t/2)]. \tag{3.55}$$

Such a wavefunction is a solution of a time-dependent Schrödinger equation of the form

$$\{\hat{H}-(\mathrm{i}\hbar\hat{\gamma}/2)\}\Psi+\frac{\hbar}{\mathrm{i}}\frac{\partial\Psi}{\partial t} = 0, \tag{3.56}$$

where the operator $\hat{\gamma}$ is defined by

$$\hat{\gamma}u_a(q) = \gamma_a u_a(q) \quad \text{and} \quad \hat{\gamma}u_b(q) = \gamma_b u_b(q). \tag{3.57}$$

We can see the meaning of each damping constant individually by calculating the probability, as a function of time, of the system being in one particular energy level. For level a, say, this is given by (3.36) as

$$\left\{\int u_a^*(q)\Psi(q, t)\mathrm{d}q\right\}^2 = aa^* \exp(-\gamma_a t). \tag{3.58}$$

In other words, we are assuming that the level decays exponentially with a lifetime given by the inverse of γ_a. We shall show in § 3.3 that the assumption of exponential damping is a valid one when the transitions from the levels are brought about by broad-band radiation, or are occurring spontaneously. For the case when the transitions between the two levels are brought about by a monochromatic radiation field, this is no longer true, and we must treat the interaction by time-dependent perturbation theory.

This we now do by considering the influence of some external time-dependent perturbation on the system, when the time-dependent Schrödinger equation becomes

$$(\hat{H}_0+\hat{H}_1)\Psi+\frac{\hbar}{\mathrm{i}}\frac{\partial\Psi}{\partial t} = 0, \tag{3.59}$$

where $\hat{H}_0$ is the Hamiltonian for the unperturbed system (but including damping) and $\hat{H}_1$ is the Hamiltonian describing the perturbation.

Following the methods of perturbation theory, we assume that the wavefunction of the perturbed system is a linear combination of the unperturbed eigenfuntions of the isolated system

$$\Psi = a(t)u_a(q)+b(t)u_b(q). \tag{3.60}$$

The constants are now allowed to be explicit functions of time and represent the time development of the system under the perturbation. (For convenience we have also included in them the harmonic and exponential time factors written out explicitly in (3.55). The point is that the constants a and b in (3.55) are now allowed to be time-dependent as well.) The above wavefunction is substituted into the time-dependent Schrödinger equation (3.11). By multiplying throughout the resulting equation by u_a^* and integrating over the whole range of the coordinates (q), we can use the orthonormal properties of the eigenfunctions (3.18), to obtain the following equation for $a(t)$,

$$\mathrm{i}\frac{\mathrm{d}}{\mathrm{d}t}a(t) = \frac{b}{\hbar}\int u_a^* \hat{H}_1 u_b \mathrm{d}q + \frac{a}{\hbar}\int u_a^* \hat{H}_1 u_a \mathrm{d}q + \frac{aE_a}{\hbar} - \frac{\mathrm{i}\gamma_a}{2}a, \tag{3.61}$$

and similarly for the coefficient $b(t)$, we obtain

$$\mathrm{i}\frac{\mathrm{d}}{\mathrm{d}t}b(t) = \frac{a}{\hbar}\int u_b^* \hat{H}_1 u_a \mathrm{d}q + \frac{b}{\hbar}\int u_b^* \hat{H}_1 u_b \mathrm{d}q + \frac{bE_b}{\hbar} - \frac{\mathrm{i}\gamma_b}{2}b. \tag{3.62}$$

3.2 *Electromagnetic transitions in a two level system*

We now consider the form of the perturbation terms in (3.61), when $\hat{H}_1$ is due to an electromagnetic field. In appendix B it is shown that the Hamiltonian for a charged particle in an electromagnetic field is given by (c.g.s. units)

$$H = \frac{1}{2m}\left(\boldsymbol{p} + \frac{e}{c}\boldsymbol{A}\right)^2 - e\phi, \tag{3.63}$$

where ϕ and $\boldsymbol{A}$ are the scalar and vector potentials associated with the applied field. We can choose the gauge of ϕ and $\boldsymbol{A}$ such that

$$\phi = 0, \qquad \nabla \cdot \boldsymbol{A} = 0, \tag{3.64}$$

and in this case, on expansion of (3.63), we obtain

$$H = \frac{p^2}{2m} + \frac{e}{2mc}(\boldsymbol{p}\cdot\boldsymbol{A} + \boldsymbol{A}\cdot\boldsymbol{p}) + \frac{e^2}{2mc^2}A^2. \tag{3.65}$$

The last term in the above is much smaller than the others, and will be neglected, so that the Hamiltonian describing the perturbation becomes

$$H_1 = \frac{e}{2mc}(\boldsymbol{p}\cdot\boldsymbol{A} + \boldsymbol{A}\cdot\boldsymbol{p}). \tag{3.66}$$

We shall now restrict ourselves to the case when the two level system we are considering arises in what is effectively a single electron atom, so that there is only one particle interacting with the electromagnetic field. This makes the subsequent workings less complex, but the results can readily be extended to the many electron case. We must now convert the Hamiltonian (3.66) into an operator by the use of the transformations discussed in § 2. Consider the perturbation term $(\boldsymbol{p} \cdot \boldsymbol{A})$ operating on one of the eigenfunctions, u_a, say, of the unperturbed system; then using the transformations we have

$$\hat{\boldsymbol{p}} \cdot (\hat{\boldsymbol{A}} u_a) = \frac{\hbar}{\mathrm{i}} \nabla \cdot (\boldsymbol{A} u_a) = \frac{\hbar}{\mathrm{i}} (\nabla \cdot \boldsymbol{A}) u_a + \frac{\hbar}{\mathrm{i}} \boldsymbol{A} \cdot (\nabla u_a), \tag{3.67}$$

and since $\nabla \cdot \boldsymbol{A} = 0$, it follows that

$$\hat{\boldsymbol{p}} \cdot \hat{\boldsymbol{A}} = \hat{\boldsymbol{A}} \cdot \hat{\boldsymbol{p}},$$

showing that the operators $\hat{\boldsymbol{p}}$ and $\hat{\boldsymbol{A}}$ commute, so that the perturbation operator can be written

$$\hat{H}_1 = \frac{\mathrm{i}e\hbar}{mc} \boldsymbol{A} \cdot \nabla. \tag{3.68}$$

We therefore have that

$$\int u_b^* \hat{H}_1 u_a \mathrm{d}q = \frac{\mathrm{i}e\hbar}{mc} \int u_b^* (\boldsymbol{A} \cdot \nabla) u_b \mathrm{d}q. \tag{3.69}$$

Optical wavelengths are considerably larger than typical atomic dimensions so we can assume that $\boldsymbol{A}$ is constant over the volume of the atomic system under consideration. This is an approximation which retains information about absorption due to the electric dipole moment but which neglects the very much smaller absorption due to the magnetic dipole moment and the electric quadrupole moment. In this approximation (3.69) becomes

$$\int u_b^* \hat{H}_1 u_a \mathrm{d}q = \frac{\mathrm{i}e\hbar}{mc} \boldsymbol{A} \int u_b^* \nabla u_a \mathrm{d}q. \tag{3.70}$$

We now require to show that

$$-\frac{\hbar^2}{m} \int u_b^* \nabla u_a \mathrm{d}q = (E_b - E_a) \int u_b^* \boldsymbol{r} u_a \mathrm{d}q. \tag{3.71}$$

In order to do this, consider the commutator

References on page 92

$$\int u_b^*[\hat{H}, \boldsymbol{r}]u_a \mathrm{d}q = \int u_b^*\{\hat{H}\boldsymbol{r} - \boldsymbol{r}\,\hat{H}\}u_a \mathrm{d}q =$$

$$= \int \boldsymbol{r}u_a(\hat{H}u_b)^* \mathrm{d}q - E_a \int u_b^* \boldsymbol{r} u_a \mathrm{d}q =$$

$$= (E_b - E_a) \int u_b^* \boldsymbol{r} u_a \mathrm{d}q. \tag{3.72}$$

Now the Hamiltonian of the unperturbed system is

$$\hat{H} = -\frac{\hbar^2}{2m}\nabla^2 + V,$$

where V is the potential function describing the interactions within the atom itself. If the coordinate operator, $\boldsymbol{r}$, commutes with V, then

$$\int u_b^* \left[-\frac{\hbar^2}{2m}\nabla^2, \boldsymbol{r} \right] u_a \mathrm{d}q = (E_b - E_a) \int u_b^* \boldsymbol{r} u_a \mathrm{d}q,$$

and therefore

$$-\frac{\hbar^2}{2m} \int u_b^* \{\nabla^2(\boldsymbol{r}u_a) - \boldsymbol{r}\nabla^2 u_a\} \mathrm{d}q = (E_b - E_a) \int u_b^* \boldsymbol{r} u_a \mathrm{d}q. \tag{3.73}$$

Making use of the well-known identity,

$$\nabla^2(\boldsymbol{r}u_a) = \boldsymbol{r}\nabla^2 u_b + 2\nabla u_b,$$

we readily obtain the required expression (3.71) from (3.73). Our final expression for the perturbation term is therefore

$$\int u_b^* \hat{H}_1 u_a \mathrm{d}q = \frac{\mathrm{i}e}{c\hbar}(E_a - E_b)\boldsymbol{A} \cdot \int u_b^* \boldsymbol{r} u_a \mathrm{d}q. \tag{3.74}$$

In equations (3.61) and (3.62) we must also consider the terms of the form

$$\int u_a^* \hat{H}_1 u_a \mathrm{d}q.$$

From simple perturbation theory, it is apparent that such terms represent shifts in the energy levels as a consequence of any permanent moment possessed by the atomic system when in the particular level concerned. If we make the dipole approximation as above, then these terms will describe shifts due to any permanent dipole moment. We shall assume from now on, that for the particular system considered, such terms are negligible, and so they will not appear in subsequent equations.

3.3 *Electromagnetic transitions (broad-band radiation)*

We begin by considering the case when the perturbing radiation field is a monochromatic plane wave

$$\boldsymbol{A} = \boldsymbol{A}_0 \cos(\omega t - \boldsymbol{k} \cdot \boldsymbol{r}). \tag{3.75}$$

Since we are not concerned about the direction of the radiation, and since its wavelength is much greater than the dimensions of the atomic system, the term $\boldsymbol{k} \cdot \boldsymbol{r}$ can be omitted, so that (3.75) becomes

$$\boldsymbol{A} = \boldsymbol{A}_0 \cos 2\pi\nu t = \boldsymbol{A}_0(\mathrm{e}^{2\pi \mathrm{i}\nu t} + \mathrm{e}^{-2\pi \mathrm{i}\nu t})/2. \tag{3.76}$$

We neglect for the present the phenomenological damping terms in (3.61) and (3.62), and assume that when the radiation field is "switched on" at $t = 0$, the atomic system is in the level a, so that

$$|a(t=0)|^2 = 1, \qquad |b(t=0)|^2 = 0. \tag{3.77}$$

If we further assume that the interaction between the radiation field and the atomic system is weak, so that the population of level b increases only slightly while the population of level remains effectively constant, (3.61) can be written

$$\mathrm{i}\frac{\mathrm{d}}{\mathrm{d}t}a(t) = \frac{aE_a}{\hbar}, \tag{3.78}$$

so that

$$a(t) = a(t=0)\exp(-\mathrm{i}E_a t/\hbar) = \exp(-\mathrm{i}E_a t/\hbar). \tag{3.79}$$

Substituting this value into (3.62), we obtain

$$\mathrm{i}\dot{b} = \frac{\mathrm{i}(E_a - E_b)}{2\hbar^2 c}\{\exp(-\mathrm{i}E_a t/\hbar)\}(\boldsymbol{A}_0 \cdot \boldsymbol{D}_{ba}) \times \times \{\exp(2\pi \mathrm{i}\nu t) + \exp(-2\pi \mathrm{i}\nu t)\} + bE_b/\hbar, \tag{3.80}$$

where

$$\boldsymbol{D}_{ba} = \int u_b^*(e\boldsymbol{r})u_a \mathrm{d}q. \tag{3.81}$$

In accordance with (3.55), we can write

$$b(t) = b_0(t)\exp(-\mathrm{i}E_b t/\hbar), \tag{3.82}$$

so that (3.80) becomes

$$\dot{b}_0(t) = \frac{2\pi\nu_{ab}}{2\hbar c}(\boldsymbol{A}_0 \cdot \boldsymbol{D}_{ba})\{\mathrm{e}^{2\pi \mathrm{i}(\nu - \nu_{ab})t} + \mathrm{e}^{-2\pi \mathrm{i}(\nu + \nu_{ab})t}\}, \tag{3.83}$$

where

$$\nu_{ab} = (E_a - E_b)/h. \tag{3.84}$$

Integrating (3.83) from $t = 0$ to $t = t$, we obtain

$$b_0(t) = -\frac{i\nu_{ba}}{2c\hbar}(\boldsymbol{A}_0 \cdot \boldsymbol{D}_{ba})\left[\frac{\exp\{2\pi i(\nu_{ba}+\nu)t\}-1}{(\nu_{ba}+\nu)} + \frac{\exp\{2\pi i(\nu_{ba}-\nu)t\}-1}{(\nu_{ba}-\nu)}\right]. \tag{3.85}$$

If $E_b > E_a$, ν_{ba} is positive, and when the frequency of the radiation, ν, lies close to $|\nu_{ba}|$, the second term in (3.85) becomes dominant. The radiation causes a transition from a to b, during which, energy is absorbed by the atomic system from the field. If, on the other hand, $E_a > E_b$, ν_{ba} is negative, and when ν lies close to $|\nu_{ba}|$, the first term becomes dominant. In this case the transition is still from a to b, but is accompanied by the emission of energy into the field from the atomic system (stimulated emission).

The probability that a transition has occurred is proportional to the probability of finding the atomic system in level b, since prior to the application of the radiation field this level was supposed unoccupied. The probability that absorption has occurred (considering only the dominant term) is therefore

$$|b(t)|^2 = \left[\frac{\nu_{ba}^2}{c^2\hbar^2}(\boldsymbol{A}_0 \cdot \boldsymbol{D}_{ba})^2\right]\frac{\sin^2\{\pi(\nu_{ab}+\nu)t\}}{(\nu_{ab}+\nu)^2}. \tag{3.86}$$

This equation refers to one frequency, ν, of the perturbing electromagnetic field, and gives the probability that radiation of this frequency has caused a transition from level a to level b after interacting with

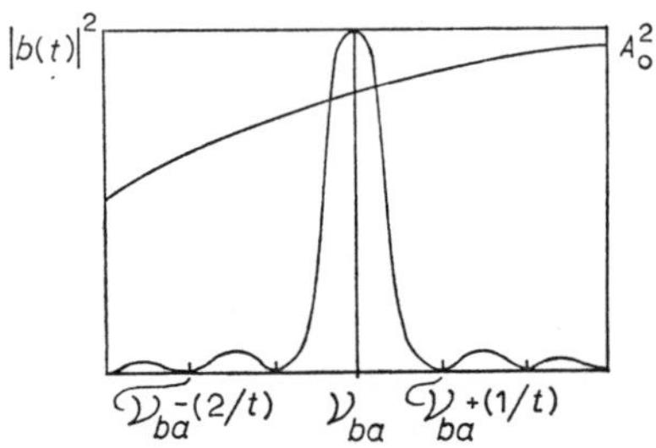

Fig. 3.1. Relationship between the probability and frequency as given by (3.86).

the atomic system for a time interval, t. Figure 3.1 shows $|b(t)|^2$ as a function of ν, on the assumption that the term in square brackets in (3.86) is effectively constant. It may be seen that those frequencies which have the greater probability of causing the transition lie within the range $\{\nu_{ba}-(1/t)\}$ to $\{\nu_{ba}+(1/t)\}$.

In deriving (3.86) we have considered a harmonic perturbation applied to the atomic system for a finite time interval. Such a perturbation is not strictly monochromatic in the sense that it has a finite bandwidth, $\Delta\nu$, because of its finite duration, t. The bandwidth and duration are related by (ch. 10 § 2)

$$\Delta\nu \cdot t \sim 1.$$

In other words, fig. 3.1 demonstrates that the electromagnetic field interacts with the atomic system only in so far as the field's bandwidth encompasses the resonance frequency, ν_{ba}, of the atomic system. If the time interval over which the field interacts with the atomic system is increased, then as may be seen from (3.86) and fig. 3.1, the probability that a transition has occurred increases in a quadratic fashion at the resonance frequency, ν_{ba}, but the frequency interval over which the interaction is significant decreases just, as the effective bandwidth of the radiation decreases. Because of the approximations we have made in deriving (3.86), namely that $|b(t)|^2$ is much less than unity, it is not possible to extend the time interval of the interaction indefinitely so as to decrease the bandwidth indefinitely. (We consider this case more fully in § 3.4 when the equations are solved without making this approximation.)

In practice, the bandwidth of radiation from most thermal sources is very much greater than that determined by the duration of its interaction with the atomic system. In other words, its spectral energy density is effectively constant over the range of frequencies covered by fig. 3.1. Also, the different frequency components may be regarded as being in random phase with respect to each other (ch. 10 § 4). In order to derive the total effect of broad-band radiation (thermal radiation) on the system, therefore, we integrate (3.86) over all frequencies and get

$$|b(t)|^2 = \left[\frac{\nu_{ab}^2}{c^2\hbar^2}(\boldsymbol{A}_0 \cdot \boldsymbol{D}_{ba})^2\right] \pi^2 t^2 \int_{-\infty}^{+\infty} \frac{\sin^2[\pi(\nu_{ab}+\nu)t]}{\pi^2(\nu_{ab}+\nu)^2 t^2}\, \mathrm{d}\nu, \qquad (3.87)$$

where we have assumed that $|\boldsymbol{A}_0|^2$, refers to unit frequency interval of the radiation, and is constant over the frequency range of significance to the integral. To evaluate this integral, we substitute $\theta = \pi(\nu_{ab}+\nu)t$ and $\mathrm{d}\nu = \mathrm{d}\theta/\pi t$ into (3.87) which then gives

$$|b(t)|^2 = \left[\frac{\nu_{ba}^2}{c^2\hbar^2}(\boldsymbol{A}_0 \cdot \boldsymbol{D}_{ba})^2\right] \pi t \int_{-\infty}^{+\infty} \frac{\sin^2\theta}{\theta^2}\, \mathrm{d}\theta.$$

References on page **92**

The value of this last integral is π, so we therefore have

$$|b(t)|^2 = \left[\frac{\nu_{ba}^2}{c^2\hbar^2}(\boldsymbol{A}_0 \cdot \boldsymbol{D}_{ba})^2\right]\pi^2 t. \tag{3.88}$$

An important feature of (3.88) is that the probability that a transition has taken place, is proportional to the time that has elapsed, and it is therefore possible to associate a time-independent transition probability with absorption brought about by weak, broad-band radiation. The value of $(\boldsymbol{A}_0 \cdot \boldsymbol{D}_{ba})^2$ is $(A_0 D_{ba} \cos\theta)^2$ where θ is the angle between the vectors $\boldsymbol{A}_0$ and $\boldsymbol{D}_{ba}$, and A_0 and D_{ba} are the magnitudes of the vectors. If the radiation field is not polarized, the direction of $\boldsymbol{A}_0$ varies randomly, and we can therefore average over θ to obtain $(A_0 D_{ba})^2/3$. Further, we can relate the magnitude of the vector potential, A_0, to the spectral energy density $\rho(\nu)$, by using (C.2), which is

$$A_0^2 = \frac{2c^2}{\pi\nu_{ba}^2}\rho(\nu_{ba}).$$

On making the above substitutions in (3.88), we finally obtain

$$|b(t)|^2 = \frac{2\pi}{3\hbar^2}D_{ba}^2\rho(\nu_{ba})t$$

which enables us to define the following transition probability per unit time interval, per unit spectral energy density,

$$B_{ab} = (2\pi/3\hbar^2)D_{ba}^2. \tag{3.89}$$

3.4 *Electromagnetic transitions with damping* (*weak signal approximation*)

We now consider solutions to (3.61) and (3.62) when the damping terms are retained, but we still assume that the interaction of the radiation with the atom is weak. The approach will differ somewhat from that in § 3.3, but will give some insight into the techniques used in Lamb's theory of the laser (discussed in ch. 9) to describe the interaction between the radiation field in the cavity and the active medium. In particular, we are now concerned with the influence of a two-level atom on a monochromatic radiation field.

As a consequence of the transitions that the radiation field induces in the atoms, it is itself modified by the interaction, either being amplified (by stimulated emission) or attentuated (by absorption). The dispersion relations, discussed in ch. 2, § 4.10, imply that the field undergoes a phase shift as well.

In order to describe the changes produced in the field, we associate with the atoms dipole moments which are induced by the field, and the resultant field is given by the sum of the dipole fields together with the original field. We restrict our attention to the scalar case where the radiation is polarized and the dipole moment is represented by a scalar quantity. The dipole moment associated with a single (two level) atom is given by

$$P = e\langle r\rangle, \tag{3.90}$$

where

$$\langle r\rangle = \int \Psi^* r \Psi \, \mathrm{d}q. \tag{3.91}$$

This latter expression is expanded in terms of the eigenfunctions of the two level system, and if it is assumed that there is no permanent dipole moment associated with the unperturbed states, we obtain

$$P = e\left(a^* b \int u_a^* r u_b \, \mathrm{d}q + a b^* \int u_b^* r u_a \, \mathrm{d}q\right). \tag{3.92}$$

We can choose the phases of u_a and u_b such that the integrals in the above are real, by taking the appropriate phase factors into the coefficients a and b. We therefore have that

$$P = D_{ab}(a^* b + a b^*), \tag{3.93}$$

where

$$D_{ab} = \int u_a^* e r u_b \, \mathrm{d}q = D_{ba}. \tag{3.94}$$

If the electric field describing the radiation field is

$$E = E_0 \cos \omega t, \tag{3.95}$$

then the associated vector potential is

$$A = \left(\frac{cE_0}{\omega}\right) \sin \omega t. \tag{3.96}$$

From (3.74), (3.95) and (3.96) we therefore have that

$$\int u_a^* \hat{H}_1 u_b \mathrm{d}q = -\mathrm{i} \cdot \frac{\omega_{ab}}{\omega} \cdot E_0 D_{ab} \sin \omega t, \tag{3.97}$$

where

$$\omega_{ab} = (E_a - E_b)/\hbar. \tag{3.98}$$

In the case when the radiation field is close to resonance (3.97) simplifies to

References on page 92

$$\int u_a^* \hat{H}_1 u_b \mathrm{d}q = -\mathrm{i}(E_0 D_{ab}) \sin \omega t. \tag{3.99}$$

The equations (3.61) and (3.62) describing the behaviour of the atom in the presence of the radiation field can therefore be written

$$\begin{aligned} \hbar \mathrm{i} \dot{a} &= -\mathrm{i} b E_0 D_{ab} \sin \omega t + a E_a - \mathrm{i}\hbar\gamma_a a/2; \\ \hbar \mathrm{i} \dot{b} &= +\mathrm{i} a E_0 D_{ab} \sin \omega t + b E_b - \mathrm{i}\hbar\gamma_b b/2. \end{aligned} \tag{3.100}$$

If we try solutions of the form

$$\begin{aligned} a(t) &= a_0(t) \exp(-\mathrm{i}E_a t/\hbar - \gamma_a t/2); \\ b(t) &= b_0(t) \exp(-\mathrm{i}E_b t/\hbar - \gamma_b t/2), \end{aligned} \tag{3.101}$$

then (3.100) reduce to

$$\begin{aligned} \hbar \mathrm{i} \dot{a}_0 &= -\mathrm{i} b_0 E_0 D_{ab} \exp\{\mathrm{i}\omega_{ab} t + (\gamma_a - \gamma_b)t/2\} \sin \omega t, \\ \hbar \mathrm{i} \dot{b}_0 &= +\mathrm{i} a_0 E_0 D_{ab} \exp\{-\mathrm{i}\omega_{ab} t + (\gamma_b - \gamma_a)t/2\} \sin \omega t. \end{aligned} \tag{3.102}$$

The sine in (3.102) is substituted as follows

$$\sin \omega t = (\mathrm{e}^{\mathrm{i}\omega t} - \mathrm{e}^{-\mathrm{i}\omega t})/2\mathrm{i}.$$

On doing this it is found that the right hand sides of (3.102) contain one term which is oscillating at a frequency $|\omega_{ab} - \omega|$ (low frequency, resonance term), and another term oscillating at a frequency $|\omega_{ab} + \omega|$ (high frequency, anti-resonance term). Over any physically significant time the anti-resonance term averages to zero, and we therefore make what is known as the "rotating wave approximation"* and neglect it from this point on. In this approximation (3.102) become

$$\begin{aligned} \dot{a}_0 &= -\mathrm{i} V b_0 \exp\{\mathrm{i}(\omega_{ab} - \omega)t + (\gamma_a - \gamma_b)t/2\}; \\ \dot{b}_0 &= -\mathrm{i} V a_0 \exp\{\mathrm{i}(\omega - \omega_{ab})t + (\gamma_b - \gamma_a)t/2\}, \end{aligned} \tag{3.103}$$

where

$$V = E_0 D_{ab}/2\hbar \qquad (V \text{ is real}). \tag{3.104}$$

We now have expressions describing the time development of the atom, and we must relate these to the polarization described by (3.90). On differentiation of (3.93) we obtain

$$\dot{P} = D_{ab}(\dot{a}^* b + a^* \dot{b} + \dot{b}^* a + b^* \dot{a}). \tag{3.105}$$

If we now substitute from (3.103) and (3.101) for the various terms in (3.105), we obtain after some re-arrangement

* This term comes from the representation of the sine by two counter-rotating vectors of frequency ω, in an Argand diagram.

$$\dot{P} = -\gamma_{ab} P/2 + 2VD_{ab}(aa^* - bb^*) \sin \omega t + i\omega_{ab} \; D_{ab} (a^*b - ab^*), \tag{3.106}$$

where

$$\gamma_{ab} = (\gamma_a + \gamma_b)/2.$$

At this stage we make the weak signal approximation by assuming that the probability of finding the atom in its initial state is independent of time; i.e. we assume that

$$aa^*(t) - bb^*(t) = aa^*(t_0) - bb^*(t_0) = \text{constant}, (\overline{N}). \tag{3.107}$$

With this approximation (3.106) is differentiated again.

After substitution from (3.106) and (3.103) into the expression thus obtained, we obtain the following final expression describing the time development of the polarization

$$\ddot{P} + \gamma_{ab}\dot{P} + (\omega_{ab}^2 + \gamma_{ab}^2/4)P = D_{ab} V\overline{N}\{2(\omega + \omega_{ab}) \cos \omega t + \gamma_{ab} \sin \omega t\}. \tag{3.108}$$

We therefore see that the polarization associated with the atom in the radiation field behave as a driven, damped, simple harmonic oscillator. The parameters describing the simple harmonic oscillator itself (i.e. its damping constant, γ_{ab}, and natural frequency ω_{ab}) are characteristic of the atom, and it is being driven by two components one in phase with the electric field of the radiation, the other in quadrature with the field. We look for steady state solutions of (3.108) of the form

$$P = P_1 \cos \omega t + P_2 \sin \omega t \tag{3.109}$$

(the component P_1 being in phase with the electric field, the component P_2 being in quadrature).

If the following approximations are made

$$(\omega_{ab} + \omega) \simeq 2\omega;$$

$$\gamma_{ab} \ll \omega_{ab},$$

then on substituting (3.109) into (3.108) and separating sine and cosine terms, we obtain

$$P_1 = \text{const.} \frac{D_{ab}^2 \cdot E_0 \cdot \overline{N} \cdot (\omega_{ab} - \omega)}{\{(\omega_{ab} - \omega)^2 + \gamma_{ab}^2\}}; \tag{3.110a}$$

$$P_2 = \text{const.} \frac{D_{ab}^2 \cdot E_0 \cdot \overline{N} \cdot \gamma_{ab}}{\{(\omega_{ab} - \omega)^2 + \gamma_{ab}^2\}}. \tag{3.110b}$$

References on page 92

3.5 *Rabi strong signal theory*

The treatments that were considered in §§ 3.3 and 3.4 were "weak signal" approximations in so far as it was assumed that the probability of finding the atom in the upper level remained constant over the period during which the radiation was interacting with the atomic system. In § 3.3 the approximation took the form of assuming that the time dependence of the coefficient describing the upper level was solely that of an imaginary exponential, see eq. (3.78). The probability of finding the atom in the lower level was then derived, and for the case of broad-band radiation, it was found that a time-independent transition probability could be associated with the transition from the upper to the lower level. In § 3.4, the influence of the atom on a monochromatic radiation field was considered in terms of the induced polarization associated with the atom. Damping was included in this treatment, but the assumption was still made that the probability of finding the atom in the upper level was time-independent, in that (aa^*-bb^*) was put equal to a constant, see eq. (3.107).

In this section we consider the case when such an approximation is no longer valid, either because the radiation field is too intense or the damping constant γ_a is too large. The Rabi strong signal theory is used to determine the time-dependence of the probability of the atom being in either the upper or lower level.

If (3.102b) is differentiated with respect to time, then on re-arrangement, we have

$$\ddot{b}_0+[\mathrm{i}(\omega_{ab}-\omega)+(\gamma_a-\gamma_b)/2]\dot{b}_0+V^2 b_0 = 0. \tag{3.111}$$

We now look for solutions of (3.111) of the form

$$b_0 = \exp[\mathrm{i}(\mu+\mathrm{i}\gamma)t], \tag{3.112}$$

where μ and γ are real parameters to be evaluated.

Substitution of (3.112) into (3.111) and separating real and imaginary parts leads to the following two equations involving these parameters

$$-\mu^2+\gamma^2-(\omega_{ab}-\omega)\mu-\gamma(\gamma_a-\gamma_b)/2+V^2 = 0; \tag{3.113a}$$

$$-2\gamma\mu-(\omega_{ab}-\omega)\gamma+\mu(\gamma_a-\gamma_b)/2 = 0. \tag{3.113b}$$

On re-arrangement, (3.113b) can be used to express γ in terms of μ

$$\gamma = \frac{\mu(\gamma_a-\gamma_b)}{2[2\mu+(\omega_{ab}-\omega)]}. \tag{3.114}$$

Substitution of (3.114) into the (3.113a) leads to the following equation for μ

$$-\mu^2+\frac{\mu^2(\gamma_a-\gamma_b)^2}{4[2\mu+(\omega_{ab}-\omega)]^2}-(\omega_{ab}-\omega)\mu+V^2-\frac{\mu(\gamma_a-\gamma_b)^2}{4[2\mu+(\omega_{ab}-\omega)]} = 0. \quad (3.115)$$

This general equation is quartic in μ, and so we now consider some special cases for which the equation may be solved.

Firstly, we treat the case when γ_a and γ_b are both zero (i.e. there is no phenomenological damping). Under this condition (3.115) becomes

$$\mu^2+(\omega_{ab}-\omega)\mu-V^2 = 0, \quad (3.116)$$

with solutions for μ

$$\left.\begin{matrix}\mu_1\\ \mu_2\end{matrix}\right\} = \frac{1}{2}(\omega-\omega_{ab})\pm[(\omega-\omega_{ab})^2+4V^2]^{\frac{1}{2}}. \quad (3.117)$$

Since γ is also zero, the general solution for $b_0(t)$ is

$$b_0(t) = A\exp(\mathrm{i}\mu_1 t)+B\exp(\mathrm{i}\mu_2 t). \quad (3.118)$$

The constants A and B are determined by the initial conditions. Since we have,

$$b_0(t=0) = 0, \quad (3.119)$$

then from (3.118) it follows that

$$A+B = 0. \quad (3.120)$$

From (3.103b) we have the relationship

$$\dot{b}_0(t=0) = -\mathrm{i}Va_0(t=0). \quad (3.121)$$

Initially the atom is in the upper level, so that the modulus of the coefficient a_0 is unity, and hence

$$a_0(t=0) = \exp(\mathrm{i}\phi_0). \quad (3.122)$$

From (3.121), (3.118) and (3.122) therefore

$$\mu_1 A+\mu_2 B = -\mathrm{i}V\exp(\mathrm{i}\phi_0). \quad (3.123)$$

From the boundary conditions (3.122) and (3.123), we obtain the following values for the coefficients A and B in (3.118)

$$\begin{aligned} A &= -V\mathrm{e}^{\mathrm{i}\phi_0}/(\mu_1-\mu_2);\\ B &= +V\mathrm{e}^{\mathrm{i}\phi_0}/(\mu_1-\mu_2). \end{aligned} \quad (3.124)$$

References on page 92

Substitution of (3.124) into (3.118) leads to the following expression for $b_0(t)$

$$b_0(t) = V e^{i\phi_0}(e^{i\mu_2 t} - e^{i\mu_1 t})/(\mu_1 - \mu_2). \tag{3.125}$$

By taking the square of the modulus of $b_0(t)$ and substituting from (3.117) for the values of μ_1 and μ_2, the following expression is obtained for the probability of the atomic system being in the lower level

$$|b(t)|^2 = |b_0(t)|^2 = \frac{4V^2 \sin^2 [\{(\omega - \omega_{ab})^2 + 4V^2\}^{\frac{1}{2}} t/2]}{\{(\omega - \omega_{ab})^2 + 4V^2\}}. \tag{3.126}$$

The probability is thus a periodic function, with a frequency at resonance ($\omega = \omega_{ab}$) determined by the strength of the interaction between the atomic system and the radiation field (i.e. by $V^2 = E_0^2 D_{ab}^2/4\hbar^2$). Further, the effective bandwidth over which monochromatic radiation of a frequency, ω, interacts with the atomic system of resonance frequency, ω_{ab}, is again determined by the strength of the interaction between the atomic system and the radiation field. The more intense the radiation, the greater is the linewidth over which it interacts with the atomic system. The behaviour of the probability function with time is illustrated in fig. 3.2(a).

We next consider the solutions to (3.113) in the presence of phenomenological damping, but in order to obtain these solutions readily, we restrict our attention to the case of resonance (i.e. $\omega = \omega_{ab}$). Under this restriction (3.113) can be written

$$\gamma^2 - \mu^2 + \gamma(\gamma_b - \gamma_a)/2 + V^2 = 0. \tag{3.127a}$$

For

$$\mu \neq 0, \qquad \gamma = -(\gamma_b - \gamma_a)/4, \tag{3.127b}$$

so that the solution for μ is

$$\left.\begin{matrix}\mu_1\\ \mu_2\end{matrix}\right\} = \pm(V^2 - \gamma^2)^{\frac{1}{2}}, \tag{3.128}$$

where γ is given by (3.127b).

The boundary conditions are evaluated in the same way as previously, and in this case lead to

$$\begin{aligned} A + B &= 0; \\ (i\mu_1 - \gamma)A + (i\mu_2 - \gamma)B &= -iVe^{i\phi_0}, \end{aligned} \tag{3.129}$$

so that the solution for $b_0(t)$ is

$$b_0(t) = V e^{i\phi_0} e^{-\gamma t}(e^{i\mu_2 t} - e^{i\mu_1 t})/(\mu_1 - \mu_2), \tag{3.130}$$

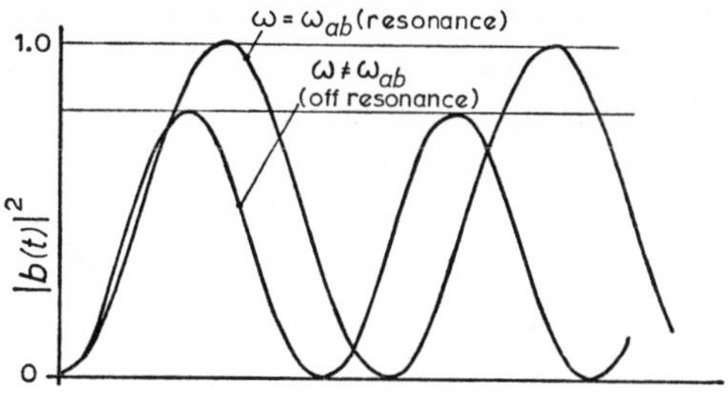

(a) Probability density for lower level as a function of time, without damping.

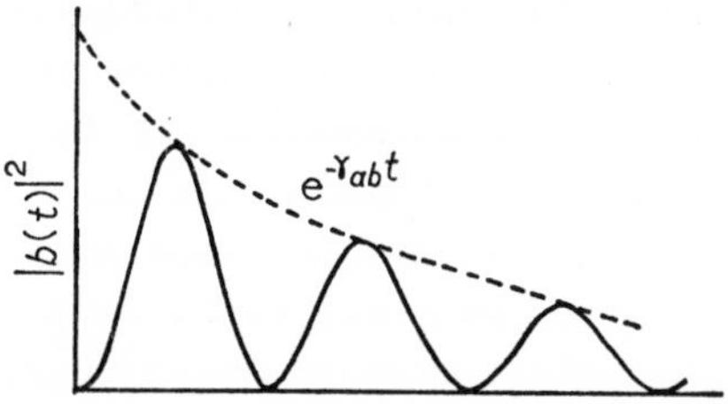

(b) Probability density for lower level as a function of time, with damping. Strong field case ($V^2 > \gamma^2$).

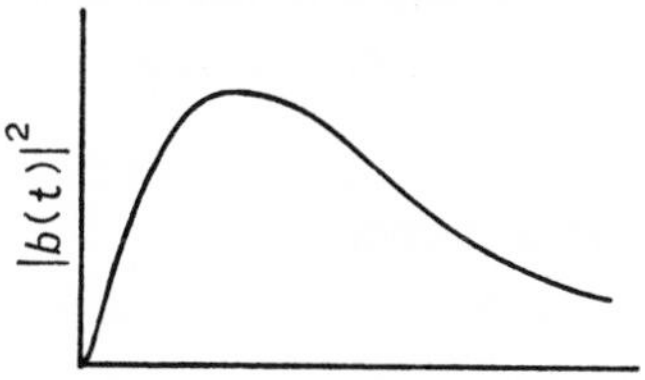

(c) As 3.2(b) but for the weak field case ($V^2 < \gamma^2$).

Fig. 3.2.

where μ_1 and μ_2 are now given by (3.128). Since μ_1 and μ_2 must, by definition, be real, the above solution, (3.130), is only valid for $V^2 > \gamma^2$. On substitution from (3.128), the probability of the atom being in the lower level may be written

$$|b(t)|^2 = |b_0(t)|^2 e^{-\gamma_b t} = \frac{V^2 e^{-\gamma_{ab} t}}{(V^2 - \gamma^2)} \sin^2 \{(V^2 - \gamma^2)^{\frac{1}{2}} t\}, \tag{3.131}$$

where γ is given by (3.127b).

References on page 92

This probability oscillates, as in the previous case, but now the envelope of the oscillation is damped at a rate determined by the damping coefficients associated with the lower and upper levels (fig. 3.2(b)).

We must also look for a solution when $V^2 < \gamma^2$. In fact this solution is obtained by putting $\mu = 0$, a case which we neglected in (3.127b). On putting $\mu = 0$ in (3.127b), and then solving and evaluating the boundary conditions as before, we obtain

$$|b(t)|^2 = \frac{V^2 e^{-\gamma_{ab}t}}{4(\gamma^2 - V^2)} \{e^{2\surd(\gamma^2 - V^2)t} + e^{-2\surd(\gamma^2 - V^2)t} - 2\}, \qquad (3.132)$$

where γ is given by (3.127b) as before. The probability function, under these conditions, after initially increasing is then exponentially damped, and does not now oscillate at all as in the previous cases, fig. 3.2(c).

The solutions that we have obtained by the use of Rabi strong signal theory for the interaction of a two level atom with monochromatic resonance radiation illustrate two important points. Firstly, under those circumstances where the rate of change of the atom due to the influence of the radiation field is large compared to the damping rates associated with the levels (i.e. $V^2 > \gamma^2$), the atom oscillates between being in one level and the other (i.e. pulsation of population between the two levels). Secondly, the interaction of the atom with the radiation broadens the linewidth over which this interaction can take place (power broadening).

4. Open quantum mechanical systems

4.1 *The density matrix*

Our particular requirement for the density matrix is to describe the interaction of a radiation field with an active medium made up of a large number of atomic systems. We have already considered, quantum-mechanically, the manner in which the individual atomic systems interact with the radiation field, on the assumption that they possess only two energy levels and that the existence of other energy levels can be taken into account by the introduction of phenomenological damping terms. In this case we were able to describe a particular atomic system by a wavefunction, once we had specified the boundary conditions.

However, when we come to consider the interaction of a large number of atomic systems with the radiation fields, it is no longer feasible to associate a wave function with the system as a whole (i.e. with the active medium).

The atomic systems are being created at different times, in one or other of their two energy levels, by excitation mechanisms which in general vary from one spatial point to another within the active medium, and which lead to a distribution of velocities amongst the systems. Unless we write down the complete set of equations describing the motion of the active medium (and this would require a detailed knowledge of its boundary conditions as well), it is not possible to associate a specific boundary condition with a particular atomic system. At the most we can only speak of a probability that within a given interval of time and within a given spatial volume, an atomic system will be created in one or other of its two energy levels, and with a velocity component within a specified range. A complete knowledge of the behaviour of the total system is not in fact usually required, since its influence on the radiation field is obtained by averaging over all the atomic systems from which it is made up. The density matrix provides a convenient method for carrying out this averaging process.

It must be stressed here that two different kinds of averaging are involved when we are dealing with the statistical properties of a quantum-mechanical system. A quantum-mechanical description is inherently statistical in the sense that even when the wavefunction for the system can be derived exactly, it is still only possible to associate probabilities with certain observables of the system. This should be clear from (3.5) which was used to find the average value of an observable of the system from a knowledge of the wavefunction. Such an average is due to the inherently statistical nature of the quantum mechanical representation, and if we can find the value of an observable to such limits, we say that we possess the maximum knowledge of the system. We now require to introduce a further averaging process, because of our lack of detailed knowledge about the system, of a type analogous to that for a purely classical system. It is when we are involved with both these types of averaging together that we use the density matrix representation. These points will become clearer in the derivation of the equations describing the density matrix that is given below.

In summary then, the density matrix representation is employed when we possess only an incomplete knowledge of the wavefunction (or probability amplitude) describing a quantum-mechanical system, and therefore require statistical concepts in order to find average values for the physical quantities in which we are interested. It is possible to introduce the density matrix representation into the description of a quantum-mechanical system in a variety of ways depending on one's viewpoint. The three main approaches have been summarized very concisely by Ter Haar [1961] as an

References on page 92

answer to the question "What is the density matrix?" and this we now quote:

"It is the quantum-mechanical counterpart of the classical distribution function (statistical point of view), or: It is the most general description of an open quantum-mechanical system, that is a system which cannot be described by a wavefunction (quantum-mechanical point of view), or, finally: It is the most convenient way to collect all parameters which are of interest for a given experimental set-up and to describe their behaviour (operational point of view)".

We now consider the general properties of the density matrix, and adopt what is essentially the statistical point of view, by introducing an ensemble to describe the physical system in which we are interested. In this section we do not specify the properties of the physical system at all; leaving until later (ch. 9 § 3) a consideration of the particular case of an active medium in a radiation field.

We suppose that there are N systems in the ensemble, each of which is described by a normalized wavefunction, Ψ^k, say. We introduce a set of orthonormal eigenfunctions, u_n, in terms of which we can expand the wavefunctions

$$\Psi^k = \sum_n a_n^k u_n \qquad (k = 1, 2, \ldots, N). \tag{3.133}$$

We now consider a particular physical observable of the system, F, and require to find its value in the ensemble representation of the system. For a particular member of the ensemble, we can find its average value from a knowledge of the wavefunction by using (3.5)

$$\langle F \rangle_k = \int (\Psi^k)^* \hat{F} \Psi^k \mathrm{d}q. \tag{3.134}$$

This average is a consequence of the inherently statistical nature of quantum-mechanics. If we now average over the ensemble, we obtain the expectation value of the observable for the system under consideration

$$\overline{\langle F \rangle} = N^{-1} \sum_{k=1}^{N} \int (\Psi^k)^* \hat{F} \Psi^k \mathrm{d}q. \tag{3.135}$$

Using (3.133), we can expand the wavefunction in terms of the chosen set of eigenfunctions, so that (3.135) becomes

$$\overline{\langle F \rangle} = N^{-1} \sum_{k=1}^{N} \sum_{m,n} (a_m^k)^* a_n^k F_{mn}, \tag{3.136}$$

where

$$F_{mn} = \int u_m^* \hat{F} u_n \mathrm{d}q. \tag{3.137}$$

We interchange the order of summation in the above, and introduce the density matrix in terms of its elements which are defined by

$$\rho_{nm} = N^{-1} \sum_{k=1}^{N} (a_m^k)^* a_n^k \tag{3.138}$$

(note that the order of the subscripts are reversed in the definition). We can then write (3.136) as

$$\overline{\langle F \rangle} = \sum_{m,n} \rho_{nm} F_{mn}, \tag{3.139}$$

or in matrix notation

$$\overline{\langle F \rangle} = \mathrm{Tr}\,(\boldsymbol{\rho}\mathbf{F}). \tag{3.140}$$

Again we stress that two types of averaging are involved in deriving the final value for the observable of the system: the former, which was introduced in (3.134), being due to the quantum-mechanical representation; the latter, (3.135), being due to our incomplete knowledge about the system, and of a kind analogous to the averaging procedures in classical statistics. It is this latter form of averaging that is carried out within the definition of the density matrix.

It can be seen from the definition (3.138), that $\boldsymbol{\rho}$ is Hermitian

$$\rho_{mn} = \rho_{nm}^* \tag{3.141}$$

and that it is normalized to unity

$$\mathrm{Tr}\,(\boldsymbol{\rho}) = N^{-1} \sum_k \sum_n (a_n^k)^* a_n^k = 1. \tag{3.142}$$

From these it readily follows that the diagonal elements are real, and that their sum is equal to unity.

The definition of the density matrix, as given above, depends on the particular representation with which we are working (i.e. the definition depends on the set of eigenfunctions that we choose for the expansion of the wavefunctions). We now consider the manner in which we can transform from one representation (with eigenfunctions u_n) to another (with eigenfunctions v_n). We can write (3.133) in terms of the new eigenfunctions as

$$\Psi^k = \sum_n a_n^k u_n = \sum_r b_r^k v_r \tag{3.143}$$

References on page 92

and note that the transformation from the one representation to the other is characterized by a unitary transformation matrix, S_{nr} (§ 2),

$$v_r = \sum_n u_n S_{nr}, \tag{3.144}$$

where

$$S^*_{nr} = S^{-1}_{rn}. \tag{3.145}$$

From (3.143) and (3.144) we have that

$$a^k_n = \sum_p S_{np} b^k_p, \tag{3.146}$$

and hence from (3.145) that

$$b^k_r = \sum_n a^k_n S^*_{nr}. \tag{3.147}$$

The transformed density matrix that we require to derive is defined by

$$\rho'_{rp} = N^{-1} \sum_k (b^k_p)^* b^k_r \tag{3.148}$$

and from (3.146) and (3.147) it can be seen that

$$\rho'_{rp} = \sum_{m,n} S^{-1}_{rm} \rho_{mn} S_{np}. \tag{3.149}$$

The transformation can be written in matrix notation as

$$\boldsymbol{\rho}' = \mathbf{S}^{-1} \boldsymbol{\rho} \mathbf{S}. \tag{3.150}$$

The density matrix transforms, therefore, in the same way as an operator. By substitution from (3.150) and using the equivalent expression for the transformation of the operator, $\hat{F}$, (3.48) it can be verified that the average value of an observable, F as defined by (3.140) is independent of the particular representation employed.

We now consider the form of the density matrix in the limiting case of it describing a system with a definite wavefunction. Such a system is said to be in a "pure state", as opposed to a "mixed state" when the wavefunction is not known with certainty. It is apparent that for a pure state all the systems within the ensemble are described by the same wavefunction, Ψ°, say, and in deriving the value of an observable, there is effectively only one averaging process, the quantum-mechanical one. With every pure state there is associated a "complete experiment" (FANO [1957]), in the sense that its result is predictable with certainty when it is performed on a system in such a state. This can be seen by remembering that with an observable of the system we associate a Hermitian operator, so that finding the complete experiment is equivalent to finding the operator which has the pure state wavefunction as an eigenfunction. The necessary and sufficient condition for ρ to describe a pure state is that it should be idempotent, that is

$$\rho^2 = \rho.$$

To prove this, consider the case when ρ has been diagonalized (and hence also ρ^2), when

$$\rho_n^2 = \rho_n$$

and therefore

$$\rho_n = 0, 1.$$

From the above and the normalization condition for the density matrix it follows that one value of ρ_n, ρ_1, say, is unity, while the rest are zero

$$\rho_1 = \sum_k |a_1^k|^2 = 1;$$

$$\rho_n = \sum_k |a_n^k|^2 = 0 \qquad (\text{for } n \neq 1),$$

and therefore

$$\Psi^k = u_1 \exp(\mathrm{i}\phi_k).$$

Thus all the systems within the ensemble are described by the same wavefunction apart from an (irrelevant) phase factor, and the "complete experiment" is described by the operator having u_1 as an eigenfunction.

In order to derive the equation of motion of the density matrix, we assume that the wavefunctions of the systems in the ensemble satisfy the Schrödinger equation (3.11)

$$\hat{H}\Psi^k + \frac{\hbar}{\mathrm{i}}\frac{\partial}{\partial t}\Psi^k = 0,$$

which can be expanded in terms of a complete, orthonormal set of eigenfunctions as

$$\frac{\partial}{\partial t}(a_m^k) = -\frac{\mathrm{i}}{\hbar}\sum_l H_{ml} a_l^k. \tag{3.151}$$

It readily follows from the Hermitian properties of $\hat{H}$ that

$$\frac{\partial}{\partial t}[(a_n^k)^* a_m^k] = \frac{\mathrm{i}}{\hbar}\sum_l [a_m^k(a_l^k)^* H_{ln} - H_{ml} a_l^k (a_n^k)^*],$$

which on averaging over the ensemble gives

$$i\hbar\dot{\rho}_{mn} = [\boldsymbol{H}, \rho]_{mn}, \tag{3.152}$$

or in matrix form

$$i\hbar\dot{\rho} = [\boldsymbol{H}, \rho], \tag{3.153}$$

where the commutator bracket has been defined previously (3.72).

4.2 *Coherence between states*

We now consider, more closely, the non-diagonal elements in the density matrix. Firstly, we note that it is impossible, by a physical measurement, to distinguish between two systems described by wavefunctions that differ only by phase factors. For example if two systems are described by wavefunctions ψ_1 and ψ_2 at a particular time, such that

$$\psi_1 = \psi_2 \exp(i\theta), \tag{3.154}$$

where θ is real, then

$$\langle F\rangle_1 = \int \psi_1^* \hat{F} \psi_1 \, dq = \int \psi_2^* \hat{F} \psi_2 \, dq = \langle F\rangle_2 ,$$

which shows that the expectation value of some observable, F, is the same for both systems. If we now express the wavefunction in terms of some set of eigenfunctions

$$\psi = \sum_n a_n u_n ,$$

we can write

$$\langle F\rangle = \sum_n \sum_m a_n^* a_m F_{nm} = \sum_n a_n^* a_n F_{nn} + \sum_{\substack{n,m \\ n>m}} \{a_n^* a_m F_{nm} + \text{c.c.}\}. \tag{3.155}$$

In the case when the u_n's are the eigenfunctions of the observable, F, we have already pointed out in § 2 that the last part of (3.155) is zero. For a general expansion, this is, of course, not so. Consider, as an example, the case when F describes a perturbation applied to the system, and the expansion is in terms of the unperturbed eigenfunctions of the system, the non-diagonal elements, F_{nm}, might be now non-zero.

If the complex coefficients are expressed in the form

$$a_n = |a_n| \cdot e^{i\phi_n}$$

it is apparent that, although the absolute values of the phases, ϕ_n, are not important, their differences are, in the case when non-diagonal elements enter into the expectation value.

Suppose that we now consider an ensemble made up of a large number (N) of identical systems of the kind described. A diagonal element of the density matrix associated with this ensemble is a real quantity of the form

$$\rho_{nn} = N^{-1} \sum_{k=1}^{N} a_n^{*k} \cdot a_n^k$$

and is the probability that an arbitrary system in the ensemble is in the state described by the eigenfunction, u_n. We can therefore regard $N\rho_{nn}$ as the "population" of this state in the ensemble.

When we consider a non-diagonal element of the density matrix

$$\rho_{nm} = N^{-1} \sum_{k=1}^{N} a_m^{*k} \cdot a_n^k = N^{-1} \sum_{k=1}^{N} |a_m| \cdot |a_n| \cdot e^{i(\phi^k{}_n - \phi^k{}_m)},$$

we see that it is the average of, in general, complex quantities. We can represent these complex quantities by vectors in the complex plane (fig. 3.3);

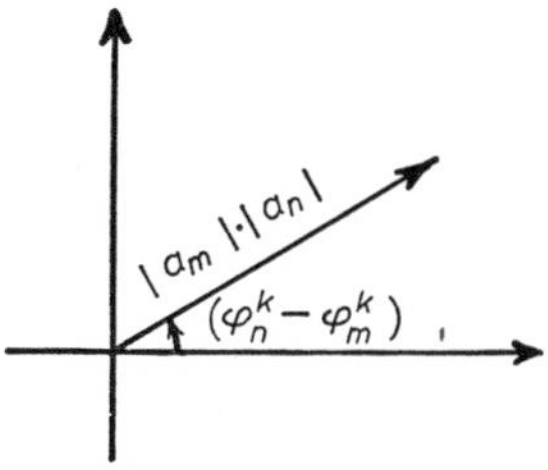

Fig. 3.3. Vector representation of the off-diagonal matrix elements.

the angle (ϕ_{nm}^k) a particular vector makes with the real axis is the phase difference between the complex coefficients, a_m and a_n, for a particular system of the ensemble, while the magnitude of the vector depends on the probabilities of the system being in one or the other of the two states.

If all values of ϕ_{nm}^k are equally probable, then the average value of $(a^{*k}_m a_n^k)$ for the ensemble will be zero, and non-diagonal terms will not occur in the density matrix. It is then possible to describe the ensemble in terms of the "populations" of the states alone. If, however, all values of ϕ_{nm}^k are not equally probable, then an average over the ensemble will yield a non-zero value for the non-diagonal elements of the density matrix. In this case we say that within the ensemble, there is "coherence" between the two states. The coherence arises because of some process that introduces the same phase difference between the states for all the systems in the ensemble. This process

References on page 92

could be coherent excitation of the states (for example, through coherent optical radiation), or the "mixing" of the states by an applied R. F. field, etc.

In ch. 9 § 3 we shall see the importance of "coherence" between states when we use Lamb's theory to describe the operation of a laser. In this theory, the wavefunction of a particular (two-level) atomic system in the active medium is expanded in terms of the unperturbed energy eigenfunctions of the system. The laser radiation field is then treated as a perturbation, and gives rise to non-diagonal terms in expressions similar to (3.155). It is these non-diagonal terms that act as the source of the field, and so the importance of their having non-zero values after averaging over the active medium is apparent.

References

Fano, U., 1957, Rev. Mod. Phys. **29**, 74.
Ter Haar, D., 1961, Rep. Prog. Phys. **24**, 304.
Weisskopf, V. and E. Wigner, 1930, Z. f. Phys. **63**, 54.

CHAPTER 4

OPTICAL RESONATORS (SIMPLE CONSIDERATIONS)

1. Laser cavities and resonators (introduction)

The purpose of a laser cavity is to direct the radiation produced by an active medium back and forth many times through the medium so that the gain by stimulated emission exceeds the losses. (Cavities used for optical pumping are not considered.) The geometry of the system chosen may result in the establishment in the cavity of standing waves, or travelling waves, or both, e.g. standing waves result from cavities constructed with plane-parallel and spherical mirrors; corner-cube reflectors give rise to travelling wave systems in the cavity. In travelling wave modes, the amplitudes of the waves travelling in opposite directions along the same path are unrelated. In standing waves there is a permanent phase and amplitude relationship between waves travelling in opposite directions.

A mode may be defined as a field distribution that reproduces itself in spatial distribution and phase, though not in amplitude as the wave travels between the reflectors of a cavity. In general, amplitude is not reproduced because of diffraction and reflection losses. Because of these unavoidable losses, a given mode has a lifetime generally defined as the time for its amplitude to decay to the fraction e^{-1} of its initial value.

Modes are classified as longitudinal and transverse. Longitudinal modes are designated according to the number of nodes along the axis of the cavity between the mirrors. Transverse modes are defined by the number of nodes in the plane of the mirror i.e. in a plane normal to the laser axis.

2. Resonant cavity

The dimensions of cavities which will oscillate in one or a few modes at microwave frequencies ($\lambda \sim 1$ cm) are of the order of one wavelength and so they are quite convenient to make and operate. At optical frequencies this is not the case and the cavities are necessarily multimode (millions).

 References on page 107

However, the possible modes may be limited by retaining high Q in one direction only. (The Q of a cavity is discussed in § 5 below.) This is conveniently done by having two parallel mirrors separated by an arbitrary distance with no walls so that the cavity is "open" and forms a Fabry-Perot resonator.

Consider a cavity with no side walls consisting of plane parallel perfect reflectors as first proposed by PROKHOROV [1958], DICKE [1958], and SCHAWLOW and TOWNES [1958] and neglect diffraction and other losses. A standing wave system is set up by radiation travelling back and forth between the mirrors. In establishing such a system the radiation must travel from A to B and arrive back at A (fig. 4.1) in the correct phase. The re-

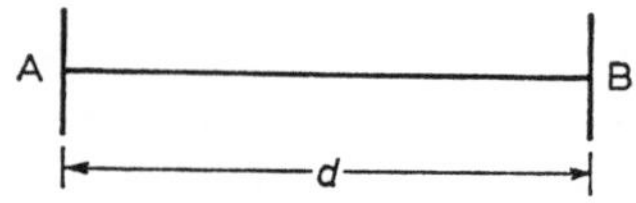

Fig. 4.1. Plane mirrors separated by distance *d*.

quirement of correct phase is only fulfilled by radiation of certain frequencies. Thus, only these frequencies can exist in the cavity and each frequency is that of a possible longitudinal mode of oscillation of the cavity. (We leave consideration of transverse modes till later). There are generally about 10^4–10^7 wavelengths between the mirrors of a laser oscillator (i.e. lasers of length in the range from a few centimetres to a few metres).

Let T be the total time to go along ABA and let τ be the period of oscillation of the light of frequency ν. If the number of wavelengths along ABA is q, then $T = q\tau$, and the number q designates the mode. Other modes are $q \pm 1$, $q \pm 2$, etc.

If c' is the velocity of the radiation in the medium of the cavity, we have

$$2d = c'T \quad \text{and} \quad \nu_q = qc'/2d,$$

where d is the distance between the mirrors.

The frequency by which modes are separated is given by

$$\nu_q - \nu_{q-1} = \delta\nu = [q-(q-1)]c'/2d = c'/2d. \tag{4.1}$$

The mode separation frequency can be derived alternatively as follows. The phase shift resulting from a wave travelling the length of the cavity in a medium of refractive index μ once from one mirror to the other at a phase velocity of c/μ during time t, is $\phi = \omega t$, where ω is the angular frequency of the radiation. Since $\omega = 2\pi\nu$ and $t = d\mu/c$, we get

$$\phi = 2\pi\nu d\mu/c.$$

The dispersion of a cavity is $\partial\phi/\partial\nu$ and is given by

$$\partial\phi/\partial\nu = 2\pi d\mu/c \tag{4.2}$$

For standing waves, neighbouring modes are given by

$$\phi = 2\pi\nu d\mu/c \quad \text{and} \quad \phi+\pi = 2\pi(\nu+\delta\nu)d\mu/c.$$

By subtraction we obtain, finally

$$\delta\nu = c/2d\mu. \tag{4.3a}$$

A cavity made from concave mirrors of equal radius of curvature separated by their radius is confocal since the mirror foci coincide. In such a cavity (fig. 4.2) standing waves may be set up by radiation along the path

Fig. 4.2. Ray path between concave mirrors.

ABCBA. As before, if T is the total time to go along ABCBA, we get $4d = c'T$. If the number of waves along ABCBA is $2q$, then $T = 2q\tau$. The frequency by which longitudinal modes are separated is thus

$$\nu_q - \nu_{(q-1)} = \delta\nu = [2q-2(q-1)]c'/4d = c'/2d. \tag{4.3b}$$

The case of transverse modes requires the more detailed treatment of the wave theory of cavities from which we obtain (ch. 6 § 1) the resonance condition to be

$$4d/\lambda = 2q+(m+n+1), \tag{4.4}$$

where q refers to the longitudinal mode and m and n to the transverse modes. From this equation the separation $\delta\nu$ between modes m and $m-1$ is $c'/4d$.

Other commonly used geometries are summarized in table 4.1.

The two essential requirements for a laser are a cavity and an active medium within the cavity, i.e. a medium which will provide the radiation and amplification necessary to establish oscillation in the cavity. A cavity is said to be "passive" when it does not contain an active medium and "active" when it does. The active material generally takes the form of a rod of luminescent material which is optically pumped, or a tube filled with gas excited by an electric discharge. The length of a cavity is dictated by that of the active medium which is 5–20 diameters for rods and 100–200 diameters for gases.

The active medium will only produce radiation and gain over a limited frequency range $\Delta\nu$. This range is determined by the width of a spectral

References on page 107

TABLE 4.1
Spherical mirror resonators

Resonator form	Mirror radii	Mirror separation
Confocal	R	R
Plane parallel	∞	any
Concentric	R	$2R$
Half confocal	R, ∞	$R/2$
Half concentric / Hemispherical	R, ∞	R
No name given	R_1, R_2	any up to R_1+R_2

line. The number of longitudinal modes possible within the range is,

$$N = \Delta\nu/\delta\nu, \tag{4.5}$$

and this is the number which will oscillate if the gain conditions (ch. 8) are sufficient. That is, the system oscillates at whatever frequencies the resonator and active medium have in common. In general, the laser gain should be well above the threshold necessary for oscillation, to obtain simultaneous oscillation of several modes. Another consideration is the type of broadening of the line (homogeneous or inhomogeneous, ch. 8 § 2).

The line-width of the 6328 Å line of neon in a He-Ne laser is due to the Doppler effect (Appendix J) and is typically about 1500 MHz (i.e. 0.02 Å). If such a laser has confocal mirrors of 1 m radius, (4.3) and (4.5) give $N = 20$. Figure 4.3 shows the modes for which oscillation is possible within the profile of the spectral line concerned provided the gain conditions are sufficient to produce oscillation. It also shows the various line widths. In the visible region, the Doppler width may be 10^2–10^3 times the natural width.

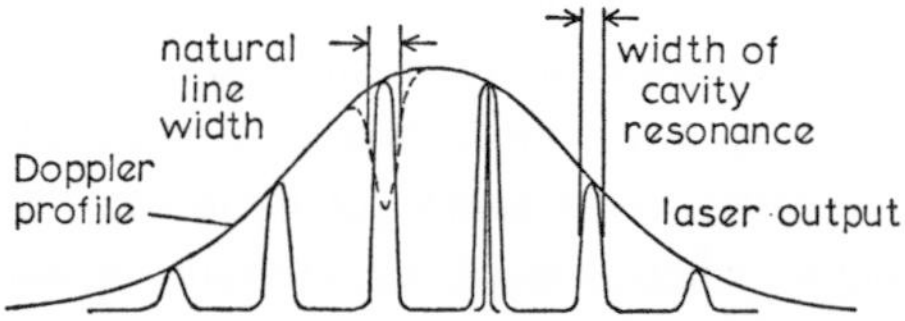

Fig. 4.3. Spectral line profiles of the various oscillations present in a laser cavity. After HERRIOTT [1962].

Each of these longitudinal modes oscillates independently and they do not compete in an inhomogeneously broadened line provided the mode separation is large compared with the natural line-width. The modes beat together, and thus beat frequencies, in addition to the discrete mode frequencies, are always present in the output of a laser unless the dimensions of the cavity are chosen so that $N = 1$. To ensure that only one longitudinal mode is present when the line-width is greater than 1500 MHz, the cavity should be less than about 10 cm if it is stabilized to the line centre, and 5 cm if it is not. When restricted to this length, the single mode output power is about 1 mW. Choice of cavity length provides an elementary means of mode selection (ch. 11 § 8). It is easy to achieve single mode operation by this means in gas lasers because the atomic line-width is generally narrow. However, solids have much broader lines and so they are generally multi-mode even at quite short lengths.

In practice, there are various cavity losses, so if a mode is not maintained by radiation from the active medium, it decays. This process is described by assigning a mean "life-time" to the radiation in the cavity (ch. 4 § 5.3). The time at which modes are established and decay is quite random and so there is no phase correlation between the various independent modes. The mode beats thus fluctuate randomly and add to the noise spectrum of the device.

These elementary considerations have ignored completely the diffraction losses which occur at each transit of the open cavity. FOX and LI [1961] and BOYD and GORDON [1961] (ch. 6) considered the important question as to whether stable modes were possible with the diffraction losses of such a cavity. Fox and Li considered a number of mirror geometries including circular plane mirrors of radius a at a separation of d (fig. 4.4)

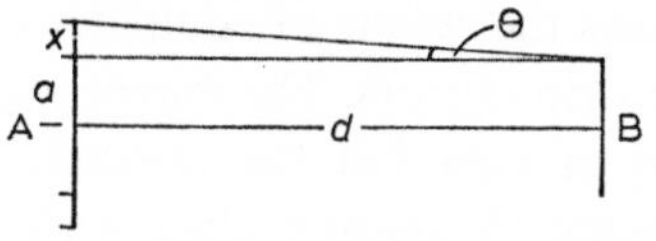

Fig. 4.4. Circular plane mirrors of radius a separated by distance d.

to determine the diffraction losses and their effect on the mode structure as energy is propagated back and forth between the mirrors. The questions they set out to answer were: (a) whether, after many transists, the relative field distribution approached a steady state; (b) how many steady states there were, if any; (c) what the losses were. A uniform plane wave, starting at one mirror, loses some energy from the peripheral regions by diffraction before it reaches the second mirror, and similarly the portion which is reflected by the second mirror loses energy before arriving back at the first mirror. As this process is repeated, the field across the wave front becomes weaker towards the edges. To calculate the field distribution across mirror B the initial wave front from the mirror A was divided into a number of

References on page 107

elementary sources and Huygens' principle was applied to each source to find the total contribution of each source to the field at each of a number of points across mirror B. The distribution obtained was then used as the initial wave front to calculate the distribution resulting when the wave reflected from B arrived back at the mirror A. This process of calculation

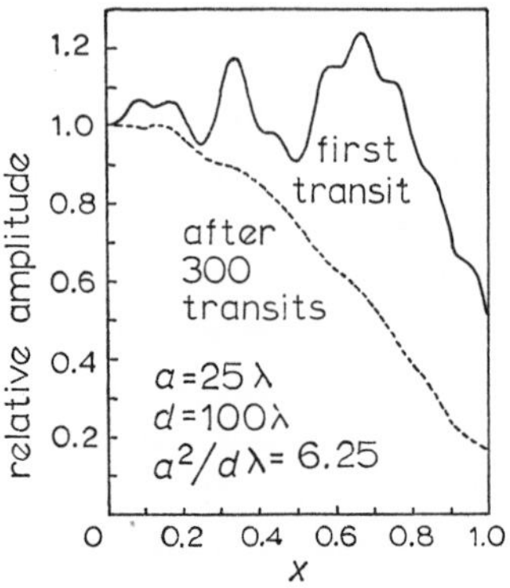

Fig. 4.5. Relative amplitude of field intensity for infinite strip mirrors. (The initially launched wave has a uniform distribution.) Reproduced from Fox and Li [1961] with permission.

was repeated for hundreds of transits and it was shown that a distribution was finally obtained which remained unchanged by further transits. Such a stable distribution of energy moving back and forth between the mirrors is a mode. Fig. 4.5 shows the results of computer calculations obtained by Fox and Li for infinite strip mirrors. The mirror geometry is not important from a practical point of view but the principle demonstrated by their calculations is fundamental. A uniform plane wave was launched from the first mirror and after the first transit the field amplitude distribution (normalized so that the amplitude at the centre is unity) was as shown in fig. 4.5. With each successive transit, the fluctuations of amplitude at a particular point x relative to the amplitude at the centre of the mirror gradually decreased until after 300 transits, they were less than 0.03 % of the final average value. The distribution reproduces itself and thus forms a stable mode.

Summarizing then, a plane wavefront was set off from one mirror and its development was followed as it travelled back and forth between the mirrors, and thus it was demonstrated that the system would eventually settle down to a stable mode. This is because it turns out that the dominant mode is characterized by fields which are considerably smaller at the edges of the mirrors than at the centres, thus making the diffraction loss orders of magnitude smaller than that predicted by considerations of plane waves.

3. Fresnel number (elementary considerations)

Consider a cavity consisting of two plane parallel circular mirrors AB of radius a separated by distance d. The energy of a parallel beam of radiation of wavelength λ impinging on A is reflected and diffracted into an angle of about λ/a (Airy disc, where about 84 % of the energy lies). The angle subtended at A by mirror B is a/d. Radiation at this angle to the resonator axis will travel the length of the cavity only once before "walking out" of the cavity. In a laser resonator, the radiation is required to travel to and fro many times between the mirrors. If the number of transits the radiation is required to make through the system is n, the maximum angle between the direction of the radiation and the resonator axis is a/nd. Thus, for small losses,

$$a/nd > \lambda/a. \tag{4.6}$$

In practice, the radius a is that of the active medium between the mirrors. The quantity $a^2/d\lambda$ will be recognized as the Fresnel number, N, given by

$$N = a^2/\lambda d. \tag{4.7}$$

When considered purely from the point of view of diffraction, the condition to be satisfied is that each mirror should intercept as much as possible of the energy originating at the other mirror. To do this, the angle subtended by each mirror at the centre of the other must be greater than the angle of the far field diffraction pattern of radiation from the other mirror. On this basis, N is approximately the number of Fresnel zones intercepted. For a system consisting of mirrors of radii a_1, and a_2 we have

$$N = a_1 a_2/\lambda d.$$

Typically, a resonator must allow a few tens or hundreds of transits, before the radiation is reduced by the various loss processes (transmission, scattering diffraction, walk out, etc.) to e^{-1} of its initial intensity. When the Fresnel number is $\sim$ 100, diffraction losses are insignificant and the system can be described quite accurately by geometric optics. Consideration should be given to diffraction losses when they become comparable with reflection losses.

Fresnel numbers are generally of order 100 for ruby lasers and similar systems and of order 10 for gas lasers, as the following two examples show.

A ruby rod has dimensions 10 cm × 1 cm dia. with reflectors at its end

References on page 107

surfaces. The wavelength of radiation in ruby at room temperature is 6943/1.76 Å where 1.76 is the refractive index of ruby. The Fresnel number given by (4.7) is ~ 625. An argon laser has a cavity length of 50 cm and a diameter of 2 mm in the active region. The wavelength is 4880 Å. The Fresnel number given by (4.7) is ~ 20.

The fractional energy, α, lost by diffraction per transit, assuming a uniform field and phase distribution, and diffraction angle $\theta = \lambda/a$ (fig. 4.4) is the ratio of the areas of the annulus of width $x (x = \theta d)$ and mirror of radius a, thus

$$2\pi a x/\pi a^2 = \lambda d/a^2 = 1/N. \tag{4.8}$$

The larger the Fresnel number, the smaller the diffraction losses. It will be shown in ch. 6 that when a standing wave is established in the cavity, the field distribution is not uniform and the diffraction losses are considerably less than predicted here. The confocal resonator has the lowest diffraction loss of all types and the fundamental mode is the mode of lowest loss.

Diffraction losses become an important consideration if the laser medium has low gain.

4. Photon representation of cavity properties

The condition for low diffraction loss from a laser cavity can be derived in terms of photons, making use of the uncertainty principle. Photons arriving at the mirrors A, B (fig. 4.4) can be located with reference to axes x, y, z, where z is the axis of the cavity. When a photon is reflected back into the cavity by mirror A, say, and we observe it, its position along the x axis has been located with an accuracy

$$\Delta x \sim a.$$

Using the uncertainty principle, this implies that its momentum in the x-direction is uncertain by an amount

$$\Delta p_x \sim h/a.$$

Therefore, the photon direction on leaving the mirror A is uncertain by an amount

$$\Delta\theta \sim \Delta p_x/p_z \sim c/a\nu \quad \text{where} \quad p_z = h\nu/c.$$

The value obtained for $\Delta\theta$ is thus just that obtained by the classical wave treatment in terms of diffraction. For diffraction losses to be a minimum, i.e. to be sure of intercepting the photon, the mirror B should be of diameter large enough to allow for the uncertainty. When the photon reaches B, its position is uncertain by the amount $d\Delta\theta$ therefore, $d\Delta\theta < a$. The condition to be fulfilled is thus

$$a^2/\lambda d > 1.$$

5. The Q of a cavity

Oscillating systems (e.g. LC resonators, microwave cavities, optical cavities, etc.) are generally described in terms of a quality factor, Q. We shall discuss the concept in terms suitable for understanding the optical cavities of lasers. Consider a cavity of dimensions such that it will support oscillations in the frequency range $d\nu_c$ about the central frequency ν_c of one of its modes. The Q of the cavity may be defined in several equivalent ways for large values of Q:

$$Q = \frac{\text{frequency at resonance}}{\text{frequency band between the half-power resonance points}}; \quad (4.9)$$

$$\text{or} \quad Q = \frac{2\pi \text{ stored oscillator energy}}{\text{energy spent per period}}; \quad (4.10)$$

$$\text{or} \quad Q = \omega \frac{\text{energy stored in the oscillator}}{\text{energy lost per second}}, \quad (4.11)$$

where $\omega = 2\pi\nu$.

When we come to consider Lamb's theory of the laser (ch. 9), we shall use a conductivity factor, σ, to describe all the losses of the cavity. By integrating (9.9) we obtain the electric field of the radiation in the cavity

$$E = E_{0n} \exp(-\sigma t/2\varepsilon_0).$$

The energy per unit volume of the cavity radiation field is therefore

$$U = E_{0n}^2 \exp(-\sigma t/\varepsilon_0).$$

Expressing (4.11) in terms of unit volume of the cavity for oscillations of angular frequency Ω_n, we get

$$Q = \frac{\Omega_n E_{0n}^2}{E_{0n}^2[1-\exp(-\sigma/\varepsilon_0)]};$$

$$Q = \Omega_n \varepsilon_0/\sigma. \quad (4.11a)$$

5.1. *Bandwidth of laser oscillator*

In a laser oscillator, the cavity mode is receiving coherent energy as well as losing energy through the various dissipation processes. In this case the

References on page 107

definition for the Q of a mode becomes

$$Q = \frac{2\pi\nu \text{ (stored energy)}}{\text{net loss of coherent energy per second}}.$$

Let $P^{(i)}_{\mathrm{coh}}$ be the rate at which coherent energy is supplied to the mode (through stimulated emission), and let $P^{(o)}_{\mathrm{coh}}$ be the rate of loss of coherent energy by transmission through the mirrors (assumed the only loss process). We then have

$$Q = \frac{2\pi\nu \text{ (stored energy)}}{P^{(o)}_{\mathrm{coh}} - P^{(i)}_{\mathrm{coh}}}.$$

Under steady state conditions, the total rate of energy input to the mode, which is made up of the coherent energy input rate (due to stimulated emission) and the incoherent energy input rate (due to spontaneous emission) must equal the total rate at which energy is dissipated by the oscillator, and therefore we have

$$P^{(i)}_{\mathrm{coh}} + P^{(i)}_{\mathrm{incoh}} = P^{(o)} \approx P^{(o)}_{\mathrm{coh}},$$

so that the above becomes

$$Q = \frac{2\pi\nu \text{ (stored energy)}}{P^{(i)}_{\mathrm{incoh}}}.$$

To calculate $P^{(i)}_{\mathrm{incoh}}$, we note that the ratio of the rates, into a given mode, of stimulated emission to spontaneous emission equals the number of photons in the mode. This number may be related to the field intensity of the mode inside the cavity, I, or the output power, $P^{(o)}$, as

$$N_p = Iad/ch\nu = P^{(o)}d/\alpha_r\, ch\nu, \tag{4.12a}$$

where a is the cross-section of the mode, d is the cavity length and α_r is the fractional transmission loss of the mirrors. The rate of stimulated emission into the mode is readily derived from (1.24) as

$$B_{mn}N_m g(\nu, \nu_{mn})Ih\nu ad/c, \tag{4.12b}$$

so that $P^{(i)}_{\mathrm{incoh}}$ is given by dividing (4.12b) by (4.12a) to obtain

$$P^{(i)}_{\mathrm{incoh}} = B_{mn}N_m g(\nu, \nu_{mn})(h\nu)^2.$$

The stored energy in the mode is $N_p h\nu$ and hence the Q of the mode is

$$Q = \frac{\nu}{\delta\nu} = \frac{2\pi\nu P^{(o)}d}{B_{mn}N_m g(\nu, \nu_{mn})(h\nu)^2\alpha_r\, c}, \tag{4.12c}$$

where $\delta\nu$ is the linewidth of the oscillator output.

At threshold for oscillation (1.28) gives

$$B_{mn} g(\nu, \nu_{mn}) h\nu N_m = \frac{\alpha_r c}{d}\left[1 - \frac{g_m N_n}{g_n N_m}\right]_{\text{th}}^{-1}.$$

Substituting this into (4.12c) and using (4.15a), we obtain,

$$\delta\nu = \frac{2\pi(\Delta\nu_c)^2 h\nu}{P^{(0)}}\left[1 - \frac{g_m N_n}{g_n N_m}\right]_{\text{th}}^{-1}. \tag{4.13a}$$

In the usual case when $N_m \gg N_n$ this simplifies to

$$\delta\nu = \frac{2\pi(\Delta\nu_c)^2 h\nu}{P^{(0)}}. \tag{4.13b}$$

The linewidth of the oscillator mode narrows as more power is coupled out of the cavity. We return to a fuller consideration of the linewidth of the laser oscillator in ch. 10 § 11.

This is the equation originally derived by GORDON et al. [1955] and by SCHAWLOW and TOWNES [1958]. Applying (4.13b) to a He-Ne laser with output 1 mW we get $d\nu \sim 5 \times 10^{-4}$ Hz if the mirrors transmit 10^{-2} at 6328 Å and are separated by 1 metre. (For calculation of $\Delta\nu_c$ see next section.) Since (10.1) shows that $\Delta\nu\Delta t = 1$, and Δt in this case is the time for which stability must be maintained, it is not possible to make a cavity dimensionally stable enough to exploit this narrow linewidth. See, for example, the classic measurements of JAVAN et al. [1962].

5.2. *Effect of losses on Q*

To obtain radiation from a laser cavity it is necessary to decouple it by using mirrors with reflectivity less than 100 %. Let the fractional power lost by reflection and diffraction from the cavity each time the radiation of a given mode of wavelength λ meets a mirror be α and let U be the total energy of the mode present in the cavity. If the mirrors are distance d apart ($d \gg \lambda$), the energy lost per second is $U\alpha c/d$, where c is the velocity of light in the cavity medium. Equation (4.11) then gives for a passive cavity,

$$Q = \frac{2\pi\nu U}{U\alpha c/d} = \frac{2\pi d}{\alpha\lambda}. \tag{4.14}$$

The Q may be increased by increasing d, but as the mirrors are separated the diffraction losses contributing to α are also increasing. The upper limit of Q is reached when the diffraction and reflection losses are about equal; at greater values of d the value of Q decreases.

The full frequency width of the cavity at half maximum response is obtained from (4.9) and (4.14),

$$\Delta\nu_c = c\alpha/2\pi d. \tag{4.15a}$$

For a 1 m cavity with 2 % loss the value of $\Delta\nu_c$ is about 1 MHz. From (4.15a) and (4.2) we get

$$(\alpha/\Delta\nu_c) = (\partial\phi/\partial\nu) \tag{4.15b}$$

The Q of a Fabry-Pérot cavity is principally determined by reflection and diffraction losses. Reflection losses result from absorption and transmission at the mirrors and diffraction losses result from the finite aperture of the mirror. We shall calculate the Q due to each of these losses separately.

Figure (4.4) shows the two mirrors of radius $2a$ separated by distance d. The radiation from mirror B will be propagated within the angle $\theta = \lambda/2a$ to the axis due to diffraction and so will miss mirror A by distance $x = \theta d$, i.e. the radiation falling per transit on the annulus between radii a, $a+x$ will be lost. If u is the energy density of the radiation field, (4.10) gives

$$Q_d = \frac{2\pi\nu\pi a^2 ud}{2\pi axuc} = \frac{2\pi a^2}{\lambda^2}. \tag{4.16}$$

If the reflection loss is α_r, we can compare the Q's given by (4.14) and (4.16)

$$\frac{Q_r}{Q_d} = \frac{2\pi d}{\alpha_r\lambda}\cdot\frac{\lambda^2}{2\pi a^2} = \frac{\lambda d}{\alpha_r a^2}.$$

If $d = 100$ cm, $\alpha_r = 10^{-2}$, $a = 1$ cm, $\lambda = 5000$ Å, $Q_r/Q_d = 0.5$.

The curves of Fox and Li [1961] show that the diffraction losses can be represented by an equation of the form

$$\alpha_d = K(\lambda d/a^2)^\beta, \tag{4.17}$$

where K and β are constants depending on the type of resonator. From (4.14) and (4.17) we get

$$\frac{1}{Q} = \frac{\lambda\alpha_r}{2\pi d} + \frac{\lambda}{2\pi d} K\left(\frac{\lambda d}{a^2}\right)^\beta.$$

Differentiating Q with respect to d shows that Q has a maximum value when

$$\alpha_r = \alpha_d(\beta-1)$$

and

$$d = \frac{a^2}{\lambda}\left(\frac{\alpha_r}{K(\beta-1)}\right)^{1/\beta}. \tag{4.18}$$

For circular plane mirrors $K \approx 0.2$, $\beta \approx 1.4$, while for confocal mirrors $K \sim 10^{-4}$, $\beta \sim 10$. Since the smallest values of α_r attainable are $\sim 0.1\%$ we see that the value of d is largely determined by a^2/λ and unpractical values of d are obtained at optical wavelengths.

5.3 *Time constant of a passive cavity*

As radiation travels back and forth within a cavity, losses occur by reflection, diffraction, etc. and so the radiation field decays. Let the radiation density be u, the velocity in the medium be c, the mirror separation d, and the losses per transit, α. The rate of change of energy density due to losses from the cavity is

$$\dot{u} = -c\alpha u/d. \qquad (4.19)$$

Solving this we get

$$u = u_0 \exp(-c\alpha t/d). \qquad (4.20)$$

The time required for the energy of the cavity to decrease to the fraction e^{-1} of its initial value is the life-time, τ_c of radiation in the cavity and is given by

$$\tau_c = d/c\alpha, \qquad (4.21)$$

e.g. when $d = 100$ cm, $\alpha = 10^{-2}$, $c = 3\times10^{10}$ cm sec^{-1}, we get $\tau_c = 3\times10^{-7}$ sec.

From (4.9), (4.20) and (4.21) we get

$$Q = \frac{2\pi}{1-\exp(-\tau/\tau_c)} \approx 2\pi\frac{\tau_c}{\tau} = 2\pi\nu\tau_c, \qquad (4.22)$$

where τ is the period of oscillation and ν is the frequency of oscillation.

For the example above, with $\nu = 4.8\times10^{14}$ Hz, we get $Q = 10^9$. Part of the loss from a cavity is due to absorption and scattering, and as the beam travels a distance l in the cavity it will decrease in intensity according to the equation

$$\mathscr{I} = \mathscr{I}_0 \exp(-\alpha_a l). \qquad (4.23)$$

The time for $\mathscr{I}_0$ to decrease to $\mathscr{I}_0/e$ is the time for the beam to travel path $l = 1/\alpha_a$, which is

$$\tau_a = 1/\alpha_a c. \qquad (4.24)$$

The corresponding Q is

$$Q_a = 2\pi\nu/\alpha_a c. \qquad (4.25)$$

In general, α_a could mean any loss process, and each one would contribute

References on page 107

a characteristic time to the cavity so that (4.21) becomes

$$\tau_{\text{loss}} = d/c\alpha_{\text{loss}}\,. \tag{4.26}$$

If all the losses, say, diffraction, reflection and absorption act simultaneously, the total decay time of the cavity is given by

$$(1/\tau_c) = (1/\tau_d)+(1/\tau_r)+(1/\tau_a) \tag{4.27}$$

and the Q is given by

$$(1/Q) = (1/Q_d)+(1/Q_r)+(1/Q_a). \tag{4.28}$$

Regarding the complete laser system defined by the mirrors as a single oscillator, and applying the relation $\Delta\nu\Delta t \sim 1$ to this oscillator, we see that Δt is τ_c. Using (4.21) we then obtain $\Delta\nu_c \sim c\alpha/d$. This agrees with (4.15).

6. Mode patterns observed

We shall consider the mode patterns of gas discharge lasers. The mode patterns of other lasers e.g. ruby, are generally distorted because of the non-uniform refractive index often present in such lasers. Gas discharge lasers generally have good quality optical components and the active medium is homogeneous and so produces undistorted modes. The patterns seen and their designations are shown in fig. 4.6. They may occur with plane, or

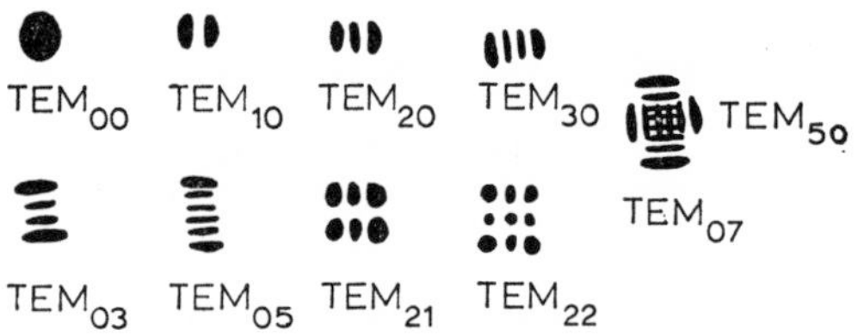

Fig. 4.6. Transverse modes and their designations.

spherical mirrors and are displayed on a plane normal to the axis of the laser beam. Modes are transverse electromagnetic, and are classified as TEM_{mnq}. (The theoretical background to modes as described by m and n is given in ch. 6.) The number of field reversals (i.e. zero intensity regions, or nodes), across the observed pattern is specified by m in the x-direction, and by n in the y-direction; the plane of polarization is in the y-direction. The TEM_{mnq} modes are "off-axis" modes. The value of q refers to the longitudinal mode specified by the number of zeros of intensity in the standing wave system along the axis of the laser and is not usually written because it is generally in the range 10^4–10^7. Longitudinal modes differ from one another only in their oscillation frequency; transverse modes differ from one another both in oscillation frequency and field distribution in a plane normal to the laser axis.

Sometimes several transverse modes may oscillate simultaneously. Figure 4.6 shows the resulting mode pattern for TEM_{50}, TEM_{07} modes. All modes are polarized in the same direction, depending upon the orientation of the Brewster windows of the laser.

Modes with axial symmetry may oscillate. These are designated TEM_{plq} where p gives the number of zeros (nodes) in the radial direction, and l gives the number of nodes in the azimuthal direction (i.e. half the number of zeros in an angular direction, see fig. 4.7). Sometimes modes are designated TEM_{lpq}. Circular symmetric modes which are marked with an asterisk are composites of two degenerate modes with $l = 1$ combining in space and phase quadrature with each other. Some typical axi-symmetric modes are given in fig. 4.7. These modes are also formed with plane, or spherical

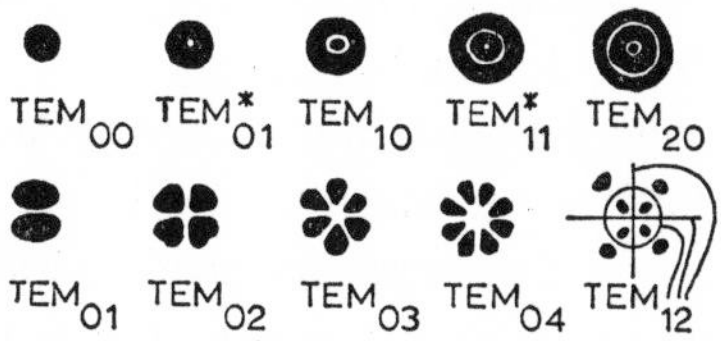

Fig. 4.7. Axi-symmetrical transverse modes and their designations.

mirrors but require rather careful alignment of the laser to obtain the axial symmetry necessary to produce them. The isolation of axi-symmetrical modes has been described by RIGROD [1963].

A linear combination of modes with the same value of $m+n$ can be found which is identical with a mode in cylindrical coordinates with $2p+l = m+n$. Both methods of description are equivalent in a perfectly symmetrical, lossless resonator. However, in a real resonator, the azimuthal symmetry is not perfect and the degeneracy under m and n intercharge is removed.

References

BOYD, G. D. and J. P. GORDON, 1961, Bell System Tech. J. **40**, 489.
DICKE, R. H., 1958, Sept., U.S. Patent 2851–652.
FOX, A. G. and T. LI, 1961, Bell System Tech. J. **40**, 453.
GORDON, J. P., H. J. ZEIGER and C. H. TOWNES, 1955, Phys. Rev. **99**, 1264.
HERRIOTT, D. R., 1962, J. Opt. Soc. Am. **52**, 31.
JAVAN, A., E. A. BALLIK and W. L. BOND, 1962, J. Opt. Soc. Am. **52**, 96.
PROKHOROV, A. M., 1958, Soviet Phys. JETP. **7**, 1140.
RIGROD, W. W., 1963, Appl. Phys. Lett. **2**, 51.
SCHAWLOW, A. L. and C. H. TOWNES, 1958, Phys. Rev. **112**, 1940.

CHAPTER 5

OPTICAL RESONATORS (GEOMETRIC THEORY)

1. Introduction

Cavities with very large Fresnel numbers ($\sim$ 100 say) have very small diffraction losses and so we can apply geometrical optics to study certain aspects. One important aspect is the effect of repeated reflections at the spherical mirrors which may form a laser cavity.

We are going to discuss the behaviour of rays in laser cavities of various geometries using paraxial ray treatment. Matrix methods will be used in the studies since these provide a very powerful technique for handling the multiple passes the rays undergo in a cavity. The aspects of matrix theory used are the representation of equations by matrices, and multiplication of matrices. The stability of a cavity is found to depend upon the eigenvalues of a matrix. A brief revision of relevant matrix theory is given in appendix E. A very good account of matrix methods as applied to optical resonators has been given by BERTOLOTTI [1964].

2. Plane mirror cavity

Consider a cavity consisting of two plane mirrors normal to the cavity axis separated by distance d as shown in fig. 5.1a. To study the behaviour of a ray reflected back and forth between the mirrors we need to know the location of the ray at one of the mirrors and the orientation of the ray. The coordinate system we shall use is shown in fig. 5.1b. The origin (fig. 5.1a) is at the mirror centre and the z-axis is the axis of the cavity.

Let a ray start from mirror 1 at $P_1(x_1, y_1)$ and leave in the direction defined by the angles $\theta_1 = \theta$ and $\phi_1 = \phi$ which the ray makes with the normal to the mirror in the xz and yz planes, respectively. The angles θ and ϕ are small so that the angles may be used instead of their tangents. The angles θ are positive when they lie on the side of the normal in the positive direction

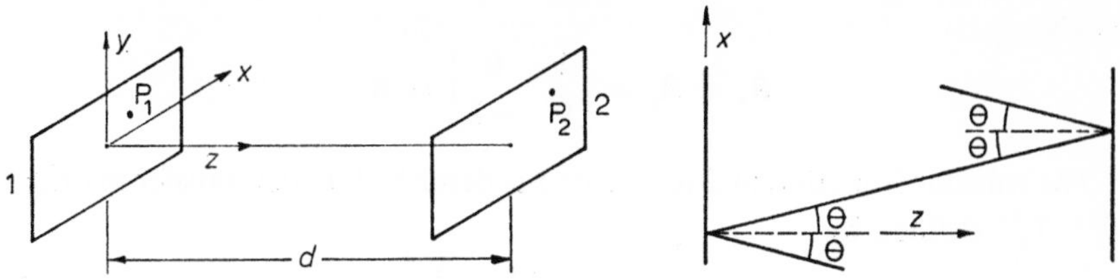

(a) (b)

Fig. 5.1. Co-ordinate system used to study a cavity consisting of two rectangular plane mirrors normal to the cavity axis separated by distance d.

of x. Similarly for ϕ in the y-direction. When the ray meets mirror 2 it will arrive at point x_2, y_2 given by

$$x_2 = x_1 + \theta d;$$
$$y_2 = y_1 + \phi d.$$

At mirror 2, the angles of arrival are $\theta_{a2} = -\theta$ and $\phi_{a2} = -\phi$. The result of the ray travelling from mirror 1 to mirror 2 is that its location at the mirror has been transferred from (x_1, y_1) to (x_2, y_2). This can be described by the transfer matrices $\boldsymbol{T}_x^{12}$ and $\boldsymbol{T}_y^{12}$, thus

$$\begin{bmatrix} x_2 \\ \theta_{a2} \end{bmatrix} = \boldsymbol{T}_x^{12} \begin{bmatrix} x_1 \\ \theta \end{bmatrix} \quad \text{and} \quad \begin{bmatrix} y_2 \\ \phi_{a2} \end{bmatrix} = \boldsymbol{T}_y^{12} \begin{bmatrix} y_1 \\ \phi \end{bmatrix},$$

where

$$\boldsymbol{T}_x^{12} = \boldsymbol{T}_y^{12} = \begin{bmatrix} 1 & d \\ 0 & -1 \end{bmatrix}.$$

No significance must be given to the location of the suffices and superscripts. Before the ray can leave mirror 2 it must be reflected and this process introduces another operation on the ray. The equations we wish to describe are

$$\begin{aligned} x_2 &= x_2 \\ \theta_{d2} &= -\theta \end{aligned} \quad \text{and} \quad \begin{aligned} y_2 &= y_2 \\ \phi_{d2} &= -\phi, \end{aligned}$$

where θ_{d2} and ϕ_{d2} are the angles at which the ray departs from the mirror. These may be represented by the reflection matrices, $\mathbf{R}_x$, $\mathbf{R}_y$ thus

$$\begin{bmatrix} x_2 \\ \theta_{d2} \end{bmatrix} = \mathbf{R}_x \begin{bmatrix} x_2 \\ \theta \end{bmatrix} \quad \text{and} \quad \begin{bmatrix} y_2 \\ \phi_{d2} \end{bmatrix} = \mathbf{R}_y \begin{bmatrix} y_2 \\ \phi \end{bmatrix},$$

References on page 130

where

$$\boldsymbol{R}_x = \boldsymbol{R}_y = \begin{bmatrix} 1 & 0 \\ 0 & -1 \end{bmatrix} = \boldsymbol{R}.$$

The transfer from mirror 2 to mirror 1 is described by the transfer matrices $\boldsymbol{T}_x^{21}, \boldsymbol{T}_y^{21}$ defined by

$$\boldsymbol{T}_x^{21} = \boldsymbol{T}_y^{21} = \begin{bmatrix} 1 & d \\ 0 & -1 \end{bmatrix}.$$

After another reflection operation the ray has made a round trip and is again leaving mirror 1 towards mirror 2. The complete operation can be represented by a matrix which has been called by Bertolotti a "round trip matrix" which for the geometry considered is

$$\boldsymbol{P} = \boldsymbol{R}\boldsymbol{T}^{12}\boldsymbol{R}\boldsymbol{T}^{21} = \begin{bmatrix} 1 & 0 \\ 0 & -1 \end{bmatrix}\begin{bmatrix} 1 & d \\ 0 & -1 \end{bmatrix}\begin{bmatrix} 1 & 0 \\ 0 & -1 \end{bmatrix}\begin{bmatrix} 1 & d \\ 0 & -1 \end{bmatrix}, \tag{5.1}$$

from which we get

$$\boldsymbol{P} = \begin{bmatrix} 1 & 2d \\ 0 & 1 \end{bmatrix}.$$

For n round trips, the matrix is

$$\boldsymbol{P}^n = \begin{bmatrix} 1 & 2nd \\ 0 & 1 \end{bmatrix}. \tag{5.2}$$

Let x_i and θ_i be the initial values of x and θ, and x_n and θ_n the values after n round trips. From (5.2) we get

$$\begin{bmatrix} x_n \\ \theta_n \end{bmatrix} = \boldsymbol{P}^n \begin{bmatrix} x_i \\ \theta_i \end{bmatrix}, \tag{5.3}$$

which gives

$$x_n = x_i + 2nd\theta_i, \qquad \theta_n = \theta_i. \tag{5.4}$$

To calculate the maximum value of θ which will allow the rays to stay within the cavity formed by mirrors of side a separated by distance d, we note that $x_n = a$ and $x_i = 0$. Then, from (5.4), we get $a = 0 + 2dn\theta_{\max}$,

$$\theta_{\max} = a/2nd.$$

3. Cavity with totally reflecting prism

A rotating 90° roof prism is often used in Q-spoiling. Also, many ruby laser units have a ruby with one end shaped as a 90° roof prism and the other

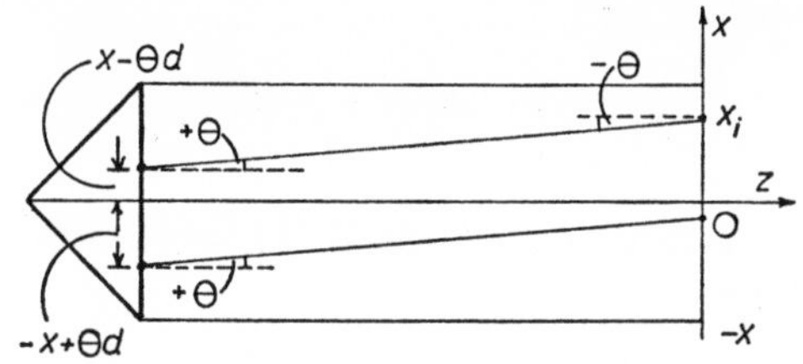

Fig. 5.2. Co-ordinate system used to study a cavity consisting of a roof prism and a plane mirror.

end plane. We may apply the matrix method to the analysis of this system. Figure 5.2 shows the cavity and the coordinate system used. The planes 1, and 2, define the planes to which the matrices refer. Considering only first order values of ϕ, the reflection matrices for the prism are

$$\boldsymbol{R}_{px} = \begin{bmatrix} -1 & 0 \\ 0 & 1 \end{bmatrix} \quad \text{and} \quad \boldsymbol{R}_{py} = \begin{bmatrix} 1 & -|x_i|\sqrt{2} \\ 0 & -1 \end{bmatrix}.$$

Rays which are in planes normal to the edge of the prism are reflected in a direction parallel to their original direction. Rays at angle ϕ with the xz plane diverge rapidly except when $|x_i| = 0$. For rays with $\phi = 0$, the round trip matrix for the plane mirror and prism is

$$\boldsymbol{M}_x = \boldsymbol{R}_x \boldsymbol{T}^{12} \boldsymbol{R}_{px} \boldsymbol{T}^{21} = \begin{bmatrix} 1 & 0 \\ 0 & -1 \end{bmatrix} \begin{bmatrix} 1 & d \\ 0 & -1 \end{bmatrix} \begin{bmatrix} -1 & 0 \\ 0 & 1 \end{bmatrix} \begin{bmatrix} 1 & d \\ 0 & -1 \end{bmatrix},$$

from which we get

$$\boldsymbol{M}_x = (-1) \begin{bmatrix} 1 & 2d \\ 0 & 1 \end{bmatrix}. \tag{5.5}$$

For n round trips, we get

$$\boldsymbol{M}_x^n = (-1)^n \begin{bmatrix} 1 & 2nd \\ 0 & 1 \end{bmatrix}. \tag{5.6}$$

By using (5.6) in the same way as (5.2) we can show that, if a is the x-dimension of the active medium, $|\theta_{\max}| = a/2nd$.

For rays with $\phi \neq 0$, the round trip matrix is

$$\boldsymbol{M}_y = \boldsymbol{R}_y \boldsymbol{T}^{12} \boldsymbol{R}_{py} \boldsymbol{T}^{21} = \begin{bmatrix} 1 & 0 \\ 0 & -1 \end{bmatrix} \begin{bmatrix} 1 & d \\ 0 & -1 \end{bmatrix} \begin{bmatrix} 1 & -\sqrt{2}|x_i| \\ 0 & -1 \end{bmatrix} \begin{bmatrix} 1 & d \\ 0 & -1 \end{bmatrix},$$

from which we get

$$\boldsymbol{M}_y = \begin{bmatrix} 1 & 2d+\sqrt{2}|x_i| \\ 0 & 1 \end{bmatrix}.$$

References on page 130

For n round trips, we get

$$\boldsymbol{M}_y^n = \begin{bmatrix} 1 & n(2d+\sqrt{2}|x_i^n|) \\ 0 & 1 \end{bmatrix}.$$

This gives $|\phi_{\max}| = b/2n[d+(|x_i+2nd\theta_i|)/\sqrt{2}]$, where b is the y-dimension of the active medium.

3.1 *Standing waves in roof-top cavities*

BOBROFF [1964] has discussed the conditions under which standing wave modes can be formed between roof-top reflectors. Instead of considering the paths of rays in the cavity as above, the phase shift is considered. For a standing wave to be established, it is required that the phase shift resulting from a round trip shall be $2\pi q$, where q is an integer. Let the cavity be formed by two roof-top reflectors facing each other distance d apart and orientated so that their ridges are normal to the optic axis but at angle θ to each other as represented in fig. 5.3. Let the x and y axes be as shown with the positive

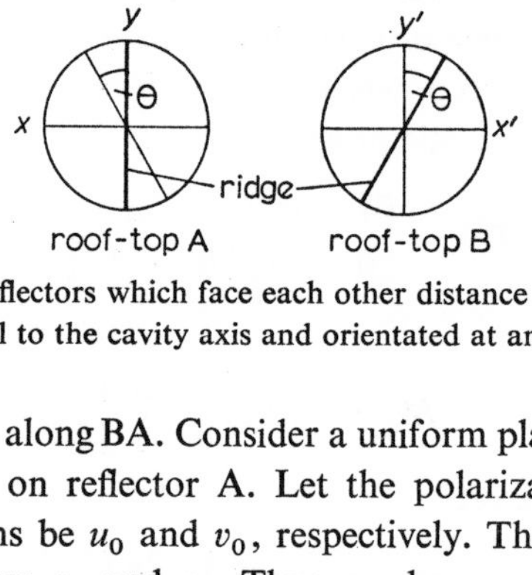

Fig. 5.3. Two roof-top reflectors which face each other distance d apart with their ridges normal to the cavity axis and orientated at angle θ.

direction of the z-axis along BA. Consider a uniform plane wave of arbitrary polarization incident on reflector A. Let the polarization components in the x- and y-directions be u_0 and v_0, respectively. The components of the reflected wave at A are u_1 and v_1. Thus, we have

$$\begin{bmatrix} u_1 \\ v_1 \end{bmatrix} = \begin{bmatrix} \exp(-i\phi_{\parallel}) & 0 \\ 0 & \exp(-i\phi_{\perp}) \end{bmatrix} \begin{bmatrix} u_0 \\ v_0 \end{bmatrix}, \tag{5.7}$$

where $\phi_{\parallel}$ and $\phi_{\perp}$ give the phase delay of the u_0 and v_0 components in traversing roof-top A. In travelling from A to B, both waves are further delayed by $\phi_0 = 2\pi d/\lambda$, where λ is the wavelength in the medium between the reflectors. We have, then

$$\begin{bmatrix} u_2 \\ v_2 \end{bmatrix} = \begin{bmatrix} \exp(-i\phi_0) & 0 \\ 0 & \exp(-i\phi_0) \end{bmatrix} \begin{bmatrix} u_1 \\ v_1 \end{bmatrix}, \tag{5.8}$$

where u_2 and v_2 are the components referred to the axes at A. The axes are now rotated through angle $(-\theta)$ so that the new components of the wave at B are parallel and perpendicular to the ridge of B. Thus, we have

$$\begin{bmatrix} u_3 \\ v_3 \end{bmatrix} = \begin{bmatrix} \cos\theta & -\sin\theta \\ \sin\theta & \cos\theta \end{bmatrix} \begin{bmatrix} u_2 \\ v_2 \end{bmatrix}; \tag{5.9}$$

reflection by B then gives

$$\begin{bmatrix} u_4 \\ v_4 \end{bmatrix} = \begin{bmatrix} \exp(-i\phi_{\parallel}) & 0 \\ 0 & \exp(-i\phi_{\perp}) \end{bmatrix} \begin{bmatrix} u_3 \\ v_3 \end{bmatrix}. \tag{5.10}$$

We now transform back to the original set of axes and obtain

$$\begin{bmatrix} u_5 \\ v_5 \end{bmatrix} = \begin{bmatrix} \cos\theta & \sin\theta \\ -\sin\theta & \cos\theta \end{bmatrix} \begin{bmatrix} u_4 \\ v_4 \end{bmatrix}. \tag{5.11}$$

Both waves now travel back to A arriving with phase delay ϕ_0, so that we get

$$\begin{bmatrix} u_6 \\ v_6 \end{bmatrix} = \begin{bmatrix} \exp(-i\phi_0) & 0 \\ 0 & \exp(-i\phi_0) \end{bmatrix} \begin{bmatrix} u_5 \\ v_5 \end{bmatrix}. \tag{5.12}$$

If the wave is to be an eigenmode of the cavity it must satisfy the self-consistency requirement that

$$\begin{bmatrix} u_6 \\ v_6 \end{bmatrix} = \begin{bmatrix} \exp(-i\phi) & 0 \\ 0 & \exp(-i\phi) \end{bmatrix} \begin{bmatrix} u_0 \\ v_0 \end{bmatrix}. \tag{5.13}$$

Substituting (5.12), (5.11), (5.10), (5.9), (5.8) and (5.7) successively into (5.13) gives

$$\left\{ \begin{bmatrix} \cos\theta & \sin\theta \\ -\sin\theta & \cos\theta \end{bmatrix} \begin{bmatrix} \exp(-i\phi_d) & 0 \\ 0 & \exp(+i\phi_d) \end{bmatrix} \begin{bmatrix} \cos\theta & -\sin\theta \\ \sin\theta & \cos\theta \end{bmatrix} \times \right.$$
$$\left. \times \begin{bmatrix} \exp(-i\phi_d) & 0 \\ 0 & \exp(+i\phi_d) \end{bmatrix} - \begin{bmatrix} \lambda & 0 \\ 0 & \lambda \end{bmatrix} \right\} \begin{bmatrix} u_0 \\ v_0 \end{bmatrix} = 0, \tag{5.14}$$

where

$$\lambda = \exp[i(2\phi_0 + 2\phi_a - \phi)]$$

with $\phi_a = (\phi_{\parallel} + \phi_{\perp})/2$ and $\phi_d = (\phi_{\parallel} - \phi_{\perp})/2$. From (5.14) we get the simultaneous equations

$$[\cos^2\theta \exp(-2i\phi_d) + \sin^2\theta - \lambda]u_0 - [\sin\theta\cos\theta\{1 - \exp(2i\phi_d)\}]v_0 = 0; \tag{5.15}$$

$$[\sin\theta\cos\theta\{1 - \exp(-2i\phi_d)\}]u_0 + [\cos^2\theta \exp(2i\phi_d) + \sin^2\theta - \lambda]v_0 = 0. \tag{5.16}$$

References on page 130

From (5.15) and (5.16) we get the determinantal equation

$$\lambda^2 - 2\lambda[\sin^2\theta + \cos^2\theta\cos 2\phi_d] + 1 = 0$$

which gives

$$\cos[\phi - 2(\phi_0 + \phi_a)] = \sin^2\theta + \cos^2\theta\cos 2\phi_d.$$

The two values given by the quadratic equation in λ may be substituted into (5.15) and (5.16) to give the two eigenpolarizations (ratios of u to v) for the cavity. The fact that solutions are obtainable which satisfy the self consistency requirement imposed, means that standing wave modes with certain polarizations do exist for cavities formed of two roof tops.

4. Spherical mirror cavities

The matrix technique may also be applied to cavities including spherical mirrors. Consider a ray incident on a spherical mirror of radius r. The radius of a concave mirror is positive, that of a convex mirror is negative.

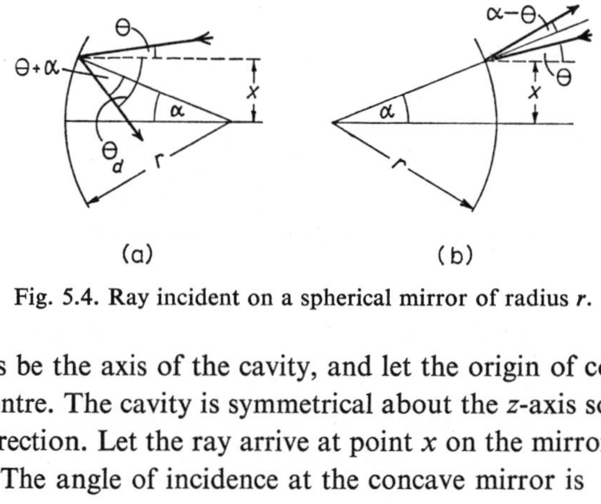

Fig. 5.4. Ray incident on a spherical mirror of radius r.

Let the z-axis be the axis of the cavity, and let the origin of coordinates be the mirror centre. The cavity is symmetrical about the z-axis so we consider only the x-direction. Let the ray arrive at point x on the mirror as shown in figs. 5.4a, b. The angle of incidence at the concave mirror is

$$\alpha + \theta \approx (x/r) + \theta$$

and at the convex mirror is

$$\alpha - \theta \approx (x/r) - \theta.$$

The angles of reflection are, respectively $(x/r)+\theta$ and $(x/r)-\theta$. Hence, reflection at the mirrors may be described by the equations,

$$\text{concave}\quad \begin{aligned} x &= x, \\ \theta_d &= -(2x/r) - \theta; \end{aligned} \qquad \text{convex}\quad \begin{aligned} x &= x, \\ \theta_d &= (2x/r) - \theta. \end{aligned}$$

These equations may be represented by reflection matrices, thus we have

$$\begin{bmatrix} x \\ \theta_d \end{bmatrix} = \boldsymbol{R}_{\text{concave}} \begin{bmatrix} x \\ \theta \end{bmatrix} \quad \text{and} \quad \begin{bmatrix} x \\ \theta_d \end{bmatrix} = \boldsymbol{R}_{\text{convex}} \begin{bmatrix} x \\ \theta \end{bmatrix},$$

where

$$\boldsymbol{R}_{\text{concave}} = \begin{bmatrix} 1 & 0 \\ -\dfrac{2}{r} & -1 \end{bmatrix} \quad \text{and} \quad \boldsymbol{R}_{\text{convex}} = \begin{bmatrix} 1 & 0 \\ \dfrac{2}{r} & -1 \end{bmatrix}. \tag{5.17}$$

If we agree that a convex radius is negative, then both reflection matrices can be represented by $\boldsymbol{R}$ (spherical) given by

$$\boldsymbol{R}_{\text{spher}} = \begin{bmatrix} 1 & 0 \\ -\dfrac{2}{r} & -1 \end{bmatrix}, \tag{5.18}$$

and can be applied to both types of spherical reflecting surface.

The transfer matrices $\boldsymbol{T}^{12}$ and $\boldsymbol{T}^{21}$ are as in (5.3), viz.

$$\boldsymbol{T}^{12} = \boldsymbol{T}^{21} = \begin{bmatrix} 1 & d \\ 0 & -1 \end{bmatrix}. \tag{5.19}$$

4.1 *Combination of plane mirror and spherical mirror*

The hemispherical cavity consists of a concave mirror and a plane mirror separated by the radius of curvature of the concave mirror. The round trip matrix for this system is

$$\boldsymbol{C}_h = \boldsymbol{R}\boldsymbol{T}^{12}\boldsymbol{R}_{\text{spher}}\boldsymbol{T}^{21} = \begin{bmatrix} 1 & 0 \\ 0 & -1 \end{bmatrix} \begin{bmatrix} 1 & r \\ 0 & -1 \end{bmatrix} \begin{bmatrix} 1 & 0 \\ -\dfrac{2}{r} & -1 \end{bmatrix} \begin{bmatrix} 1 & r \\ 0 & -1 \end{bmatrix},$$

from which we get

$$\boldsymbol{C}_h = -1 \begin{bmatrix} 1 & 0 \\ \dfrac{2}{r} & 1 \end{bmatrix}.$$

For n round trips, we have

$$\boldsymbol{C}_h^n = (-1)^n \begin{bmatrix} 1 & 0 \\ \dfrac{2n}{r} & 1 \end{bmatrix}. \tag{5.20}$$

We shall now consider the question of stability of this system. Let x_i and θ_i

References on page 130

be the initial values, and x_n and θ_n the values after n round trips, of the distance from the mirror centres and the angle the ray makes with the cavity axis, respectively.

The final matrix is

$$\begin{bmatrix} x_n \\ \theta_n \end{bmatrix} = \mathbf{C}_h^n \begin{bmatrix} x_i \\ \theta_i \end{bmatrix} = (-1)^n \begin{bmatrix} 1 & 0 \\ \dfrac{2n}{r} & 1 \end{bmatrix} \begin{bmatrix} x_i \\ \theta_i \end{bmatrix}.$$

Hence, we get

$$x_n = (-1)^n x_i;$$

$$\theta_n = (-1)^n \{(2n/r)x_i + \theta_i\}.$$

This shows that if $x_i = 0$, then $x_n = 0$ and $\theta_n = (-1)^n \theta_i$; that is, the rays do not walk out of the cavity and so the cavity is stable for this condition. If $x_i \neq 0$, the values of θ_n can range up to a/r where a is the radius of the mirror aperture. The limiting condition is

$$|\theta_{\max}| \leqslant a/r.$$

Therefore, the number of complete round trips the cavity can maintain when $x_i \neq 0$ is given by

$$(2n/r)x_i + \theta_i \leqslant a/r. \tag{5.21}$$

[*Example.* The number of complete round trips in a hemispherical cavity with mirrors 2 cm diameter of radius of curvature 1 metre if $\theta_i = 0$ and $x_i = 1$ mm is 5.]

Let us now examine the more general case of a plane mirror and a concave mirror separated by distance d. The round trip matrix is

$$\mathbf{C}_g = \begin{bmatrix} 1 & 0 \\ 0 & -1 \end{bmatrix} \begin{bmatrix} 1 & d \\ 0 & -1 \end{bmatrix} \begin{bmatrix} 1 & 0 \\ -\dfrac{2}{r} & -1 \end{bmatrix} \begin{bmatrix} 1 & d \\ 0 & -1 \end{bmatrix}.$$

Multiplying out, we get

$$\mathbf{C}_g = \begin{bmatrix} 1-(2d/r) & 2d-(2d^2/r) \\ -2/r & 1-(2d/r) \end{bmatrix}. \tag{5.22}$$

Applying this matrix to determine the value of x and θ after one round trip we get

$$\begin{aligned} x &= x_i(1-2d/r) + \theta_i(2d-2d^2/r), \\ \theta &= -(2x_i/r) + \theta_i(1-2d/r). \end{aligned} \tag{5.23}$$

Inspection of (5.23) shows that x is independent of θ_i for $d = r$. The equations also show that x is a minimum when $d = r/2$. When this value for d is substituted in (5.22), we get,

$$\mathbf{C}_{r/2} = \begin{bmatrix} 0 & r/2 \\ -2/r & 0 \end{bmatrix},$$

and for n round trips, we have

$$\mathbf{C}_{r/2}^{2n} = (-1)^n \begin{bmatrix} 1 & 0 \\ 0 & 1 \end{bmatrix}. \qquad (5.24)$$

Since this is independent of x and θ, the cavity is stable against walk-out for all rays originating at the centre of the plane mirror with $\theta_i \leqslant a/d = 2a/r$.

4.2 *Combination of two spherical mirrors*

The round trip matrix for concave mirrors of radius r, distance d apart is given by

$$\mathbf{C}_c = \begin{bmatrix} 1 & 0 \\ -2/r & -1 \end{bmatrix} \begin{bmatrix} 1 & d \\ 0 & -1 \end{bmatrix} \begin{bmatrix} 1 & 0 \\ -2/r & -1 \end{bmatrix} \begin{bmatrix} 1 & d \\ 0 & -1 \end{bmatrix},$$

from which we get

$$\mathbf{C}_c = \begin{bmatrix} 1-2d/r & 2d-2d^2/r \\ -(4/r)+(4d/r^2) & 1+(4d^2/r^2)-(6d/r) \end{bmatrix}. \qquad (5.25)$$

Inspection shows that x is independent of θ_i when $d = r$ i.e. the confocal case. In this case, the round trip matrix becomes, for one round trip,

$$\mathbf{C}_{\text{confocal}} = \begin{bmatrix} -1 & 0 \\ 0 & -1 \end{bmatrix},$$

and, for n round trips,

$$\mathbf{C}_{\text{confocal}}^{n} = (-1)^n \begin{bmatrix} 1 & 0 \\ 0 & 1 \end{bmatrix}, \qquad (5.26)$$

which shows the confocal case is very stable, being independent of x and θ.

4.3 *General conditions of stability*

Having calculated $\mathbf{R}_n$, the matrix for n round trips we may then use it to deduce whether or not the cavity is stable. By stable we mean that the values of $|x|$, $|y|$, $|\theta|$ and $|\phi|$ do not become greater than their initial values during any number of round trips. The possibility of attaining this condition with

References on page 130

a given cavity may be examined by inspecting the elements of $\mathbf{R}_n$. This may most easily be done by diagonalizing $\mathbf{R}_n$, for then we are only concerned with the diagonal elements, the others being zero. To diagonalize $\mathbf{R}_n$ we use the transformation matrix $\mathbf{A}$, thus,

$$\mathbf{A}^{-1}\mathbf{R}_n\mathbf{A} = \mathbf{B}, \tag{5.27}$$

where $\mathbf{B}$ is the diagonalized matrix $\mathbf{R}_n$, given by

$$\mathbf{B} = \begin{bmatrix} \lambda_1 & 0 \\ 0 & \lambda_2 \end{bmatrix}. \tag{5.28}$$

The diagonal elements λ_1, λ_2 are the eigenvalues of $\mathbf{R}_n$ and are given by

$$\lambda^2 - \alpha\lambda + \beta = 0, \tag{5.29}$$

where α is the sum of the diagonal elements of $\mathbf{R}_n$ (the trace, or spur) and β is the determinant of $\mathbf{R}_n$. Both the trace and the determinant are invariant, i.e. they do not change when a matrix is transformed. We also know from the theory of quadratic equations that

$$\lambda_1 + \lambda_2 = \alpha \quad \text{and} \quad \lambda_1\lambda_2 = \beta. \tag{5.30}$$

To maintain $|x|$, $|y|$, $|\theta|$ and $|\phi|$ at values at or below their initial value, it is necessary for $|\lambda_1|$ and $|\lambda_2|$ to be less than 1. Thus a cavity is stable if the round trip matrix $\mathbf{R}_n$ is such that

$$\text{Trace } \mathbf{R}_n \leqslant 2 \quad \text{and} \quad \text{Det } \mathbf{R}_n \leqslant 1. \tag{5.31}$$

These are necessary conditions but they are not sufficient. However, we shall not discuss this aspect since the treatment is adequate for the cavities we have studied. (A fuller discussion is given by Bertolotti.)

We shall now determine the condition of stability of a cavity with two concave mirrors of radii r_1 and r_2 at a distance d apart. The round trip matrix is

$$\mathbf{R} = \begin{bmatrix} 1 & 0 \\ -2/r_1 & -1 \end{bmatrix} \begin{bmatrix} 1 & d \\ 0 & -1 \end{bmatrix} \begin{bmatrix} 1 & 0 \\ -2/r_2 & -1 \end{bmatrix} \begin{bmatrix} 1 & d \\ 0 & -1 \end{bmatrix}.$$

Multiplying out, we get

$$\mathbf{R} = \begin{bmatrix} 1-2d/r_2 & 2d-2d^2/r_2 \\ -r_1+(4d/r_1r_2)-(2/r_2) & -(4d/r_1)+(4d^2/r_1r_2)-2dr_2+1 \end{bmatrix}.$$

For stability we require that

$$\text{Trace } \mathbf{R} = 2 - \frac{4d}{r_2} - \frac{4d}{r_1} + \frac{4d^2}{r_1 r_2} \leqslant 2.$$

If we add 2 to both sides of this equation, divide through by 4 and then let

$$g_1 = 1 - \frac{d}{r_1} \quad \text{and} \quad g_2 = 1 - \frac{d}{r_2}, \tag{5.32}$$

we get

$$g_1 g_2 \leqslant 1. \tag{5.33}$$

This is the stability condition derived in ch. 7 § 4 and in § 5.1 of this chapter.

5. Cavity considered as a lens sequence

In studying the geometry of repeated reflections at the spherical mirrors of a laser cavity using ray theory we are not concerned about the direction of the rays but are concerned only with their convergence or divergence. On this basis, a lens produces the same effect as a spherical mirror and we can replace the mirrors with a periodic sequence of lenses in which there is one lens for each reflection†. The axis of cylindrical symmetry is the line of centres of the lenses. Such a sequence of lenses forms an optical trans-

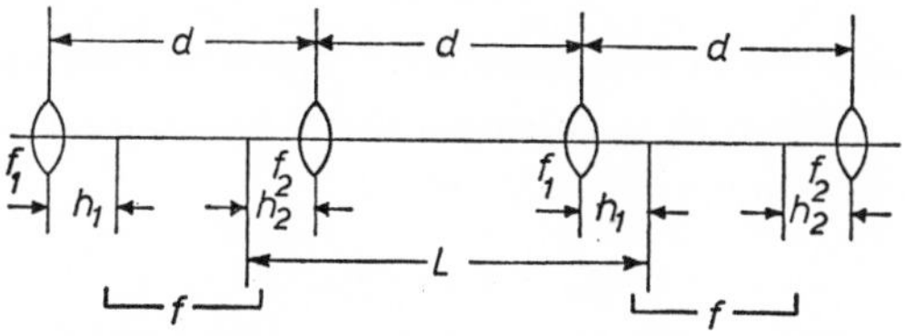

Fig. 5.5. Ray in a cavity considered as a sequence of lenses.

mission line. If the mirrors are of equal curvature and aperture, the lenses are of equal focal length, f, and equal aperture radius a. Let the lenses be separated by distance d as shown in fig. 5.5. (Mirrors of unequal curvature can be represented by a sequence of lenses of focal length f_1, and f_2 in which each pair can be combined to form one equivalent lens of focal length f.) Let the ray leave the nth lens at distance x_n from the axis at angle θ_n to the axis. We wish to determine a condition (if any) for which $x_k < a$ $(k = 1, 2, 3, \ldots)$

† The advantage of this approach is that the ray travels in one direction, thus avoiding complications due to sign conventions.

References on page 130

for all values of k; i.e. the ray does not miss any lens. If this condition can be fulfilled, by a suitable choice of scale, we can arrange that $|x_k| < 1$. As the ray travels through the system it intersects the plane of each lens at a distance x_k from the axis. When we come to use the matrix representation to describe the path of a ray through the lens sequence (§ 5.2), we shall see that the location x_n of the ray at lens n, can be represented by an nth order polynomial of some function of the system parameters d and f. At this stage all we need to note is that n gives both the order of the polynomial and the location of the particular lens considered in the sequence of lenses. We now seek a polynomial which can describe the characteristics of a stable cavity, viz. $|x_k| < 1$.

The Chebyshev polynomials are defined by (COURANT and HILBERT [1953])

$$T_n(y) = \cos(n \cos^{-1} y), \tag{5.34}$$

where $n = 1, 2, 3, \ldots$; they have the property that, of all polynomials of degree n, $T_n(y)$ has the smallest maximum value in the interval $-1 \leqslant y \leqslant 1$ i.e. it deviates least from zero. (Proof of this is given in appendix I.) This is just the sort of characteristic we wish for the ray we consider. By trigonometry we have

$$\cos(n+1)\phi + \cos(n-1)\phi = 2 \cos n\phi \cos \phi,$$

therefore

$$T_{n+1}(y) + T_{n-1}(y) = 2yT_n(y). \tag{5.35}$$

We now consider the lens sequence, fig. 5.5, and note that the rays' distance from the axis is unchanged in passing through a lens, but that its angle with respect to the axis changes by $-x/f$; we get

$$x_n = x_{n-1} + \theta_{n-1} d; \tag{5.36}$$

$$x_{n+1} = x_n + \theta_n d; \tag{5.37}$$

$$\theta_n = \theta_{n-1} - (x_n/f). \tag{5.38}$$

Combining (5.36), (5.37) and (5.38) we get

$$x_{n+1} + x_{n-1} = [2 - (d/f)]x_n. \tag{5.39}$$

Comparing (5.35) and (5.39) we see that if they are to both represent the same stable condition, $T_{n+1}(y)$ and $T_{n-1}(y)$ represent x_{n+1} and x_{n-1}, respectively, and $2y = 2 - (d/f)$ showing that the $T_n(y)$ are polynomials in $(1 - d/2f)$. Since $y = 1 - (d/2f)$ and $|y| \leqslant 1$, we obtain the stability condition

$$0 \leqslant d/f \leqslant 4. \tag{5.40}$$

5.1 *Stability condition*

Consider a resonator with mirrors of radius of curvature r_1 and r_2, respectively. The focal lengths of the equivalent thin lenses are $f_1 = r_1/2$ and $f_2 = r_2/2$ and the resonator can be represented by a sequence of such lenses separated by d, the mirror spacing, as shown in fig. 5.6. It is well

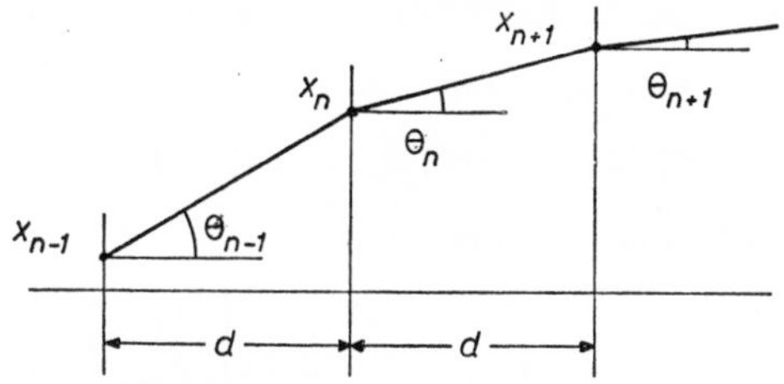

Fig. 5.6. Lens sequence equivalent to cavity with spherical mirrors of unequal focal length.

known that a pair of thin lenses of spacing d, can be represented by a single thick lens of focal length given by

$$\frac{1}{f} = \frac{1}{f_1} + \frac{1}{f_1} - \frac{d}{f_1 f_2}. \tag{5.41}$$

The principle planes of the equivalent thick lens are located at distances $h_1 = d(f/f_2)$ and $h_2 = d(f/f_1)$ from the corresponding lenses. The "effective" spacing between the equivalent thick lenses may be defined as

$$L = d + h_1 + h_2. \tag{5.42}$$

Equations (5.41), (5.42) and the expressions for h_1 and h_2 may be combined to give

$$\frac{L}{f} = d\left[\frac{2}{f_1} + \frac{2}{f_2} - \frac{d}{f_1 f_2}\right]. \tag{5.43}$$

For a sequence of thick lenses, (5.40) becomes

$$0 \leqslant L/f \leqslant 4. \tag{5.44}$$

From the stability condition (5.44) and (5.43) the boundaries of the stable regions may be shown to be given by

$$0 \leqslant \left(\frac{d}{r_1} - 1\right)\left(\frac{d}{r_2} - 1\right) \leqslant 1. \tag{5.45}$$

This stability condition is derived in ch. 7 § 4 by considering a Gaussian

References on page 130

beam propagating in a cavity. An unstable cavity which should be noted is that for which $r_1 < d < r_2$.

5.2 *Matrix treatment of thin lens sequence*

Consider a paraxial ray propagating through a sequence of thin lenses separated by distance d. As in § 5, let the ray leave the nth lens at distance x_n from the axis at angle θ_n. Rewriting (5.36) and (5.38) we have

$$x_n = x_{n-1} + \theta_{n-1} d; \tag{5.46}$$

$$\theta_n = [1-(d/f)]\theta_{n-1} - (x_{n-1}/f). \tag{5.47}$$

In matrix form (5.46) and (5.47) become

$$\begin{bmatrix} x_n \\ \theta_n \end{bmatrix} = \begin{bmatrix} 1 & d \\ -(1/f) & 1-(d/f) \end{bmatrix} \begin{bmatrix} x_{n-1} \\ \theta_{n-1} \end{bmatrix}. \tag{5.48}$$

The square matrix of (5.48) and its inverse are of the form

$$\begin{bmatrix} A & B \\ C & D \end{bmatrix}, \begin{bmatrix} D & -B \\ -C & A \end{bmatrix} (AD-BC)^{-1}. \tag{5.49}$$

Because light paths are reversible (reciprocity), the determinant is unity

$$AD-BC = 1. \tag{5.50}$$

This is a useful check on the matrix products representing many operations in an optical system. The matrix for the full sequence of n lenses is

$$\begin{bmatrix} x_n \\ \theta_n \end{bmatrix} = \begin{bmatrix} 1 & d \\ -(1/f) & 1-(d/f) \end{bmatrix}^n \begin{bmatrix} x_0 \\ \theta_0 \end{bmatrix}. \tag{5.51}$$

Kogelnik and Li [1966] give the following value for the nth power of a square matrix whose determinant is unity:

$$\begin{bmatrix} A & B \\ C & D \end{bmatrix}^n = \frac{1}{\sin\phi} \cdot \begin{bmatrix} A\sin n\phi - \sin(n-1)\phi & B\sin n\phi \\ C\sin n\phi & D\sin n\phi - \sin(n-1)\phi \end{bmatrix}, \tag{5.52}$$

where

$$\cos\phi = (A+D)/2.$$

Equation (5.52) follows from (E.34a) and (E.36).

6. Stability diagram

Fox and Li [1963] have shown that the stability of resonators consisting of two mirrors distance d apart and of radii of curvature r_1 and r_2 (positive for concave mirrors and negative for convex mirrors) can be represented graphically by choosing coordinate axes g_1 and g_2 where $g_1 = 1-d/r_1$ and $g_2 = 1-d/r_2$ as shown in fig. 5.7. Alternatively, and equivalently, BOYD and KOGELNIK [1962] chose the axes d/r_1 and d/r_2. Each point on the diagram represents a particular resonator geometry. Points within the shaded area represent stable systems. By stable we mean that the system is not subject to the type of loss which can be predicted by geometric optics alone. That is, if the mirrors are regarded as a periodic focussing system, a ray within

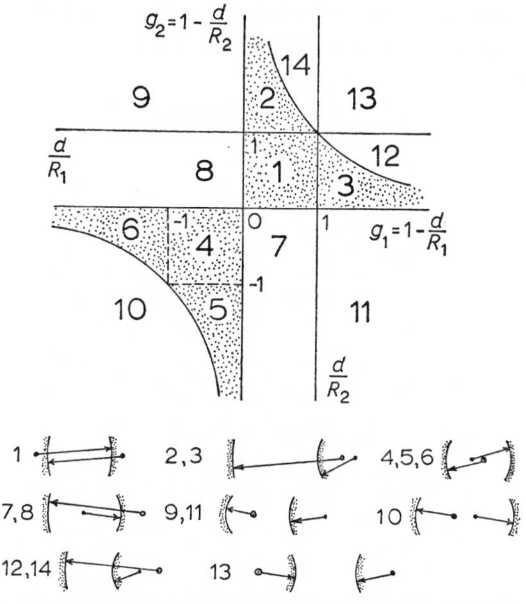

Fig. 5.7. Stability diagram for cavities with spherical mirrors.
Stable cavities are in the shaded regions.
The cavity configurations and corresponding regions of the stability diagram are numbered.

a stable resonator remains within the resonator after any number of round trips but a ray in an unstable resonator walks out. The boundaries between the stable and unstable regions are the axis and the hyperbolae defined by $g_1 g_2 = 1$.

The confocal resonator with identical mirrors lies at the origin and it may be seen that slight deviations (due to manufacturing tolerances, say) makes the system unstable and may incur serious losses. For this reason, it is better to avoid this region.

References on page 130

In addition to geometric losses, both stable and unstable resonators are subject to diffraction losses and it turns out that these losses are very much greater for unstable geometries than for the stable resonators. This is because the fields of the modes in stable resonators remain concentrated near the resonator axis whereas the fields in unstable resonators move progressively towards the periphery of the mirrors where the diffraction losses are greater.

Thus, it is more appropriate to describe the regions of the stability diagram as low-loss and high-loss. The sharp boundaries of the purely geometric case are approached by systems with large Fresnel numbers but for systems with lower Fresnel numbers, in which the diffraction losses are more significant, the boundaries between the regions are less well defined.

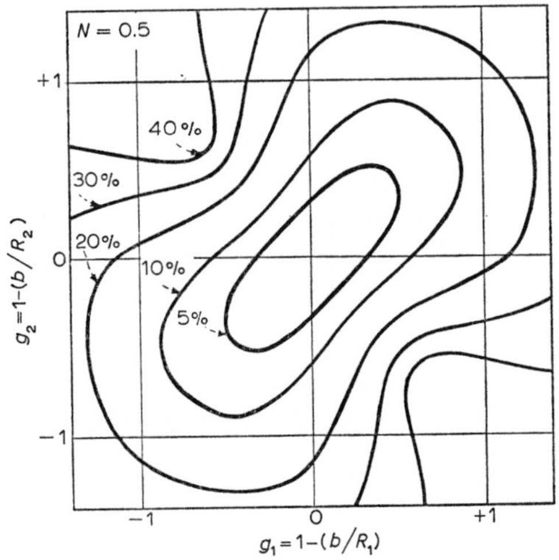

Fig. 5.8. Contour plot of the computed loss function for the dominant mode of a family of infinite-strip curved-mirror interferometers ($N = 0.5$) (reproduced from A. G. Fox and T. Li, Proc. I.E.E.E. **51**, 80 (1963) with permission).

Fox and Li [1963, 1964] have used their iterative technique to calculate the dominant TEM_0 mode and its loss function for a pair of curved infinite-strip mirrors of unequal radii of curvature but of the same width. Their results are shown in fig. 5.8 for $N = 0.5$. The contour lines are equi-loss lines giving the average power loss per transit calculated from the total losses of one round trip. The average is taken because the losses at the two mirrors are not the same when their radii of curvature are not equal. The loss at the origin is 1.9 %. As the Fresnel number increases the contour lines move to the boundaries between high and low loss regions defined by geometry.

A geometrical treatment of losses has been given by Siegman whose results are discussed below in § 7.

7. Geometrical losses from unstable cavities

It has been shown in § 6 that when diffraction losses are neglected ($N = \infty$), a resonator consisting of mirrors of radius R_1 and R_2 separated by distance d can be described by a single point in the $g_1 g_2$ plane, where $g_1 = (1-d/R_1)$ and $g_2 = (1-d/R_2)$. The stability of the resonator may be determined by its location on this diagram, the stable region being defined by the condition $0 \leqslant g_1 g_2 \leqslant 1$. If a resonator lies well within the stable region the low order modes have very small diffraction losses. Near the boundaries of the stable region the diffraction losses increase.

The radiation field after reflection by a mirror is distributed according to the diffraction process outlined in ch. 6. The second mirror does not have infinite aperture and so some of the radiation is lost from the cavity. Thus losses can be described generally as "diffraction" losses. However, it is convenient to consider cavities from the point of view of geometric optics, in which case, rays which miss the mirrors may be designated geometric losses. Cavities which lie in the unstable region have geometrical losses which connect smoothly with the diffraction losses of stable resonators. Unstable cavities and their geometric losses have been studied by SIEGMAN [1965] whose work we now follow. Figure 5.9 shows some examples of unstable cavities.

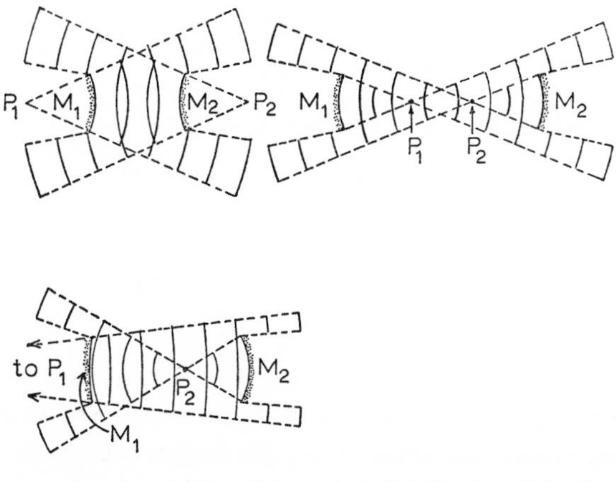

Fig. 5.9. Some examples of unstable cavities and mode behaviour (after SIEGMAN [1965]).

Assume that diffraction losses are negligible (i.e. Fresnel number is large), so that the mirrors are uniformly illuminated. Considering the geometry shown in fig. 5.10, we also assume that the wave leaving mirror 1 is spherical and diverging from a point P_1 which is not necessarily the centre of curvature of mirror 1. Some of this wave is reflected by mirror 2 and the rest misses it.

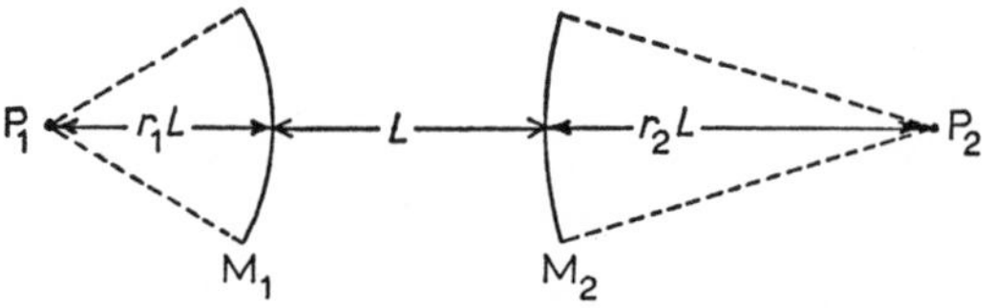

Fig. 5.10. Geometry used to analyse the unstable optical cavity (after SIEGMAN [1965]).

The reflected wave appears to come from point P_2 and reaches mirror 1 where it is again reflected. For self-consistency, the points P_1 and P_2 are required to be images of each other in the appropriate mirrors. The sign convention to be used with the dimensionless distances r_1 and r_2 is that the distances are positive if the point concerned is outside the cavity and negative if it is inside (i.e. between the two mirrors). The mirror radius R is negative if the centre concerned is outside the cavity and positive if it is inside. Since P_1 and P_2 must image each other upon reflection in the respective spherical surfaces, we have, by elementary geometrical optics,

$$\frac{1}{r_1} - \frac{1}{r_2+1} = -\frac{2L}{R_1} = 2(g_1-1);$$

$$\frac{1}{r_2} - \frac{1}{r_1+1} = -\frac{2L}{R_2} = 2(g_2-1).$$

Simultaneous solution of these two equations gives

$$r_1 = \frac{\pm[1-(g_1g_2)^{-1}]^{\frac{1}{2}}-1+g_1^{-1}}{2-g_1^{-1}-g_2^{-1}}; \qquad r_2 = \frac{\pm[1-(g_1g_2)^{-1}]^{\frac{1}{2}}-1+g_2^{-1}}{2-g_1^{-1}-g_2^{-1}}. \tag{5.53}$$

To decide which sign to take, consider the effect of a small displacement δr_1 of P_1 from the self consistent point. If the negative value of the square root is taken, successive values of δr_1 get larger, the centre of curvature of the spherical wave moves further from the self-consistent position, that is, the system is unstable. If the positive value is taken, the successive values of δr_1 get smaller; the system is stable.

The method of studying the behaviour of cavities by considering spherical wavefronts in the simple geometrical way described above and shown in fig. 5.9 gives a very useful and simple physical picture which retains some of the character of the diffraction process while simplifying the calculations.

The energy loss per transit is that which misses the mirror as shown in fig. 5.9. Let $\Gamma_{1,2}$ be the fraction of the energy leaving one mirror which is reflected by the other. The fractional energy remaining after a complete round-trip is $\Gamma^2 \equiv \Gamma_1 \Gamma_2$. If the gain per pass of a laser is G then the threshold for laser oscillation is given by

$$G^2 \Gamma_1 \Gamma_2 = G^2 \Gamma^2 = 1. \tag{5.54}$$

Consider parallel strip mirrors of infinite length, and widths $2a_1$ and $2a_2$ respectively. From elementary geometry the fractional energies reflected are

$$\Gamma_1 = \frac{r_1 a_2}{(r_1+1)a_1}, \qquad \Gamma_2 = \frac{r_2 a_1}{(r_2+1)a_2}. \tag{5.55}$$

For the round trip, we have

$$\Gamma^2_{\text{strip}} = \Gamma_1 \Gamma_2 = \frac{r_1 r_2}{(r_1+1)(r_2+1)}. \tag{5.56}$$

It is important to note that for unstable resonators, the losses are independent of mirror sizes. SIEGMAN [1965] has found this to be true for arbitrary and asymmetrically shaped mirrors as well, provided that the mirror surfaces extend past the centre line of the resonator on both sides. The physical reason for this is that reduction of one mirror reduces the cone angles of the waves in both directions proportionately, so that the relative transverse sizes stay the same, and the fractional power losses stay the same. Substituting the values of r_1 and r_2 from (5.53) into (5.56) we may write Γ^2 in terms of g_1, g_2, thus

$$\Gamma^2_{\text{strip}} = \pm \frac{1-[1-(g_1 g_2)^{-1}]^{\frac{1}{2}}}{1+[1-(g_1 g_2)^{-1}]^{\frac{1}{2}}}. \tag{5.57}$$

The positive sign is valid for g-values in the first and third quadrants of fig. 5.7; the negative sign is valid for the other two quadrants.

Figure 5.11 gives the geometrical losses per transit for half symmetric resonators ($g_1 = 1$, $g_2 = g$) and symmetric resonators ($g_1 = g_2 = g$).

References on page 130

The losses involved are seen to be such that some high gain lasers can operate with such cavities.

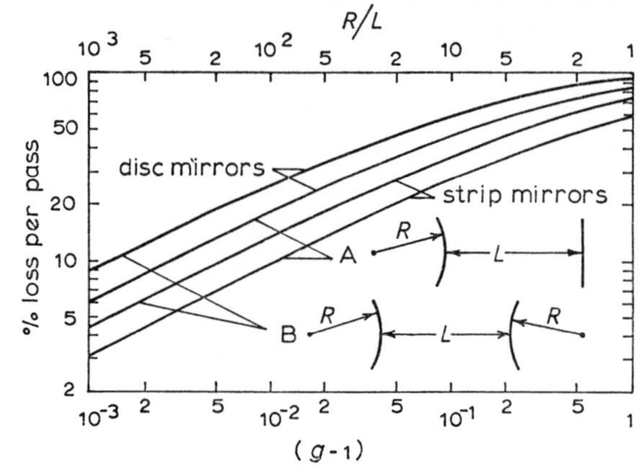

Fig. 5.11. Loss per pass for strip and disc mirrors forming half symmetric and symmetric unstable cavities (after SIEGMAN [1965]).

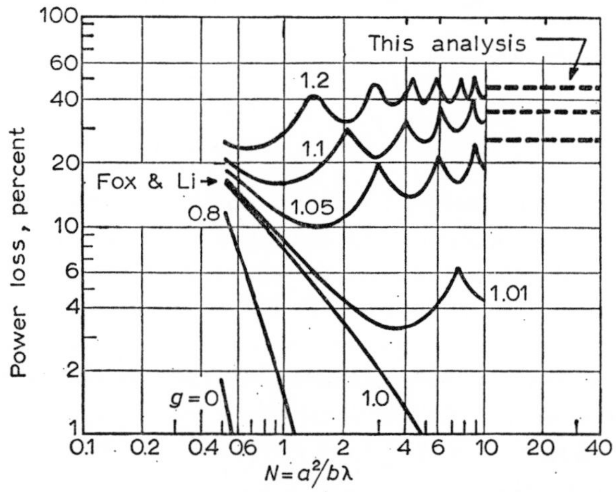

Fig. 5.12. Power loss per transit versus Fresnel number N for the dominant mode of a family of infinite-strip curved mirror cavities. The simple geometrical approach of Siegman is compared with results of the more detailed analysis of Fox and LI (reproduced from A. E. SIEGMAN, Proc. I.E.E.E. **53**, 277 (1965) with permission).

FOX and LI [1963, 1964] have used their iterative technique to study stable and unstable resonators. For mirrors of geometries corresponding to points lying along the line $g_1 = g_2$ of fig. 5.7 they have obtained results which are compared with those of Siegman in fig. 5.12. It is seen that at large Fresnel numbers, Siegman's method for calculating the geometrical losses agrees well with the more detailed results of Fox and Li.

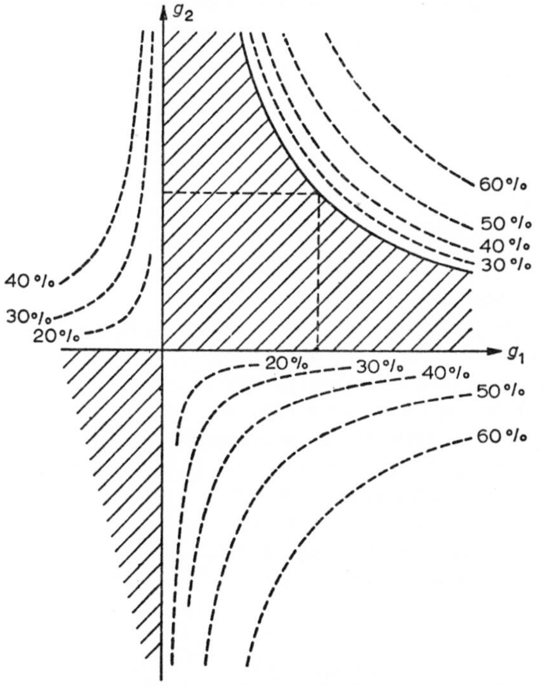

Fig. 5.13. Equiloss contours on the stability diagram for cavities with infinite-strip mirrors. Disc mirrors have substantially larger losses (reproduced from A. E. SIEGMAN, Proc. I.E.E.E. **53**, 277 (1965) with permission).

Siegman has calculated loss contours for the unstable regions of the stability diagram (fig. 5.13) for infinite strip mirrors.

Contours of equal loss in the g plane may be obtained from (5.57); we get

$$g_1 g_2 = \frac{(1 \pm C)^2}{(1 \pm C)^2 - (1 \mp C)^2},$$

where $C = \Gamma^2_{\text{strip}}$. Hyperbolae for several values of the average loss per transit are plotted in fig. 5.13.

Siegman has verified by experiments with ruby that the theory outlined above is good to first-order for the lowest order mode in an unstable resonator. A useful feature of this type of resonator is that the output of the

References on page 130

laser may consist of the geometrical losses. This type of output coupling is generally called "diffraction" coupling even though, in this case, it is primarily of geometric origin rather than diffraction.

The cavity dimensions and mirror reflectivities of many lasers (e.g. ruby) are such that transverse modes of quite high order can oscillate before the diffraction losses become greater than the reflection losses. Geometrical losses provide a useful means of discriminating against these high order modes.

References

BERTOLOTTI, M., 1964, Nuovo Cimento **32**, 1242.

BOBROFF, D. L., 1964, Appl. Opt. **3**, 1485.

BOYD, G. D. and H. KOGELNIK, 1962, Bell Syst. Tech. J. **41**, 1347.

COURANT, R. and D. HILBERT, 1953, Methods of Mathematical Physics, Vol. I (Interscience Publishers, Inc., New York) pp. 87–88.

FOX, A. G. and T. LI, 1963, Proc. I.E.E.E. **51**, 80.

FOX, A. G. and T. LI, 1964, Modes in a maser interferometer with curved mirrors. In: Grivet, P. and N. Bloembergen, eds., Quantum Electronics III (Columbia University Press, New York) pp. 1263–1270.

KOGELNIK, H., 1965, Bell. Syst. Tech. J. **44**, 455.

KOGELNIK, H. and T. LI, 1966, Appl. Opt. **5**, 1550.

SIEGMAN, A. E., 1965, Proc. I.E.E.E. **53**, 277.

CHAPTER 6

OPTICAL RESONATORS (WAVE THEORY)

Whilst many properties of laser cavities have been explained by the treatments given in chs. 4 and 5, we have not yet explained those characteristics which depend on the wave theory of the cavity radiation field. Perhaps the most obvious characteristic to be explained is the mode patterns observed (ch. 4 § 6). We shall now consider laser cavities from the point of view of the scalar theory of electromagnetic radiation as applied to Fresnel diffraction. Background theory is presented in appendix F.

1. Confocal multimode resonators

Understanding of this fundamental aspect of lasers began with the classic papers of Fox and Li [1961] and Boyd and Gordon [1961]. These papers are most important and form the basis for much of the subsequent work on cavity analysis. We shall follow closely the treatment of Boyd and Gordon and use their nomenclature.

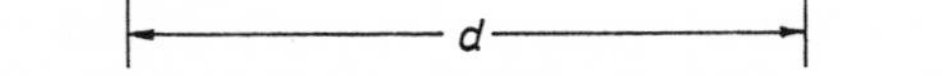

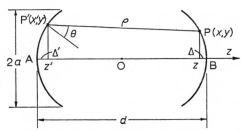

Fig. 6.1. Confocal cavity with spherical square mirrors of side $2a$ and radius of curvature equal to the mirror separation d.

Consider the confocal resonator with square spherical mirrors (side $2a$) shown in fig. 6.1. The medium of the cavity is passive, i.e. non-amplifying. The mirror apertures, radii of curvature and separation are much larger

 References on page 152

than the wavelength. Assume that the radiation field is linearly polarized in the y-direction and that it travels back and forth between the mirrors of the resonator with very little loss at each traversal so that the decay constant of the cavity (ch. 4 § 5.3) is much greater than a period of oscillation of the field and the system may be regarded as being in a steady state. Let us start with the radiation field distributed over mirror A and consider its reflection at B and arrival back at A. At A the radiation field distributed over the mirror surface is E_A, at B it is E_B, when it arrives back at A we shall assume that it is a constant fraction of E_A (because of losses on the round trip). The field at one mirror is expressed in terms of the field at the other by the Kirchhoff-Fresnel diffraction equation which then leads to an integral equation, the eigenfunctions of which give the mode patterns and the eigenvalues give the resonance condition. Let the field just after reflection at mirror A be given by $E_0 f_m(x')g_n(y')$ where E_0 is a constant amplitude factor and $f_m(x')$ and $g_n(y')$ are the radiation field variations over the aperture. The subscripts m and n are integers referring to the x- and y-directions respectively. In a resonator, only certain field distributions are possible and the equations reveal that these are specified by the values of the integers m, n, i.e. the values of the functions $f_m(x')$, $g_n(y')$ depend upon m and n; the values m, n, designate the mode. The field at point $\mathrm{P}(x, y)$ on the surface of the other mirror (B) may be calculated by summing over contributions from the differential Huygens sources at all points $\mathrm{P}'(x', y')$. The result is the Kirchhoff-Fresnel equation (appendix F),

$$E = \int_{S'} \frac{ik(1+\cos\theta)}{4\pi\rho} \,\mathrm{e}^{-ik\rho}\, E_0 f_m(x')g_n(y')\mathrm{d}S', \tag{6.1}$$

where ρ is the distance between P and P′, θ is the angle between the line PP′ and the normal to the reflector at P′, and $k = 2\pi/\lambda$ is the propagation constant of the medium between the mirrors. The electromagnetic waves in the cavity depart from the direction of the z-axis by small angles only, so the electric field in the xz-plane is very nearly zero. Hence, we are investigating transverse electromagnetic modes, TEM. We shall consider the case in which the mirror spacing, d is much greater than the mirror dimensions $2a$. This condition makes $\theta \to 0$. Some of the radiation leaving A is lost by diffraction, the rest reaches B. The field distribution of normal modes or eigenfunctions of resonators reproduce themselves within a constant factor (due to diffraction losses) over each successive mirror. To determine whether this is possible in the present case we investigate the consequences of the requirement that the equation $E = E_1 f_m(x)g_n(y)$ is fulfilled, where $E_1 = \sigma_m \sigma_n E_0$. The proportionality factor $\sigma_m \sigma_n$ is dual to take care of the x- and y-directions and is generally complex, giving both amplitude and phase changes. (We shall use the factor to obtain the condition for resonance and to calculate the energy loss per reflection due to diffraction effects.) Equation (6.1) becomes

$$E = E_0 \sigma_m \sigma_n f_m(x) g_n(y) = \int_{-a}^{+a} \int_{-a}^{+a} \frac{2\mathrm{i}k}{4\pi\rho} \mathrm{e}^{-\mathrm{i}k\rho} E_0 f_m(x') g_n(y') \mathrm{d}x' \mathrm{d}y'. \tag{6.2}$$

Therefore, we have

$$\sigma_m \sigma_n f_m(x) g_n(y) = \int_{-a}^{+a} \int_{-a}^{+a} \frac{\mathrm{i}k}{2\pi\rho} \mathrm{e}^{-\mathrm{i}k\rho} f_m(x') g_n(y') \mathrm{d}x' \mathrm{d}y'.$$

For small aperture mirrors, $d \approx \rho$. From the point of view of magnitudes it is quite acceptable to make this approximation. However, the phase term,

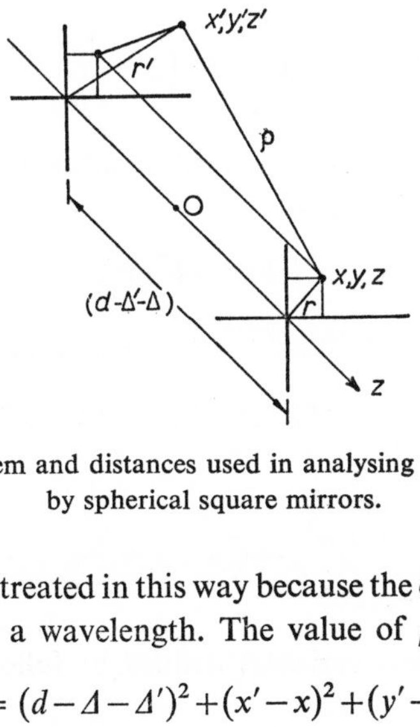

Fig. 6.2. Co-ordinate system and distances used in analysing the confocal cavity formed by spherical square mirrors.

exp $(-\mathrm{i}k\rho)$, cannot be treated in this way because the distances of importance here are a fraction of a wavelength. The value of ρ (fig. 6.2) is given by

$$\rho^2 = (d - \Delta - \Delta')^2 + (x' - x)^2 + (y' - y)^2. \tag{6.3}$$

Simple geometry shows that for mirrors of radius of curvature, b,

$$\Delta = b - (b^2 - r^2)^{\frac{1}{2}} \approx r^2/2b$$

and

$$\Delta' = b - (b^2 - r'^2)^{\frac{1}{2}} \approx r'^2/2b.$$

References on page 152

These two equations may be combined with (6.3) to give

$$\rho = d - \frac{xx'+yy'}{d} - \frac{d-b}{2bd}(x^2+x'^2+y^2+y'^2) + \text{neglected terms.} \tag{6.4}$$

We shall consider the case for which $d = b$, the confocal case. The condition under which terms other than those given may be neglected may be obtained as follows. The greatest possible difference between ρ and b occurs when $x = a$, $y = a$, $x' = -a$, $y' = -a$. Then we have

$$\rho^2 = b^2 + 4a^2,$$

which gives

$$\frac{\rho}{b} = \left(1 + \frac{4a^2}{b^2}\right)^{\frac{1}{2}}.$$

Since this is an alternating convergent series, the error involved in taking only a few terms is not greater than the first neglected term, hence if we write the phase term of (6.2) thus,

$$\exp(-ik\rho) = \exp\left[-ik\left(b + \frac{4a^2}{b} - \frac{4a^4}{b^3}\right)\right],$$

the condition to be fulfilled for the third member of the bracket to be negligible is $4ka^4 \ll b^3$, which gives

$$(a^2/\lambda b) \ll (b^2/a^2). \tag{6.5}$$

Provided this condition is fulfilled we may write, more generally,

$$\exp(-ik\rho) = \exp[-ikb\{1-(xx'+yy')/b^2\}].$$

Hence, from (6.2) we get,

$$\sigma_m \sigma_n f_m(x) g_n(y) = \int_{-a}^{+a}\int_{-a}^{+a} \frac{ik}{2\pi b}\, e^{-ikb} e^{ikxx'/b} e^{ikyy'/b} f_m(x') g_n(y') dx' dy'. \tag{6.6}$$

Introducing dimensionless variables defined as follows

$$X \equiv x\sqrt{c}/a, \quad Y \equiv y\sqrt{c}/a, \quad c \equiv a^2k/b = 2\pi N \tag{6.7}$$

(N is the Fresnel number, $a^2/b\lambda$) and with $F_m(X) \equiv f_m(x)$ etc., (6.6) then becomes,

$$\sigma_m \sigma_n F_m(X) G_n(Y) = \frac{i}{2\pi} e^{-ikb} \int_{-\sqrt{c}}^{+\sqrt{c}}\int_{-\sqrt{c}}^{+\sqrt{c}} F_m(X') e^{iXX'} G_n(Y') e^{iYY'} dX' dY'. \tag{6.8}$$

SOOHOO [1963] investigated the general non-confocal case with mirrors of equal curvature, b; he used (6.4) as above to obtain the non-confocal form of (6.8),

$$\sigma_m \sigma_n F_m(X) G_n(Y) =$$

$$= \frac{\mathrm{i}}{2\pi} \mathrm{e}^{-\mathrm{i}kd} \int_{-\sqrt{c}}^{+\sqrt{c}} F_m(X') \exp(\mathrm{i}XX') \exp\left[\mathrm{i}\frac{d-b}{2b}(X^2+X'^2)\right] \mathrm{d}X' \times$$

$$\times \int_{-\sqrt{c}}^{+\sqrt{c}} G_n(Y') \exp(\mathrm{i}YY') \exp\left[\mathrm{i}\frac{d-b}{2b}(Y^2+Y'^2)\right] \mathrm{d}Y',$$

where $c = a^2k/d$. Soohoo solved this equation numerically. We shall continue with the confocal case.

Let $\chi_m \chi_n = \sigma_m \sigma_n / \mathrm{i}\mathrm{e}^{-\mathrm{i}b}$ and write (6.8) as a pair of equations (we can do this because there are no cross products), thus

$$F_m(X) = \frac{1}{\sqrt{2\pi\chi_m}} \int_{-\sqrt{c}}^{+\sqrt{c}} F_m(X') \mathrm{e}^{\mathrm{i}XX'} \mathrm{d}X' \tag{6.9}$$

and

$$G_n(Y) = \frac{1}{\sqrt{2\pi\chi_n}} \int_{-\sqrt{c}}^{+\sqrt{c}} G_n(Y') \mathrm{e}^{\mathrm{i}YY'} \mathrm{d}Y'. \tag{6.10}$$

Since these equations are identical in form, we shall consider only (6.9). This is a homogeneous Fredholm equation of the second kind with $F_m(X')$ as the kernel (see appendix G, eq. G.2).

The value of c for a He-Ne laser operating at 6328 Å with a confocal cavity of 1 metre and usable mirror dimension $a = 0.2$ cm is about 40. High values of c are typical of most laser systems. For this case and near the axis of the laser, we may obtain an approximate solution of (6.9) by comparing (6.9) with (I.37)† which shows that (6.9) is satisfied by putting

$$F_m(X) = \exp(-X^2/2) H_m(X) \tag{6.11}$$

and

$$\chi_m = \mathrm{i}^m. \tag{6.12}$$

The eigenfunctions $F_m(X)$ give the distribution of field (i.e. the mode patterns ch. 4 § 6) at the mirror surface. The losses and resonance condition may be obtained from the eigenvalues. The phase condition to be fulfilled by the

† By allowing the limits of integration to approach infinity, we assume that contributions to the integrals from points remote from the axis are negligible; i.e. the modes are confined closely to the resonator axis.

cavity at resonance is that the round-trip phase shift must be $2\pi q$, where q is an integer. The phase condition may be obtained from

$$\sigma_m \sigma_n = \chi_m \chi_n \mathrm{i} e^{-ikb}. \tag{6.13}$$

From (6.12) and (6.13), we get

$$\sigma_m \sigma_n = \mathrm{i}^{m+n+1} e^{-ikb}. \tag{6.14}$$

Remembering that $\mathrm{i} = \exp(\mathrm{i}\pi/2)$, we may extract the phase information from (6.14) to obtain the condition

$$2|(m+n+1)(\pi/2) - kb| = 2\pi q,$$

where the factor 2 on the LHS arises because there are two reflections. (We consider the field immediately after reflection at mirror A.) The resonance condition is, therefore

$$(4b/\lambda) = 2q + (m+n+1). \tag{6.15}$$

The cavity will support resonances only for values of λ which satisfy this equation. If $4b/\lambda$ is odd, $(m+n)$ must be even, and vice versa. Since many combinations of m, n, q, can yield the same integer, there is considerable degeneracy at a given value of λ; e.g. increasing $(m+n)$ by 2 and decreasing q by 1 gives the same value for $[2q+(1+m+n)]$. Thus, the TEM_{00q} mode has the same frequency as the $\mathrm{TEM}_{02(q-1)}$ mode. This degeneracy is removed if the spacing is not confocal, but other degeneracies appear at certain other spacings.

The complete specification of a mode is TEM_{mnq}, where $m, n = 0, 1, 2, 3, \ldots$ refer to variations in the x, y-directions and thus specify transverse modes, and $q \sim 10^5$–10^7 specifies the number of half-wavelengths in the z-direction. Other aspects of the modes TEM_{mnq} are discussed in ch. 4 § 6.

The fractional energy loss per transit due to diffraction is

$$\alpha = 1 - |\sigma_m \sigma_n|^2 = 1 - |\chi_m \chi_n \mathrm{i} e^{-ikb}|^2 = 1 - |\chi_m \chi_n|^2. \tag{6.16}$$

The diffraction losses for various TEM_{mnq} modes are shown in fig. 6.3 for confocal and plane parallel resonators. Diffraction losses increase rapidly with increasing values of m, n, i.e. with increase of transverse mode order. Any variation there may be due to longitudinal mode order is entirely negligible. The diffraction loss for the plane-parallel case as calculated for the simple case described in ch. 4. § 3 is also shown. The eigenfunctions (6.11) may be written

$$f_m(x) = \text{constant} \times H_m[x(2\pi/b\lambda)^{\frac{1}{2}}] \exp(-\pi x^2/b\lambda). \tag{6.17}$$

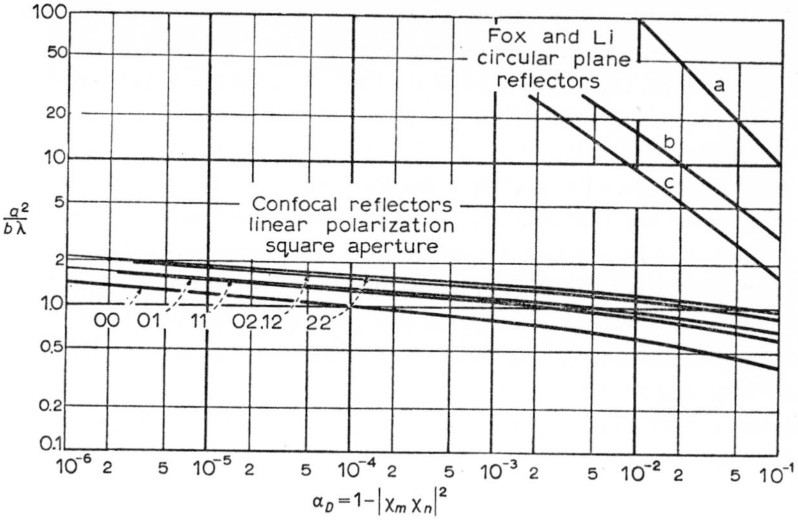

Fig. 6.3. Diffraction losses for confocal and plane-parallel resonators (reproduced from G. D. Boyd and J. P. Gordon, Bell System Tech. J. **40**, 489 (1961) with permission).

a – Fraunhofer diffraction angles (simple calculation, ch. 4, § 3).
b – Circular polarization, symmetric mode
c – Linear polarization, symmetric mode

This states that the eigenfunction (i.e. the mode shape) near the axis of the laser is equal to a normalization constant times the product of a Hermite polynomial of order m and a Gaussian. (For mirrors of circular section, the result is a similar product involving a Laguerre polynomial and a Gaussian; eq. (6.42).) The eigenfunctions of the modes are Hermite-Gaussian functions only for the case of $N = \infty$, i.e. infinite aperture (low loss modes). For mirrors with finite apertures, the approximation is good if the energy of the mode is concentrated within the mirror apertures. The normalization constant depends upon the values of m and n and so contains information about the mode structure.

More accurately, the eigenfunctions and eigenvalues of (6.9) used by Boyd and Gordon are those given by Slepian and Pollak [1961],

$$F_m(c, \eta) \propto S_{0m}(c, \eta), \quad \chi_m = (2c/\pi)^{\frac{1}{2}} i^m R^{(1)}_{0m}(c, 1), \qquad m = 0, 1, 2, \cdots, \tag{6.18}$$

where $S_{0m}(c, \eta)$ and $R^{(1)}_{0m}(c, 1)$ are, respectively, the angular and radial wave functions in prolate spheroidal coordinates as defined by Flammer [1957], and where $\eta = X/\sqrt{c} = x/a$. Since the eigenfunctions are real for real η (inspection of the functions given by Flammer shows this), the

References on page 152

fields of the modes at reflecting surfaces are of constant phase. Equation (6.18) leads to the same resonance condition as was obtained from (6.14), viz. (6.15). There is an infinite number of eigenfunctions and corresponding eigenvalue solutions to (6.9) for any given value of c. Flammer gives some values of these functions for $c \leqslant 5$. According to Flammer, the expansions he has tabulated of the prolate spheroidal wave functions describing the eigenfunctions are tractable only for values of c from zero to about ten. The range of c over which the expansions are usable varies with n for a given m. For larger values of c, an asymptotic expansion of the spheroidal wave function is necessary. Equation (6.17) is a good approximation for the paraxial condition but gets progressively worse as the point considered moves away from the mirror centre.

Remembering that we started out with the product $F_m(X)G_n(Y)$ we get, finally, for the Gaussian,

$$\exp\{-(x^2+y^2)\pi/b\lambda\}. \tag{6.19}$$

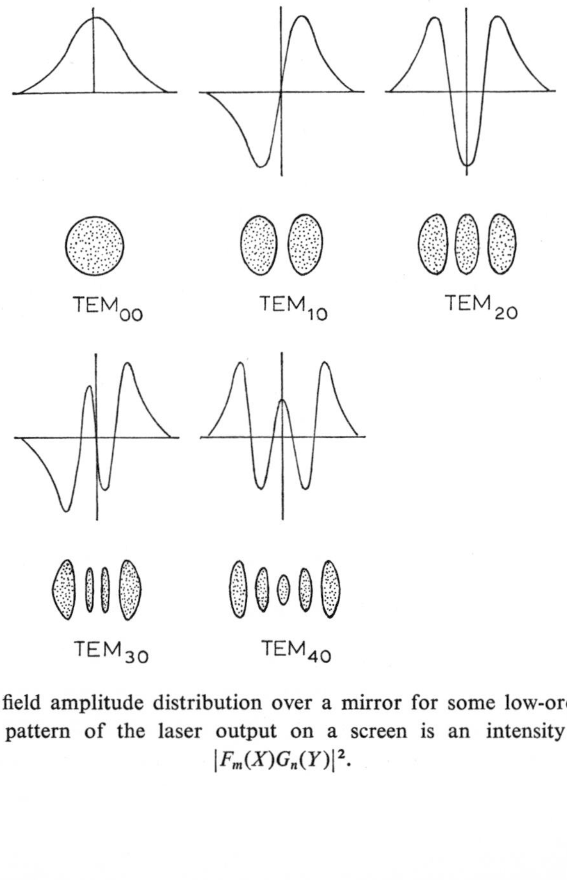

Fig. 6.4. The field amplitude distribution over a mirror for some low-order transverse modes. The pattern of the laser output on a screen is an intensity distribution, $|F_m(X)G_n(Y)|^2$.

Figure 6.4 shows the field distribution† over a mirror for some low-order transverse modes. The fundamental mode, $\mathrm{TEM}_{00}(m = 0,\ n = 0)$ is a Gaussian because $H_0 = 1$. The higher order modes are the Hermite polynomials times the same Gaussian. The negative values of $E(x)$ mean that there is a phase reversal. For example the mode appears as shown in fig. 6.4. The phase across the pattern reverses in going from one spot to the other. The question of patterns and mode designation is taken up more fully in ch. 4 § 6. A simple demonstration of the 180° phase shift between the neighbouring lobes of a given mode pattern may be given by using Young's double source technique in which the sources are located by pinholes. When the two holes are placed in one lobe, a set of interference fringes is obtained. When one hole is placed in one lobe and the other in the lobe next to it, the bright fringes become dark and vice-versa indicating a phase reversal.

The region around the mirror centre defined by the radius at which the intensity falls to 1/e of its value at the centre is, by (6.19), of radius $w = w_s$ where $w^2 = x^2 + y^2$, and

$$w_s = (b\lambda/\pi)^{\frac{1}{2}}. \tag{6.20}$$

It should be noted that the radius of this region, called the "spot-size", or "beam-radius", where most of the energy is concentrated, is independent of the mirror dimensions. Increasing the mirror size only reduces the diffraction losses; it does not affect the spot-size.

2. Confocal resonator fields

Using the same techniques as described above, § 1, we may investigate the field distribution at an arbitrary point within the resonator cavity or outside the cavity. The field at any point inside is essentially that of a standing wave (made up of two waves travelling in opposite directions), the field outside is that of a travelling wave of amplitude reduced by the transmission coefficient of the mirror. If the point we consider is located at x, y, z where z is measured from the resonator centre, we can define a dimensionless parameter

$$\xi \equiv z/(b/2) = 2z/b.$$

Within the resonator $0 < \xi < 1$. Outside the resonator $\xi > 1$. The field at the point is the travelling wave field due to waves originating at one of the

† Remember that the pattern of the laser output on a screen is an intensity distribution $|F_m(X)G_n(Y)|^2$.

References on page 152

mirrors. As before, the field, for large values of c (i.e. large Fresnel number, N, (6.7)) may be approximated by Hermite-Gaussian functions for rectangular mirrors and by Laguerre-Gaussian functions for circular mirrors. Continuing the discussion of square section mirrors, the travelling wave field obtained by BOYD and GORDON [1961] at a point in a confocal resonator resulting from waves originating at one of the mirrors is given by

$$\frac{E(x,y,z)}{E_0} = \left(\frac{2}{1+\xi^2}\right)^{\frac{1}{2}} \frac{\Gamma(m/2+1)\Gamma(n/2+1)}{\Gamma(m+1)\Gamma(n+1)} \exp\left\{-\frac{kw^2}{b(1+\xi^2)}\right\} \times$$
$$\times H_m\left\{X\left(\frac{2}{1+\xi^2}\right)^{\frac{1}{2}}\right\} H_n\left\{Y\left(\frac{2}{1+\xi^2}\right)^{\frac{1}{2}}\right\} \times$$
$$\times \exp\left[-\mathrm{i}\left\{k\left(\frac{b}{2}(1+\xi)+\frac{\xi}{1+\xi^2}\frac{w^2}{b}\right)-(1+m+n)\left(\frac{\pi}{2}-\phi\right)\right\}\right], \tag{6.21}$$

where $w^2 = x^2+y^2$, and $\tan\phi = (1-\xi)/(1+\xi)$. The phase factor in square brackets describes variation of intensity in the z-direction.

The transverse standing wave in the cavity is obtained by replacing the exponential phase function in (6.21) by the corresponding sine function.

The constant phase surface may be obtained from the phase function. A point on this surface is x, y, z and the surface cuts the axis at $0, 0, z_0$. If we neglect the small variation of ϕ due to variation in z we obtain for the surface of constant phase,

$$k\left\{\frac{b}{2}(1+\xi)+\frac{\xi}{1+\xi^2}\frac{w^2}{b}\right\} = k\left\{\frac{b}{2}(1-\xi_0)\right\}.$$

Hence we get

$$z-z_0 \approx -\frac{\xi}{1+\xi^2}\frac{w^2}{b} \tag{6.22}$$

and, close to the axis, the constant phase surface is spherical. The radius of curvature of this surface is

$$b' = \left|\frac{1+\xi^2}{2\xi}\right| b. \tag{6.23}$$

Differentiation of (6.23) with respect to ξ shows that the surfaces of minimum radius of curvature are at $z = \pm b/2$ i.e. the mirror surfaces. Figure 6.5 shows the mirrors and some equiphase spherical surfaces of the cavity within the confines of the envelope of the TEM_{00q} mode where the field distribution falls to $1/e$ of its axial value.

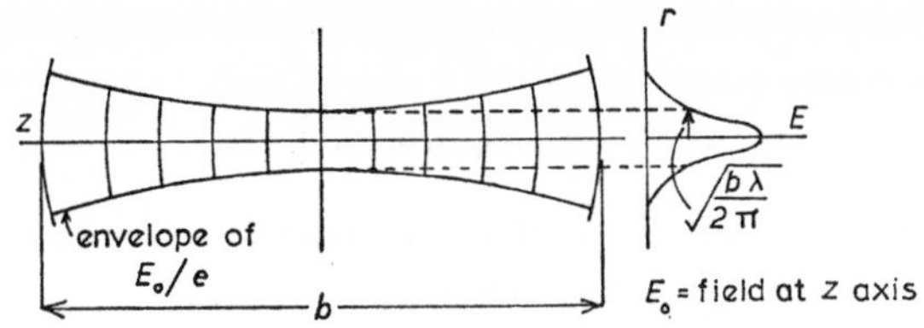

Fig. 6.5. Concave spherical mirrors and some equiphase spherical surfaces within the confines of the envelope of the TEM_{00q} mode, where the field amplitude distribution falls to $(1/e)$ of its axial value.

The field distribution at the arbitrary point considered is shown by (6.21) to fall to $1/e$ at a radius given by

$$w_s = \{\lambda b(1+\xi^2)/2\pi\}^{\frac{1}{2}}. \tag{6.24}$$

The smallest beam radius is at $\xi = 0$, i.e. at the focal plane of the confocal system. For a laser operating at about 6000 Å with a confocal cavity of 1 metre the beam radius at the focal plane is about 3×10^{-2} cm. At the mirrors, it is $\sqrt{2}$ times larger.

In terms of beam radius (spot size), equation (6.21) may be written, for the field amplitude

$$\frac{E(x, y, z)}{E_0} = \text{const.}\,\frac{1}{w_s}\exp\left(-\frac{w^2}{w_s^2}\right) H_m\left(\sqrt{2}\,\frac{x}{w_s}\right) H_n\left(\sqrt{2}\,\frac{y}{w_s}\right). \tag{6.21a}$$

3. Non-confocal concave resonator

We have seen that the field of a confocal cavity consists of spherical equiphase surfaces. Mirrors of the same curvature may replace any of these surfaces without altering the electromagnetic field within the new cavity from what it was in the old. The particular cavity formed when a plane mirror is located at O, the centre of the cavity, is called "hemispherical". The wavefronts of equal curvature are symmetrically spaced about the plane through O. When mirrors of equal curvature are placed symmetrically with respect to the plane at $z = 0$, the resulting mode is also symmetrical. However, mirrors could be located asymmetrically and asymmetrical modes result; this is discussed below. The frequencies at which the new cavity will resonate depend on the phase condition to be satisfied. Alternatively, we may regard a cavity of length d where the mirrors are not separated by their radius of curvature b' provided $b' \geqslant d/2$ as having an equivalent confocal cavity with mirror separation and radius of curvature b_{eq}. Putting $\xi = d/b_{eq}$ into (6.23) we get

$$b_{eq}^2 = 2db' - d^2. \tag{6.25}$$

This quadratic equation shows that two mirror separations d_1 and d_2 are possible for a given b_{eq} and b'. Thus

$$d_{1,2} = b' \pm (b'^2 - b_{eq}^2)^{\frac{1}{2}}. \tag{6.26}$$

Thus spot-size at the reflectors is given by

$$w_s' = \left(\frac{d\lambda}{\pi}\right)^{\frac{1}{2}} \left[2\frac{d}{b'} - \left(\frac{d}{b'}\right)^2\right]^{-\frac{1}{4}}. \tag{6.27}$$

This shows that for a given spacing between the mirrors, the spot-size is minimum for confocal resonators, i.e. $b' = d$.

Summarizing the nomenclature, we have

- $d \equiv$ Spacing of mirrors.
- $b \equiv$ Radius of curvature (mirror spacing as well for the confocal case).
- $b' \equiv$ Radius of curvature of a constant phase surface. It is also used for the radius of curvature of the mirrors of a nonconfocal resonator.
- $b_{eq} \equiv$ Radius of curvature of mirrors forming the confocal cavity equivalent to the actual cavity.

We may use the equivalent confocal resonator concept to calculate the spot-size of the fundamental mode of a non-confocal system. The spot-size at the centre when the mirrors are of equal curvature is obtained from (6.24) and is given by

$$w_0 = \left(\frac{b_{eq}\lambda}{2\pi}\right)^{\frac{1}{2}}. \tag{6.28}$$

The spot-size at distance $d/2$ from the centre is given by

$$w_s' = w_0(1 - d^2/b_{eq}^2)^{\frac{1}{4}}. \tag{6.29}$$

Figure 6.5 shows the equiphase surfaces symmetrical about the plane, $z = 0$, which results from symmetrical placing of the mirrors about this plane.

3.1 *Mode separation and degeneracy*

The condition for resonance of a nonconfocal system may be obtained from the phase angle term of (6.21) which is in square brackets. As in deriving (6.15), the round-trip phase shift is $2\pi q$. The condition finally obtained is

$$\frac{4d}{\lambda} = 2q+(1+m+n)\left(1-\frac{4}{\pi}\tan^{-1}\frac{b_{eq}-d}{b_{eq}+d}\right). \tag{6.30}$$

When $b_{eq} \approx d$ (6.30) becomes

$$\frac{4d}{\lambda} = 2q+(1+m+n)\left(1-\frac{2}{\pi}\left(1-\frac{d}{b}\right)\right). \tag{6.31}$$

Comparing (6.30) and (6.15) we see that 4 × (mirror spacing)/λ is necessarily an integer in the confocal case but is not in the nonconfocal resonator. The important consequence of this is that, unlike the confocal resonator, the modes of the nonconfocal resonator are not degenerate in $(m+n)$, though other degeneracies may appear. The mode separation is given by

$$\Delta\left(\frac{1}{\lambda}\right) = \frac{1}{4d}\left[2\Delta q+\Delta(m+n)\left(1-\frac{4}{\pi}\tan^{-1}\frac{b_{eq}-d}{b_{eq}+d}\right)\right]. \tag{6.32}$$

4. Resonators with cylindrical symmetry

4.1 *Circular mirrors*

The analysis presented in this chapter has been based on Cartesian coordinates. In practice most mirrors have circular section and, for these cylindrical coordinates are more convenient and appropriate. As before (ch. 6. § 1), we consider mirrors A and B so that in cylindrical coordinates, (6.1) for the field at mirror B becomes

$$E = \int_0^a\int_0^{2\pi} \frac{i}{2\lambda}\frac{e^{ik\rho}}{\rho}\left(1+\frac{b}{\rho}\right) E_0 f_{lp}(r_1, \phi_1) r_1\,d\phi_1\,dr_1 .$$

The field just after reflection at mirror A is given by $E_0 f_{lp}(r_1, \phi_1)$ where E_0 is a constant amplitude factor and $f_{lp}(r_1, \phi_1)$ is the radiation field variation over the aperture. The subscripts l, p are integers designating the mode and refer to the radial and azimuthal directions, respectively. Before the radiation arrives back at A, it loses a fraction κ_{lp} of its amplitude. Hence, the self-consistency condition for the radiation at A becomes

$$\kappa_{lp} f_{lp}(r_2, \phi_2) = \int_0^a\int_0^{2\pi} \frac{i}{2\lambda}\frac{e^{ik\rho}}{\rho}\left(1+\frac{b}{\rho}\right) f_{lp}(r_1, \phi_1) r_1\,d\phi_1\,dr_1 .$$

This may be simplified by referring to fig. 6.1 and noting that

$$\rho = (b_1^2+r_1^2+r_2^2-2r_1 r_2 \cos(\phi_1-\phi_2))^{\frac{1}{2}}$$

and

$$b_1 = b - \Delta_1 - \Delta_2,$$

and that for confocal spherical mirrors

$$\Delta_i = b - (b^2 - r_i^2)^{\frac{1}{2}}, \qquad i = 1, 2.$$

If b/a is large, then we get

$$\Delta_i \approx r_i^2/2b,$$

and finally, provided $(a^2/b\lambda) \ll (b/a)^2$ (as in (6.5)) we obtain

$$\kappa_{lp} f_{lp}(r_2, \phi_2) =$$
$$= \frac{\mathrm{i}e^{-ikb}}{\lambda b} \int_0^a \int_0^{2\pi} f_{lp}(r_1, \phi_1) \exp\left[\mathrm{i}k(r_1 r_2/b)\cos(\phi_1 - \phi_2)\right] r_1 \,\mathrm{d}\phi_1 \,\mathrm{d}r_1 .$$

We shall assume that the azimuthal variation is sinusoidal so that

$$f_{lp}(r_i, \phi_i) = R_{lp}(r_i) \exp(-\mathrm{i}l\phi_i), \qquad i = 1, 2. \tag{6.33a}$$

We then obtain

$$\kappa_{lp} R_{lp}(r_2) \exp(-\mathrm{i}l\phi_2) =$$
$$= \frac{k\mathrm{i}e^{-ikb}}{b} \int_0^a \left\{ \frac{1}{2\pi} \int_0^{2\pi} \exp\left[\mathrm{i}k\left(\frac{r_1 r_2}{b}\right)\cos(\phi_1 - \phi_2) - \mathrm{i}l\phi_1\right] \mathrm{d}\phi_1 \right\} R_{lp}(r_1) r_1 \,\mathrm{d}r_1 .$$

Using (M.9) and identifying n with l, we get

$$\kappa_{lp} R_{lp}(r_2) = \frac{k\mathrm{i}^{l+1} e^{-ikb}}{b} \int_0^a J_l\left(\frac{kr_1 r_2}{b}\right) R_{lp}(r_1) r_1 \,\mathrm{d}r_1 . \tag{6.34a}$$

Multiplying both sides by $\sqrt{r_2}$ and letting $\gamma_{lp} = \kappa_{lp} \exp(-\mathrm{i}kb)$ we get

$$\gamma_{lp} R_{lp}(r_2)\sqrt{r_2} = \int_0^a K_{lp}(r_1, r_2) R_{lp}(r_1) \sqrt{r_1} \,\mathrm{d}r_1$$

and, similarly

$$\gamma_{lp} R_{lp}(r_1)\sqrt{r_1} = \int_0^a K_{lp}(r_1, r_2) R_{lp}(r_2) \sqrt{r_2} \,\mathrm{d}r_2 ,$$

where

$$K_{lp}(r_1, r_2) = \frac{\mathrm{i}^{l+1} k}{b} J_l\left(\frac{kr_1 r_2}{b}\right) \sqrt{r_1 r_1} .$$

4.2 *Spherical annular mirrors*

Cavities with output coupling apertures are important for lasers operating in the infra-red region. The eigenmodes of a symmetric cylindrical confocal laser resonator and their perturbation by the output coupling apertures have been analysed by McCumber whose treatment we now follow. For convenience, we shall also use his nomenclature. Figure 6.6 shows the cavity considered. The annular mirrors are identical and totally reflecting between radii a_0 and a_m. The eigenmodes of a cavity with no coupling holes (i.e. $a_0 = 0$) are derived first and then the results of introducing coupling holes are given.

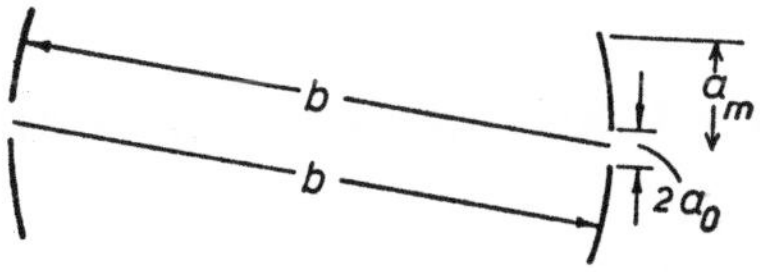

Fig. 6.6. Confocal cavity formed by spherical annular mirrors.

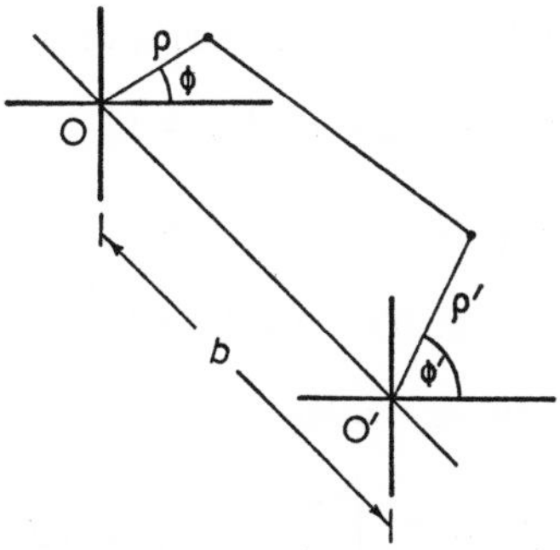

Fig. 6.7. Co-ordinate system and distances used in analysing confocal cavity formed by spherical annular mirrors.

The field amplitude at the mirrors for a typical mode may be described in terms of radial and angular coordinates (ρ, ϕ) as in fig. 6.7 and integers, l, p thus

$$F_{lp}(\rho, \phi) = f_{lp}(\rho) \exp(-il\phi). \qquad (6.33b)$$

Equation (6.33b) is simply (6.33a) rewritten. The integer p describes the angular (azimuthal) dependence of the modes and l describes the radial dependence. Since the system is symmetric and the mirrors identical, the field at one mirror must be a constant multiple of that at the other.

References on page 152

As in ch. 6 § 1 we use the scalar formulation of Huygens' principle together with the self-consistency requirement to get

$$\kappa_{lp} f_{lp}(\rho) = \frac{2\pi}{b\lambda} \int_{a_0}^{a_m} \rho' J_l \left(\frac{2\pi\rho\rho'}{b\lambda} \right) f_{lp}(\rho') \mathrm{d}\rho', \tag{6.34b}$$

where $J_l(z)$ is the Bessel function of order $|l|$. Equation (6.34b) may be obtained from (6.34a). The eigenvalue κ_{lp} gives the diffraction loss, α, per pass for the (lp) mode, thus,

$$\alpha = 1 - |\kappa_{lp}|^2. \tag{6.35}$$

The phase of the eigenvalue determines the resonant wavelength, λ, (cf. ch. 6 § 1).

$$\lambda = 4\pi b[(l+1)\pi - 2 \operatorname{Arg} \kappa_{lp} - 2\pi n]^{-1}, \tag{6.36}$$

where n is any integer.

We define the Fresnel number

$$N(\rho) \equiv r^2 \equiv \rho^2/\lambda b \tag{6.37}$$

and describe the hole and the mirror in terms of it, thus

$$N_0 = r_0^2 = a_0^2/\lambda b, \quad N_m = r_m^2 = a_m^2/\lambda b.$$

As in ch. 6 § 1 we define a new function thus

$$g_{lp}(r) \equiv f_{lp}(r\sqrt{(\lambda b)}). \tag{6.38}$$

Equation (6.34b) may now be written

$$\kappa_{lp} g_{lp}(r) = 2\pi \int_{r_0}^{r_m} r' J_l(2\pi r r') g_{lp}(r') \mathrm{d}r'. \tag{6.39}$$

McCumber has expanded the Bessel-function kernel as a power series

$$J_l(z) = \left(\frac{z}{2}\right)^l \sum_{m=1}^{\infty} \frac{(-1)^{m-1}}{(m+l-1)!(m-1)!} \left(\frac{z}{2}\right)^{2(m-1)} \tag{6.40}$$

which is truncated after M terms. The integral eigenvalue equation (6.39), was then reduced to an M-dimensional-matrix eigenvalue equation and solved numerically, by matrix-diagonalization, for Fresnel numbers in the range $0.6 \leqslant N_m \leqslant 2.0$.

When there is no aperture, $N_0 = 0$, the power lost by diffraction per transit (6.35) is given in fig. 6.8.

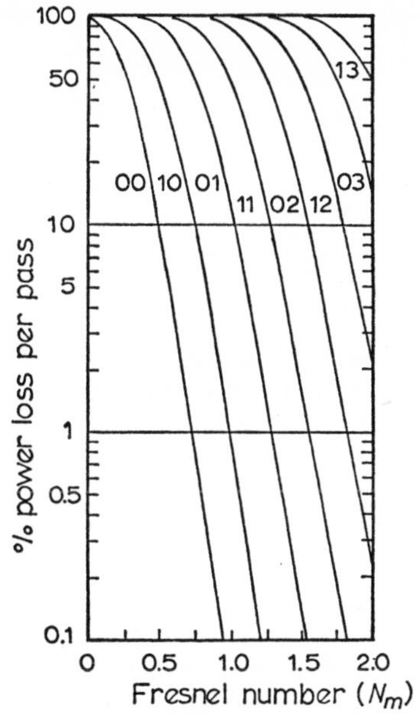

Fig. 6.8. Power loss by diffraction per transit between spherical circular mirrors for low loss modes with no output coupling hole and small values of mirror Fresnel number N_m (after McCumber [1965]).

The plots of field intensity $g_{lp}^2(r)$ against r obtained by McCumber are given in figs. 6.9–6.12. They show that the higher order modes have more intensity lying outside the reflecting mirrors than do the lower order modes. Since the radiation intensity distributed outside the mirror diameter is lost we see that the power loss per transit increases as the mode order increases.

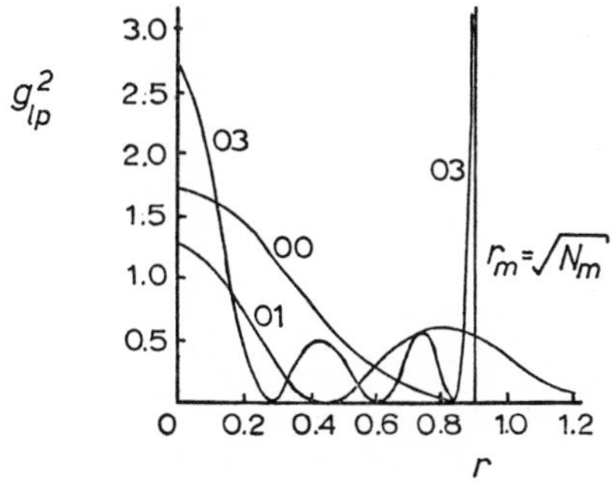

Fig. 6.9. Field intensity distribution versus r for 00, 01, and 03 modes with $N_m = 0.8$ and $N_0 = 0$ (after McCumber [1965]).

References on page 152

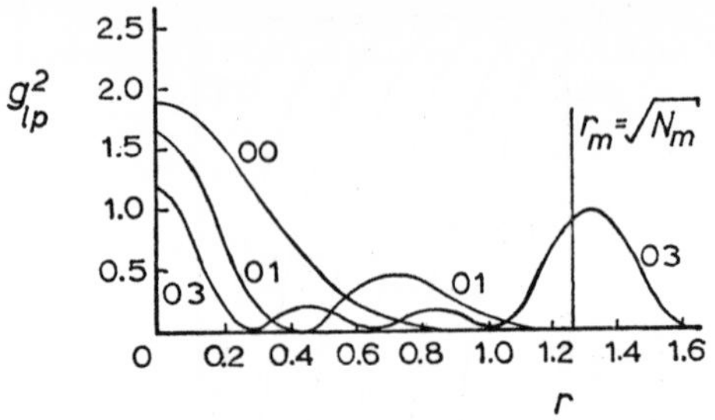

Fig. 6.10. Field intensity distribution versus r for 00, 01, and 03 modes with $N_m = 1.6$ and $N_0 = 0$ (after McCUMBER [1965]).

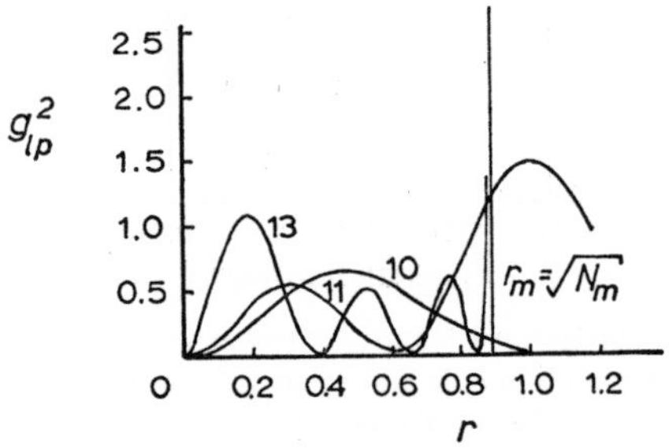

Fig. 6.11. Field intensity distribution versus r for 10, 11, 13 modes with $N_m = 0.8$ and $N_0 = 0$ (after McCUMBER [1965]).

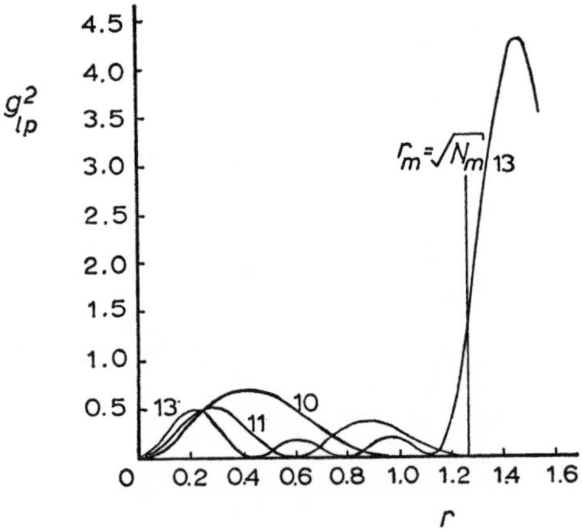

Fig. 6.12. Field intensity distribution versus r for 10, 11, and 13 modes with $N_m = 1.6$ and $N_0 = 0$ (after McCUMBER [1965]).

Only the angular independent modes ($l = 0$) have non-zero intensity at $r = 0$. This indicates that the $l = 0$ modes will be more sensitive to a coupling hole located at $r = 0$ than are the $l \neq 0$ modes.

The eigenvalues and eigenfunctions for infinite mirrors without apertures ($N_m \to \infty$, $N_0 = 0$) are

$$\kappa_{lp} = (-1)^p \tag{6.41}$$

and

$$g_{lp}(r) = \left[\frac{2p!}{(l+p)!}\right]^{\frac{1}{2}} (2\pi r^2)^{l/2} \mathrm{e}^{-\pi r^2} L_p^l(2\pi r^2), \tag{6.42}$$

where $L_p^l(z)$ is the associated Laguerre polynomial,

$$L_p^l(z) = \frac{\mathrm{e}^z z^{-l}}{p!} \frac{\mathrm{d}^p}{\mathrm{d}z^p} (\mathrm{e}^z z^{p+l}) = \sum_{m=0}^{p} \frac{(p+l)!(-z)^m}{(p-m)!(l+m)!m!}. \tag{6.43}$$

The sign of an eigenvalue determines the phase.

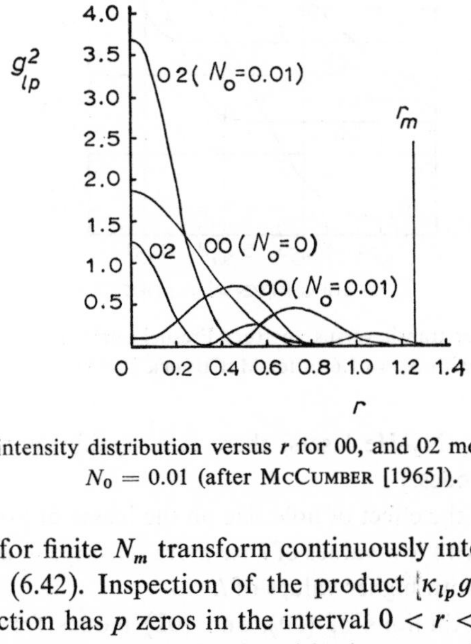

Fig. 6.13. Field intensity distribution versus r for 00, and 02 modes with $N_m = 1.6$ and $N_0 = 0.01$ (after McCumber [1965]).

The results for finite N_m transform continuously into the solutions given by (6.41) and (6.42). Inspection of the product $\kappa_{lp} g_{lp}(r)$ shows that the amplitude function has p zeros in the interval $0 < r < r_m$. We refer to the interpretation of modes of confocal cavities in terms of l and p in ch. 4 § 6.

Having shown how some of the lower order eigenmodes are obtained together with some of their characteristics, for $N_0 = 0$, we consider the effects of a hole of radius a_0 in the mirror, and centred on the resonator axis so that $N_0 \neq 0$, fig. 6.13. Designation of the modes has now to be related

References on page 152

to those obtained with zero diameter hole already studied. The $N_0 \neq 0$ modes are identified by the indices that those modes would carry if they deformed continuously from $N_0 = 0$. The power loss per transit of the $N_0 = 0$ modes with $p \geqslant 0$ increases with increase of p. This is not so with the $N_0 \neq 0$ modes, the higher order modes, may, or may not have the greater power losses.

McCumber has shown that the circular holes cause mode mixing but they do not mix modes with different values of l. Mode mixing is strongest among modes with the same p parity $(-1)^p$. (If the resonators have dissimilar mirrors where only one mirror has an output-coupling hole, as is the

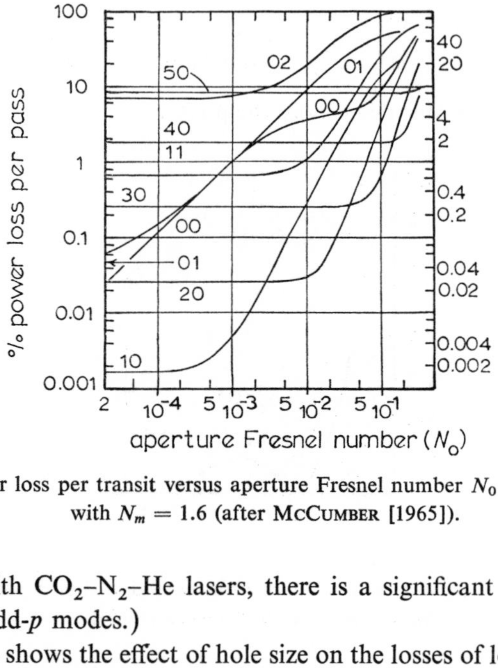

Fig. 6.14. Power loss per transit versus aperture Fresnel number N_0 for low loss modes with $N_m = 1.6$ (after McCumber [1965]).

usual case with CO_2–N_2–He lasers, there is a significant mixing between even-p and odd-p modes.)

Figure 6.14 shows the effect of hole size on the losses of low order modes. For given p, modes with low values of l are more sensitive to a small aperture than are the modes with higher values of l.

The intensity distributions of (00), and (02) modes for $N_m = 1.6$ and $N_0 = 0.01$ and the same modes with $N_0 = 0$ are compared in fig. 6.13. The intensity of the (00) mode is considerably reduced in the region of $r = 0$ by the presence of the hole but the intensity of the (02) mode is increased. This increase in intensity over the hole diameter implies that more of the (02) mode can escape through the hole. Thus we expect greater losses from the (02) mode when there is a hole at the mirror centre. Figure 6.14 shows that this is, indeed, the case for values of $N_0 \gtrsim 10^{-3}$. The power losses per transit for some modes for fixed hole size are shown in

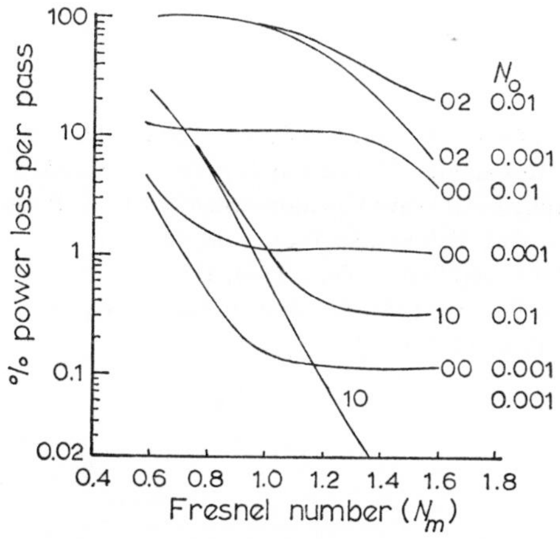

Fig. 6.15. Power loss per transit versus mirror Fresnel number N_m for some low order modes for various aperture Fresnel numbers N_0 (after McCumber [1965]).

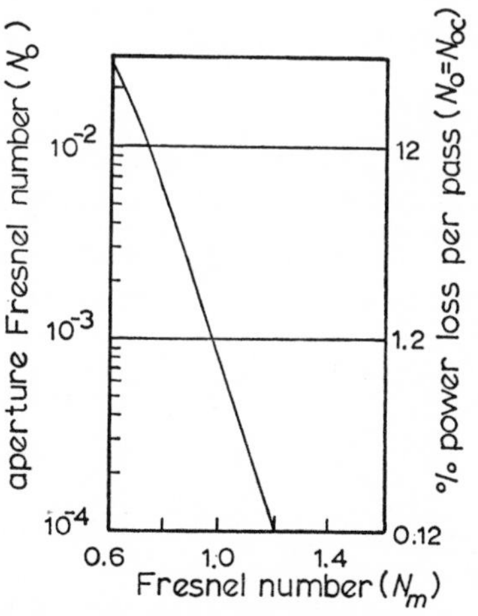

Fig. 6.16. Critical aperture Fresnel number N_{0c} for which diffraction losses of (00) mode equal those of (10) mode versus the mirror Fresnel number N_m. The loss per transit when $N_0 = N_{0c}$ is also shown for each value of N_m (after McCumber [1965]).

fig. 6.15. The aperture for which the losses of the (00) mode are equal to those of the (10) mode is designated N_{0c} and is of interest because,

References on page 152

provided other conditions are not changed, a laser will oscillate in the (00) mode for $N_0 < N_{0c}$ and in the (10) mode for $N_0 > N_{0c}$, or in some other mode if N_0 is very large. Figure 6.16 shows the relationship between N_{0c} and N_m.

References

BOYD, G. D., and J. P. GORDON, 1961, Bell System Tech. J. **40**, 489.

BOYD, G. D., and H. KOGELNIK, 1962, Bell System Tech. J. **41**, 1347.

FLAMMER, C., 1957, Spheroidal Wave Functions (Stanford Univ. Press, Palo Alto, Calif.).

FOX, A. G. and T. LI, 1961, Bell System Tech. J. **40**, 453.

MCCUMBER, D. E., 1965, Bell System Tech. J. **44**, 333.

SLEPIAN, D. and H. O. POLLAK, 1961, Bell System Tech. J. **40**, 43.

SOOHOO, R. F., 1963, Proc. I.E.E.E. **51**, 70.

CHAPTER 7

GAUSSIAN BEAMS

1. Introduction

In chapter 6 we used the scalar theory of the diffraction of light to show that the electric field distribution of the fundamental mode (TEM_{00}) of oscillation of a stable cavity is a Gaussian†. Looked at from the point of view of Fox and Li (ch. 4 § 2) the beam travelling back and forth between the mirrors eventually settles down with a Gaussian distribution. The reason for this may be found in the fact that each reflection produces a Fourier transformation of the distribution and the Fourier transform of a Gaussian is a Gaussian.

The starting point for the analysis of chapter 6 is the Kirchhoff-Fresnel equation (6.1) which, in turn, is based upon the scalar theory of electromagnetic radiation in which we represent the electric field (or the magnetic field) by a single scalar function u as described in appendix F. This is equivalent to considering a single component of the electric field, i.e. a beam plane polarized with the electric vector in the x-direction (say), E_x. The scalar theory leads to the scalar wave equation (Helmholtz)

$$\nabla^2 u + k^2 u = 0, \tag{7.1}$$

where $k = 2\pi/\lambda$ is the propagation constant in free space. We shall use (7.1) as our starting point and shall show that a Gaussian beam is a solution of the equation. In so doing we shall derive an equation which describes the propagation of the beam in terms of certain "beam parameters". We then relate these beam parameters to the boundary conditions imposed by a cavity, and show how they are transformed when the beam intercepts an optical component (such as a lens). We shall use these results to show how cavities can be matched to one another. A thorough review of laser beams and resonators has been given by Kogelnik and Li [1966].

† Plane-parallel and concentric mirrors have intensity distributions transverse to the beam axis which are sine or cosine functions.

References on page 175

2. Propagation in free space

In this section we derive the parameters which describe the propagation of a Gaussian beam in an unbounded medium. The reader will be familiar with the propagation characteristics of plane waves and spherical waves. In a plane wave (i.e. a plane phase front) the amplitude u_0 of the electric field is independent of the location of the point considered in the xy-plane and is independent of z, the location of the plane considered. Similarly, the amplitude of the electric field in a spherical wave is uniform over any spherical surface which can be drawn about the centre of propagation but depends upon the distance of the surface from the centre. A Gaussian beam generated by a laser differs from the plane waves and spherical waves described above in some important aspects. We have seen in ch. 6 § 2 that the waves close to the laser axis (z-axis) are approximately spherical, with the approximation getting worse as the distance from the axis increases. Inspection of (6.23) shows that the centre from which the spherical wavefront appears to originate is not fixed as it is in the case of a conventional spherical wave system, and, finally, the amplitude of the electric field is a Gaussian function of distance from the z-axis, (6.21).

Equation (7.1) is known to be satisfied by plane waves and spherical waves; we now seek the requirements to be fulfilled by a Gaussian beam if it is to be a solution. We have seen above that the amplitude of a Gaussian beam is a function of x, y, and z and thus, we may assume a solution of the form

$$u = u_0 \chi(x, y, z) \exp(-ikz), \tag{7.2}$$

where χ is the unknown amplitude function. At this stage (7.2) could apply to other waveforms than the Gaussian and we may find eventually that the function χ contains phase information. To determine the form of $\chi(x, y, z)$ we substitute (7.2) into (7.1) and simplify the equation obtained by noting that for a laser beam it is reasonable to assume that $\chi(x, y, z)$ is a slowly varying function of z (because of the beam directivity) so that $\partial^2\chi/\partial z^2$ may be neglected: we get

$$\frac{\partial^2\chi}{\partial x^2} + \frac{\partial^2\chi}{\partial y^2} - 2ik\frac{\partial\chi}{\partial z} = 0. \tag{7.3}$$

We now assume a solution which has Gaussian characteristics and determine the conditions to be fulfilled. Such a solution could be of the form

$$\chi = \exp\left[F_2(z) - \frac{x^2+y^2}{F_1(z)}\right]. \tag{7.4}$$

The function $F_1(z)$ describes the variance (i.e. beam width) and may be expected to accommodate two possibilities. The first is that of a change in beam width with distance along the laser axis (z); and the second is that of a phase shift in the xy-plane associated with the changing curvature of the wave front caused by the changing width. The function $F_2(z)$ is to describe both the change in amplitude at the axis of the beam (where $x^2+y^2 = 0$) due to its changing width, and an additional phase shift as the beam propagates. Substitution of (7.4) into (7.3) gives the equation

$$(x^2+y^2)\left[\frac{2}{F_1^2(z)} - \mathrm{i}k\frac{F_1'(z)}{F_1^2(z)}\right] - \left[\frac{2}{F_1(z)} + \mathrm{i}kF_2'(z)\right] = 0, \tag{7.5}$$

where a prime denotes differentiation with respect to z.

For the above to be satisfied for all values of x and y, both of the square bracketed expressions must be identically zero. Solving the equations and integrating we get the following two relations

$$F_1(z) = (A+2z/\mathrm{i}k), \tag{7.6}$$

where A is a constant (possibly complex) determined by $F_1(z = 0)$; and

$$F_2(z) = -\ln\left(z+\frac{\mathrm{i}Ak}{2}\right) + B, \tag{7.7}$$

where B is a complex constant determined by $F_2(z = 0)$. We have therefore shown that, with the above assumptions, a Gaussian beam is an acceptable solution of the wave equation. The expressions for $F_1(z)$ and $F_2(z)$ describe the way in which this beam propagates. We now rewrite these functions in a form which provides a clearer insight into the nature of the beam.

We consider, firstly, the function $F_1(z)$. By choosing the origin of the z-axis accordingly, we can make the constant A real, and so write

$$F_1(z) = (A^2-4z^2/k^2)/(A+2\mathrm{i}z/k). \tag{7.8}$$

The real part of the above expression leads to an amplitude term in the expression for χ, (7.4) of the form

$$\exp\left[-A(x^2+y^2)/(A^2+4z^2/k^2)\right]. \tag{7.9}$$

For a fixed value of z, therefore, $(A^2+4z^2/k^2)/A$ is a measure of the distance from the z-axis in the xy-plane before the beam has fallen to a fraction $1/\mathrm{e}$ of its value at the axis (it is twice the variance of the Gaussian), and it further describes how this "beam width" varies with distance along

References on page 175

the axis of propagation. (The solution is only physically meaningful for $A > 0$.) At $z = 0$, the beam has a minimum width of radius w_0 given by

$$w_0^2 = A; \tag{7.10}$$

this is known as the "beam waist". It is emphasised that at this stage the value of w_0 is arbitrary. When the beam is subject to the boundary conditions of a cavity, the value of w_0 and its position are fixed. At distance z along the axis of propagation the beam has a width given by

$$w^2(z) = w_0^2[1+(2z/kw_0^2)^2]. \tag{7.11}$$

The parameter w is known as the beam radius or "spot size". The imaginary part of $F_1(z)$ leads to a phase term in the expression for χ of the form

$$\exp\left[-\frac{2iz(x^2+y^2)/k}{w_0^4\{1+(2z/w_0^2k)^2\}}\right]. \tag{7.12}$$

A wave-front is a surface of constant phase, and hence from (7.2), (7.4) and (7.12) it is given by

$$kz+\frac{k(x^2+y^2)}{2z[1+(w_0^2k/2z)^2]} = \text{constant}, \tag{7.13}$$

where we have assumed the phase term introduced by $F_2(z)$ varies so slowly with z that it can be neglected here.

Making use of the symmetry of the wave front, we can obtain its radius of curvature by simple differentiation of (7.13) after putting $y = 0$. At the point of intersection of the wave front with the z axis, its radius of curvature is

$$R(z) = z[1+(w_0^2k/2z)^2]. \tag{7.14}$$

At the beam waist ($z = 0$), the wave front is that of a plane wave, while for large z it approaches that from a point source located at the origin.

We now consider the term $F_2(z)$ given by (7.7) which we rewrite in the form

$$F_2(z) = -\ln[(z^2+k^2w_0^2/4)^{\frac{1}{2}}e^{i\phi}]+B, \tag{7.15}$$

where

$$\phi = \tan^{-1}(kw_0^2/2z). \tag{7.16}$$

This leads to the following term in the expression for χ given by (7.4), $(w_0/w)\exp(-i\phi)$. We have derived the conditions to be fulfilled by a Gaussian beam if it is to satisfy the scalar wave equation (7.1). In doing so we have obtained the following important beam parameters,

$$\phi = \tan^{-1}(kw_0^2/2z), \tag{7.17}$$

$$w^2(z) = w_0^2[1+(2z/kw_0^2)^2], \tag{7.18}$$

$$R(z) = z[1+(w_0^2k/2z)^2]. \tag{7.19}$$

The angle of divergence, θ, of the Gaussian beam as z becomes very large (fig. 7.1) is obtained from $\tan\theta = w/R \sim \theta$ which gives

$$\theta = 2/kw_0 . \tag{7.20}$$

The beam parameters $R(z)$ and $w(z)$ are the same for modes of all orders.

The parameters associated with the Gaussian beam are summarized in (fig. 7.2).

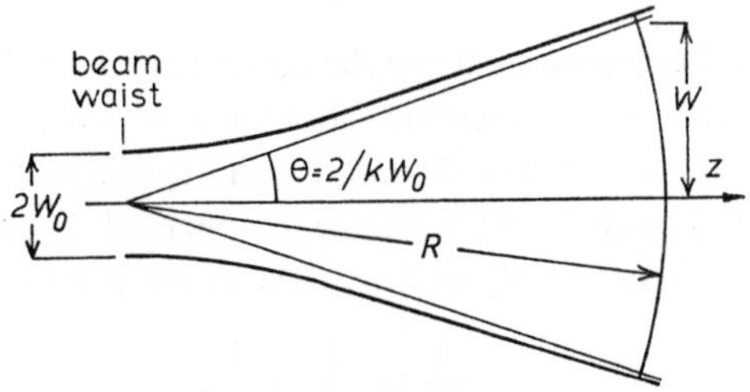

Fig. 7.1. In the far field, the Gaussian beam approaches a spherical wave front diverging from a point on the axis at the beam waist.

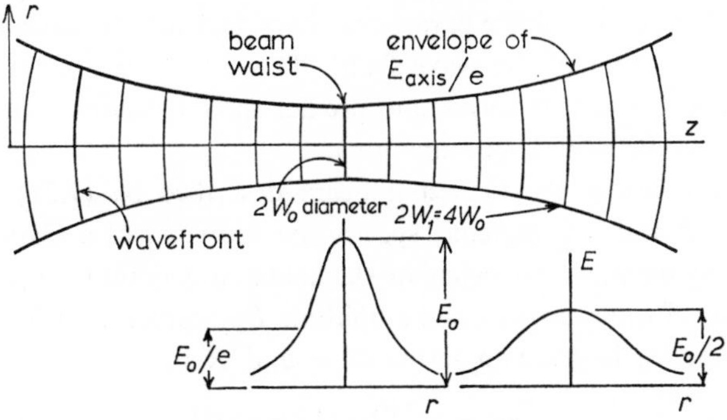

Fig. 7.2. The intensity distribution, wave fronts, and envelope of a Gaussian beam.

The complete expression for the field distribution E_x of a Gaussian is

$$E_x = E_{x0}\left(\frac{w_0}{w}\right)\exp\left\{-\mathrm{i}(kz+\phi)-(x^2+y^2)\frac{\mathrm{i}k}{2}\left[\frac{1}{R}-\frac{2\mathrm{i}}{kw^2}\right]\right\}. \tag{7.21}$$

The reason for expressing this equation in this particular form will be explained in the next section (§ 3).

3. Transformation by a lens

Before we consider the effect on Gaussian beams of optical components, such as lenses, we shall briefly review some of the familiar features of spherical waves and the effect of such components on them. Consider spherical waves propagating from a stationary point source. The radii of any two phase fronts separated by distance z along the axis of propagation normal to the phase fronts are related by

$$R_2 = R_1 + z. \tag{7.22}$$

Throughout this chapter we shall use the convention that R is positive if the wave front appears convex when viewed from $z = +\infty$. When a spherical wave passes through a thin lens of focal length f (positive if the lens causes a plane wave front to be concave), it retains its spherical shape but its radius of curvature, R_1, is altered to R_2 as described in the equation

$$\frac{1}{R_2} = \frac{1}{R_1} - \frac{1}{f}. \tag{7.23}$$

Thus, to determine the effect of propagation, or transformation by a lens on a spherical wave front, we simplify the problem considerably by considering only the effects on the single parameter characterizing the wave, namely, the radius of curvature of the wave front. This simple treatment enables us to manipulate spherical waves and predict their behaviour with adequate accuracy (geometrical approximation).

We now consider the Gaussian beam described by (7.21). The beam parameters R and w completely specify the geometry of a Gaussian beam, and knowing them, we can calculate the field at any point (x, y, z) in a beam of radiation of wave length λ and amplitude E_{x0} using (7.21). The minimum beam width, w_0 is given in terms of w and R by

$$w_0^2 = w^2/[1+(\lambda R/\pi w^2)^2]$$

and w_0, in turn, gives ϕ. As with the spherical wave, for most purposes we do not want to know the field but only the beam parameters R and w and how they are transformed by propagation or by optical components. We shall consider the effects of propagation in free space and through a thin lens. In both these processes the Gaussian character of the beam is retained.

As in the case of a spherical wave, we should like to have just one parameter to manipulate but the complication is that a Gaussian beam is characterized by two parameters. However, the quantity in square brackets in (7.21) contains both parameters conveniently separated by being respectively in the real and imaginary parts of the quantity. This suggests that we represent the terms in square brackets by a single term, thus

$$\frac{1}{q} = \frac{1}{R} - \frac{2i}{kw^2}, \tag{7.24}$$

where q is called the complex beam parameter. This is found to be a successful substitution with q playing for the Gaussian wave, a role similar to that of the radius of curvature of a spherical wave as shown below. Sometimes

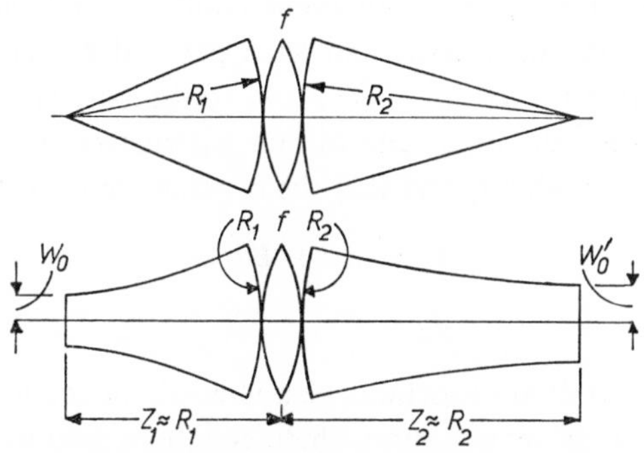

Fig. 7.3. Comparison of the effects of a lens on a spherical wave and on a Gaussian wave.

q is called the complex radius of curvature. Figure 7.3 summarizes the analogy between the propagation of spherical and Gaussian waves.

For a freely propagating Gaussian wave, if the value of the complex radius of curvature at one point is q_1, and at a point distance z away, it is q_2 then (7.18), (7.19) and (7.24) show that

$$q_2 = q_1 + z. \tag{7.25}$$

We now consider the transformation of a Gaussian beam as it passes through a lens, by investigating the effect of the lens on each beam parameter in turn. If the lens is thin, then the width of the beam does not alter as the beam passes through the lens, and therefore the beam width immediately to the left of the lens, w_1, is equal to the beam width immediately to the right of the lens, w_2; thus, we have

$$w_2 = w_1. \tag{7.26}$$

The curvature of the wave-front of the Gaussian beam in the paraxial approximation is spherical (ch. 6 § 2) and transforms in exactly the same way as a spherical wave, as given in (7.23) where R_1 and R_2 are the radii of curvature before and after the lens respectively.

From equation (7.23) and (7.26) the equation describing the transformation of a beam by a thin lens of focal length f is given by

$$\frac{1}{q_2} = \frac{1}{q_1} - \frac{1}{f}, \tag{7.27}$$

where q_1 and q_2 are the input and output parameters, respectively.

3.1 *Ray transfer matrix*

Consider an optical system with input (suffix 1) and output (suffix 2) planes normal to the optic axis, and let a paraxial ray cut the planes at distance x_1, x_2 from the axis at angle θ_1, θ_2 to the axis. To obtain a general relationship between the input and output parameters, we assume that x_2 and θ_2 depend on x_1 and θ_1 and that the dependence is of the form,

$$x_2 = Ax_1 + B\theta_1,$$

$$\theta_2 = Cx_1 + D\theta_1,$$

where A, B, C, and D are functions characteristic of the particular optical system considered. In matrix form, these equations become

$$\begin{bmatrix} x_2 \\ \theta_2 \end{bmatrix} = \begin{bmatrix} A & B \\ C & D \end{bmatrix} \begin{bmatrix} x_1 \\ \theta_1 \end{bmatrix},$$

where the $ABCD$ matrix is the ray transfer matrix.

Example 1

The ray transfer matrix for a ray travelling an optical distance d between two planes in a homogeneous medium is given by

$$\begin{bmatrix} 1 & d \\ 0 & 1 \end{bmatrix}.$$

Example 2

In a thin lens, the input and output planes coincide and the ray transfer matrix is given by

$$\begin{bmatrix} 1 & 0 \\ -1/f & 1 \end{bmatrix}.$$

Example 3

The ray transfer matrix for an input plane distance d from the output plane located at a thin lens may be obtained by multiplication of the ray transfer matrices of examples 1 and 2 above, thus we get

$$\begin{bmatrix} 1 & d \\ 0 & 1 \end{bmatrix} \begin{bmatrix} 1 & 0 \\ -1/f & 1 \end{bmatrix} = \begin{bmatrix} 1-d/f & d \\ -1/f & 1 \end{bmatrix}.$$

Reversal of the ray direction results in an interchange of the diagonal elements of the ray matrix.

3.2 *Application to Gaussian beams*

It will be remembered that the rays used in geometrical optics are normal to the wave front. If the waves are spherical with radius of curvature R, we have

$$x = R\theta$$

for paraxial rays. From the above equations the radius of curvature at the output plane is related to that at the input by

$$R_2 = \frac{AR_1+B}{CR_1+D}. \tag{7.27a}$$

It is shown in ch. 7 § 3 that the complex parameter q of a Gaussian beam is formally equivalent to the radius of curvature, R, of a spherical wave. Therefore, we may apply the transformation (7.27a) to find the effect of the optical system on a Gaussian beam and get

$$q_2 = \frac{Aq_1+B}{Cq_1+D}. \tag{7.27b}$$

Equation (7.27b) is called the *ABCD* law by Kogelnik [1965].

4. Cavity boundary conditions

Consider the Gaussian beam as the mode of a laser cavity formed from two spherical mirrors (radii of curvature R_A, R_B) distance d apart. We assume that the cavity is of infinite aperture so that diffraction effects at the mirror surfaces can be neglected. A mode is defined as a self-consistent field configuration, and if it is to be represented by a beam travelling backwards and forwards within the cavity, then this demands that the beam parameters be unchanged after a round trip.

References on page 175

A useful way of visualizing a cavity for problems of this sort is to 'unfold' it by replacing the cavity (from the point of view of the calculation) by a series of lenses (ch. 5 § 5). The focal lengths of the lenses are determined by the radii of curvature of the mirrors that they replace, and their spacing by the separation of the mirrors. We can then treat the problem as one of a beam propagating through a periodic lens sequence.

Suppose that the Gaussian beam has a complex beam parameter, q_1 immediately after leaving mirror A (lens A). Then, when it reaches mirror B, the complex beam parameter is

$$q_2 = q_1 + d.$$

After reflection by mirror B (or transmission through lens B), we have

$$\frac{1}{q_3} = \frac{1}{q_1+d} - \frac{2}{R_B}.$$

On reaching mirror A again (this time travelling in the opposite direction)

$$q_4 = q_3 + d.$$

After reflection by mirror A,

$$q_5 = \frac{R_B R_A(2d+q_1) - 2R_A d(d+q_1)}{R_B(R_A-2d) - 2(R_A+R_B)(q_1+d) + 4d(q_1+d)}. \tag{7.28}$$

Since the beam has now undergone a complete round trip inside the cavity, the condition for a self-consistent field configuration (and hence for the Gaussian to be a mode) is

$$q_5 = q_1. \tag{7.29}$$

From (7.28) and (7.29) we obtain a quadratic equation for the beam parameter

$$dR_A(R_B-d)(1/q_1)^2 + 4d(R_B-d)(1/q_1) + 2(R_A+R_B-2d) = 0,$$

which has the solutions

$$\frac{1}{q_1} = -\frac{1}{R_A} \pm \frac{i}{R_A}\left[\frac{(R_A+R_B-2d)R_A}{(R_B-d)d} - 1\right]^{\frac{1}{2}}. \tag{7.29a}$$

Since all real Gaussian beams must have a non-zero width, the complex beam parameter must have a non-zero imaginary part (which by definition contains information about the beam width). Therefore, we get

$$(R_A+R_B-2d)R_A/(R_B-d)d > 1.$$

We could also write

$$(R_A+R_B-2d)R_B/(R_A-d)d > 1,$$

since our choice of mirror for launching the beam is arbitrary. With algebraic manipulation, the above two expressions can be combined into the following

$$0 < \left(1-\frac{d}{R_A}\right)\left(1-\frac{d}{R_B}\right) < 1. \tag{7.30}$$

We have therefore found the limits on the cavity parameters for it to be capable of supporting a stable mode. This equation is very important.

If we define $g_1 \equiv (1-d/R_A)$ and $g_2 \equiv (1-d/R_B)$ and then plot g_2 against g_1 as shown in fig. 5.7, each point in the plane defines a particular cavity configuration. The limit of the right hand inequality in (7.30) defines a hyperbola in this plane, and further, the inequality is satisfied for all points lying below the upper curve and above the lower curve. The left hand inequality is satisfied as long as g_1 an g_2 are of the same sign. Therefore, the area of the plane for a stable cavity configuration is as shown shaded in the stability diagram. Fig. 5.7 shows diagramatically the various cavity configurations corresponding to the numbered points in the stability diagram. The quantities g_1 and g_2 also play an important part in the geometric theory of the cavity (ch. 5 § 4.3).

We now continue to investigate the beam parameters of the Gaussian mode of a cavity. It is clear from (7.23) that the radii of curvature of the Gaussian beam at the mirrors must match the respective mirror radii of curvature. The required width of the Gaussian at the mirrors is given by the imaginary part of (7.24) and after some algebraic manipulation is seen to be (from 7.29a)

$$w_A^4 = \frac{4R_A}{k}\,\frac{(R_B-d)d}{(R_A-d)(R_A+R_B-d)}. \tag{7.31}$$

For problems associated with the matching of cavities, etc., it is convenient to define the Gaussian mode in terms of the position of its waist (either inside or outside the cavity) and the waist diameter. These parameters can be derived by remembering that at the waist, the wavefront is that of a plane wave (infinite radius of curvature), and consequently the complex beam parameter is purely imaginary. If z_{wA} is the distance of the waist from mirror A, then we have

$$z_{wA} = \mathscr{R}(q_A). \tag{7.32}$$

References on page 175

Using this relation, the position and diameter of the waist is found to be

$$z_{wA} = d(R_B - d)/(R_A + R_B - 2d) \tag{7.33}$$

and

$$w_0^4 = \left(\frac{2}{k}\right)^2 \frac{d(R_A - d)(R_B - d)(R_A + R_B - d)}{(R_A + R_B - 2d)^2}. \tag{7.34}$$

We have, therefore, deduced the beam parameters (in two forms) of the Gaussian mode of a cavity in terms of the cavity parameters. Knowing the beam parameters, we can use (7.33) and (7.34) to follow the propagation of a Gaussian beam both inside and outside a cavity. We shall use the above relations when we come to consider the matching of cavities by optical components in § 5.

For the cavity to resonate, we must add a further condition to those derived above. This is that the phase shift when the beam travels from one mirror to the other must be an integral multiple of π, for only in this case will a field be set up inside the cavity with a well defined phase structure (a standing wave). Using (7.17) and the phase term of (7.21) for the phase shift per transit, along the z-axis we have the following condition for resonance

$$kd + \tan^{-1}[kw_0^2/2z_{wA}] + \tan^{-1}[kw_0^2/2(d - z_{wA})] = \pi(p+1), \tag{7.35}$$

where p is an integer which is equal to the number of nodes of the standing wave that is set up.

The values of k which satisfy the above equation determine the frequencies of the longitudinal modes of the cavity (which are designated by their different p values). Using expressions (7.33) and (7.34) for z_{wA} and w_0^2, it is apparent that the second two terms on the left hand side of (7.35) are independent of k so that the frequency separation of successive longitudinal modes (fundamental beat frequency) depends only on the mirror separation, and is given by

$$\nu_0 = c/2d. \tag{7.35a}$$

Substituting for z_{wA} and w_0^2 using equations (7.33) and (7.34) and using (7.35a) then on re-arranging terms, (7.35) can be written

$$\frac{\nu}{\nu_0} = (p+1) + \frac{1}{\pi}\cos^{-1}\left[\left(1 - \frac{d}{R_A}\right)\left(1 - \frac{d}{R_B}\right)\right]^{\frac{1}{2}}, \tag{7.36}$$

where we have used the relationship

$$\tan^{-1} x + \tan^{-1} y = \tan^{-1}\{(x+y)/(1-xy)\} \qquad \text{for} \quad xy < 1.$$

Notice that the cavity stability conditions appear again here, since the square root term is only real and its modulus less than unity if (7.30) is satisfied. The square root is given the sign of one of its component terms.

5. Matching of cavities

Now that we are able to derive the Gaussian beam parameters from known cavity parameters, we consider matching the Gaussian mode of one cavity to that of another by the use of an optical component such as a lens. Problems of this nature arise when, for example, we wish to use a passive cavity as an interferometer to investigate the longitudinal mode structure of the radiation emerging from a laser cavity (ch. 4 § 2). In general, when a beam from a mode of one system is injected into another system the mode parameters of the two systems should be matched. If they are not matched, a single mode from the one system (e.g. laser) will couple to several modes of the other system and mode conversion occurs.

There are several ways of approaching the matching problem depending on the form of the known parameters. For example, we could work in terms of the beam waist and calculate the focal length and position of the lens required, using the criterion that the lens transforms the position and magnitude of the beam waist of one cavity into that of the other. Alternatively, we could work in terms of the spot size at the mirrors when the requirement is that the lens transforms the mode spot size and radius of curvature (equal to the mirror radius of curvature) for one cavity into those for the other.

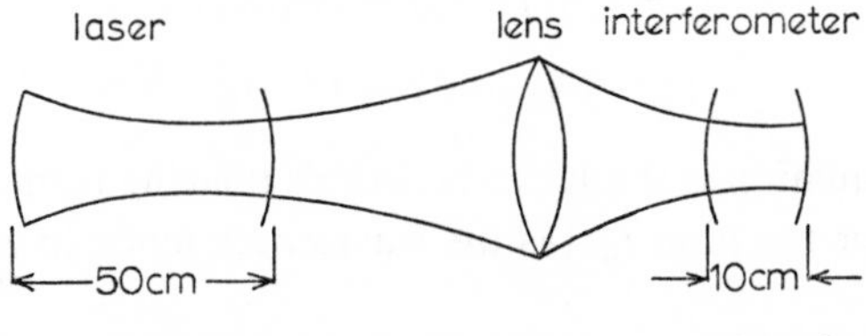

Fig. 7.4. Matching the Gaussian beam from a laser cavity to that of an interferometer.

We consider in more detail the matching of two cavities when the positions and diameters of their respective beam waists are known (fig. 7.4). Suppose that for one cavity the waist diameter is w_1 and its distance from the matching lens is z_1 (w_1 depends only on the cavity parameters, while z_1 depends also on the distance from one of the cavity mirrors to the lens). Let the corresponding values for the other cavity be w_2 and z_2 respectively.

References on page 175

At the beam waists the corresponding complex beam parameters (q_1 and q_2) are both imaginary (since the radius of curvature of the beam at these points is infinite), and are given by

$$q_1 = i\pi w_1^2 k/2, \qquad q_2 = i\pi w_2^2 k/2. \tag{7.37}$$

Using the transformation equations (7.25) and (7.27) for the complex beam parameters we get

$$q_2 = \left[\frac{1}{q_1+z_1} - \frac{1}{f}\right]^{-1} + z_2. \tag{7.38}$$

Since both q_1 and q_2 are imaginary, we can split the above equation into real and imaginary parts

$$(z_1-f)(z_2-f) = f^2 + q_1 q_2; \tag{7.39}$$

$$q_2(f-z_1) - q_1(f-z_2) = 0. \tag{7.40}$$

We can find the value of z_2, say, for a lens of a given focal length, by eliminating z_1, to obtain the following quadratic equation

$$z_2^2 - 2fz_2 - q_2^2 - f^2\left(\frac{q_2}{q_1} - 1\right) = 0. \tag{7.41}$$

It is apparent that the above only has a real solution for z_2 if

$$f > f_0, \quad \text{where} \quad f_0 = (|q_1| \cdot |q_2|)^{\frac{1}{2}} = w_1 w_2 k/2. \tag{7.42}$$

Equations (7.39) and (7.40) can be written in the form

$$(f-z_1)/(f-z_2) = (w_1/w_2)^2; \tag{7.43}$$

$$(z_1-f)(z_2-f) = f^2 - f_0^2. \tag{7.44}$$

The second equation is similar to Newton's imaging formula in geometric optics, except for the term f_0^2 (as the wavelength tends to zero, f_0 tends to zero also).

We conclude this section by giving a numerical example of cavity matching which involves the use of most of the formulae that we have derived in this and earlier sections on the Gaussian beam.

A 50 cm laser cavity (lm. mirrors) is to be matched to a 10 cm interferometer cavity (lm. mirrors) at 5000 Å (fig. 7.4).

The beam waists are positioned at the centres of their respective cavities, and are of magnitudes given by (7.34).

Laser: $w_1 = 1.7 \times 10^{-2}$ cm.

Interferometer: $w_2 = 1.05 \times 10^{-2}$ cm.

The minimum focal length of the lens used for matching is given by inequality (7.42), $f > 11.2$ cm.

If a lens of focal length 50 cm is chosen, expressions (7.43) and (7.44) enable the positions of the centres of the laser and interferometer cavities to be determined

$$z_1/50 = 1 \pm 1.58; \qquad z_2/50 = 1 \pm 0.6.$$

Laser: $z_1 = 130$ cm.

Interferometer: $z_2 = 80$ cm.

The output mirror of the laser must be positioned 105 cm to one side of the lens, the interferometer input mirror being 75 cm to the other side, for the two cavities to be matched to each other.

Laser mirrors usually consist of a multilayer dielectric structure on a fused silica substrate. The substrate acts as a plano-concave lens and consequently the parameters of the emerging beam differ from those of the beam inside the cavity (the latter being determined by the cavity parameters). In calculations connected with a practical system it is necessary to take into consideration the lens action of the mirrors of both the laser and interferometer cavities.

5.1 *Effects of cavity mismatch*

When the output of a laser is to be injected into another resonator or a transmission line such as a sequence of lenses, the question arises as to the modes which the injected beam will excite in the new system if the two systems are not matched. A single mode of the laser may couple to several modes of the new system. Coupling between the modes of the laser beam and the modes of the system into which it is injected is the concern of this section. The theory to be presented is due to KOGELNIK [1964]. In the previous section (§ 5), we were concerned with matching systems so that a mode of one system excited the same mode in the system into which it was injected.

For a rectangular geometry, the approximate transverse (x, y) field distribution of a TEM_{mn} mode is given by $\psi_m(x)\psi_n(y)$ where

$$\psi_m(x) = \left[\left(\frac{2}{\pi}\right)^{\frac{1}{2}} \frac{1}{2^m w m!}\right]^{\frac{1}{2}} \cdot H_m\left(\sqrt{2}\,\frac{x}{w}\right) \cdot \exp\left[-\frac{x^2}{w^2} - ik\frac{x^2}{2R}\right], \tag{7.45}$$

with a similar equation for $\psi_n(y)$. The field amplitudes are normalized for

References on page 175

unit power, thus

$$\int_{-\infty}^{+\infty} \psi_m(x)\psi_m^*(x)\mathrm{d}x = 1, \tag{7.46}$$

where the asterisk denotes a complex conjugate and $H_m(x)$, $H_n(y)$ are Hermite polynomials. These mode functions are approximate and in this form are a complete and orthogonal set. The discussion to follow is only valid for situations in which the approximation is good, viz. large aperture mirrors or lenses, and low values of m and n.

Consider a reference plane at the left boundary of the system into which a mode is to be injected from the left. Let the incoming mode have beam parameters $\bar{w}$ and $\bar{R}$ to the left of the plane, and w and R to the right. Waves travelling towards an observer at the right are of positive radius of curvature if they have a convex phase front as seen by that observer. Let a $\mathrm{TEM}_{\bar{m}\bar{n}}$ mode of unit power enter from the left. If the system and incoming beam are not matched, this mode will excite a set of TEM_{mn} modes in the system with fields $C_{\bar{m}\bar{n}mn}\psi_m\psi_n$, where $C_{\bar{m}\bar{n}mn}$ is the coupling coefficient introduced by Kogelnik. For a characteristic impedance of unity, the power in the TEM_{mn} mode is $C_{\bar{m}\bar{n}mn} \cdot C^*_{\bar{m}\bar{n}mn}$.

Equating the field distributions at the reference plane, we have

$$\bar{\psi}_{\bar{m}}(x) \cdot \bar{\psi}_{\bar{n}}(y) = \sum_m \sum_n C_{\bar{m}\bar{n}mn}\psi_m(x) \cdot \psi_n(y). \tag{7.47}$$

Since fields in direction x are independent of those in direction y, the coupling coefficient can be expressed as a product in much the same way as was the proportionality factor $\sigma_m\sigma_n$ (ch. 6 § 1); thus, we get

$$C_{\bar{m}\bar{n}mn} = c_{\bar{m}m} \cdot c_{\bar{n}n}. \tag{7.48}$$

Equation (7.47) can be written as two equations; we shall consider the one for the x-direction, multiply it by $\psi_m^*(x)$ and integrate, thus

$$\int_{-\infty}^{+\infty} \bar{\psi}_{\bar{m}}(x) \cdot \psi_m^*(x)\mathrm{d}x = \int_{-\infty}^{+\infty} \sum_l c_{\bar{m}l}\psi_l(x)\psi_m^*(x)\mathrm{d}x.$$

Using (7.46) and the orthogonality of the mode functions we get

$$c_{\bar{m}m} = \int_{-\infty}^{+\infty} \psi_{\bar{m}}(x) \cdot \psi_m^*(x)\mathrm{d}x. \tag{7.49}$$

A similar equation may be derived for $c_{\bar{n}n}$. Combining (7.45) and (7.49), we get

$$c_{\bar{m}m} = \left(\frac{2}{\pi\bar{w}w2^{(\bar{m}+m)} \cdot \bar{m}!m!}\right)^{\frac{1}{2}} \int_{-\infty}^{+\infty} H_{\bar{m}}(x\sqrt{\alpha})H_m(x\sqrt{\beta}) \exp(-qx^2)\mathrm{d}x, \tag{7.50}$$

where α, β and q are defined as follows:

$$\alpha = 2/\overline{w}^2, \qquad \beta = 2/w^2, \tag{7.51}$$

$$q = \frac{1}{\overline{w}^2} + \frac{1}{w^2} + \frac{ik}{2}\left(\frac{1}{\overline{R}} - \frac{1}{R}\right). \tag{7.52}$$

Inspection of (7.50) and the Hermite polynomials (appendix H) shows that if $\overline{m}+m$ is odd, then the integrand is an odd function of x and we get $c_{\overline{m}m} = 0$. This shows that there is no coupling between odd and even modes.

Integration of (7.50) (ERDELYI et al. [1954]) for both $\overline{m}$ and m even, where $\overline{m} = 2\nu$ and $m = 2\mu$ gives

$$c_{\overline{m}m} = A_+ \left(\frac{2}{\overline{w}wq}\right)^{\frac{1}{2}} \cdot \frac{(2\mu+2\nu)!}{(\mu+\nu)!(2\mu!2\nu!)^{\frac{1}{2}}}; \tag{7.53}$$

and for $\overline{m}$ and m odd, where $\overline{m} = 2\nu+1$ and $m = 2\mu+1$, we get

$$c_{\overline{m}m} = A_- \left(\frac{2}{\overline{w}wq}\right)^{\frac{3}{2}} \frac{(2\nu+2\mu+1)!}{(\nu+\mu)!\{(2\nu+1)!(2\mu+1)!\}^{\frac{1}{2}}}, \tag{7.54}$$

where

$$A_\pm = (-\tfrac{1}{2})^{\mu+\nu} \cdot \frac{(q-\alpha)^\nu (q-\beta)^\mu}{q^{(\nu+\mu)}} \cdot F\left[-\nu, -\mu; -\nu-\mu\pm\tfrac{1}{2}, \frac{q(q-\alpha-\beta)}{(q-\alpha)(q-\beta)}\right]$$

and F is the hypergeometric series defined by

$$F(a, b; c, z) = 1 + \frac{ab}{c} z + \frac{a(a+1)b(b+1)}{c(c+1)\cdot 2!} z^2 + \cdots.$$

Coupling coefficients for some low order modes are given below:

$$c_{00} = (2/\overline{w}wq)^{\frac{1}{2}}, \quad \text{(a)} \qquad c_{11} = c_{00}^3, \quad \text{(e)}$$

$$c_{02} = \frac{c_{00}}{\sqrt{2}}\left(\frac{\overline{w}}{w} c_{00}^2 - 1\right), \quad \text{(b)} \qquad c_{13} = \sqrt{3} c_{00}^2 c_{02}, \quad \text{(f)}$$

$$c_{20} = \frac{c_{00}}{\sqrt{2}}\left(\frac{w}{\overline{w}} c_{00}^2 - 1\right), \quad \text{(c)} \qquad c_{31} = \sqrt{3} c_{00}^2 c_{20}, \quad \text{(g)}$$

$$c_{22} = (1/c_{00})(c_{00}^6 + c_{02}c_{20}), \quad \text{(d)} \qquad c_{33} = c_{00}(c_{00}^6 + 3c_{02}c_{20}). \quad \text{(h)} \tag{7.55}$$

For imperfect matching, $|c_{00}| < 1$ and, therefore, $|c_{11}| = |c_{00}^3| < |c_{00}|$. This shows that the fundamental mode is less sensitive to mismatch than the mode of next higher order.

References on page 175

The fraction κ of power coupled from the fundamental mode (TEM_{00}) of one system into the fundamental of the other is given by

$$\kappa = (c_{00} \cdot c_{00}^*)^2. \tag{7.56}$$

From (7.55a) and (7.52) we get

$$\kappa = \frac{4}{\left(\dfrac{w}{\bar{w}} + \dfrac{\bar{w}}{w}\right)^2 + \left(\dfrac{\pi \bar{w} w}{\lambda}\right)^2 \left(\dfrac{1}{\bar{R}} - \dfrac{1}{R}\right)^2}.$$

This may be re-written as

$$\kappa = \frac{4}{\left(\dfrac{w}{\bar{w}} + \dfrac{\bar{w}}{w}\right)^2 + \rho^2 \dfrac{\bar{w}^2}{w^2}}, \tag{7.57}$$

where Kogelnik interprets $\bar{w}/w$ as the mismatch in beam radius and

$$\rho^2 = \sigma^2 \left(\frac{R}{\bar{R}} - 1\right)^2. \tag{7.58}$$

as the mismatch between the curvatures of the two phase fronts. The factor $\sigma = \pi w^2/\lambda R$ is called the "system parameter". For symmetric, nonconfocal resonators with mirrors distance d apart and radius of curvature R, we have

$$\sigma = \left(2\frac{R}{d} - 1\right)^{-\frac{1}{2}}. \tag{7.59}$$

When the systems are perfectly matched, $\bar{w} = w$ and $\bar{R} = R$. For a cavity with a plane input mirror ($R = \infty$), we have $\rho = \pi w^2/\lambda \bar{R}$, as may be seen by combining (7.58) and the expression for σ. Equation (7.59) is plotted in fig. 7.5 which shows that systems with values of d less than the confocal value are relatively insensitive to mismatch in phase front curvature.

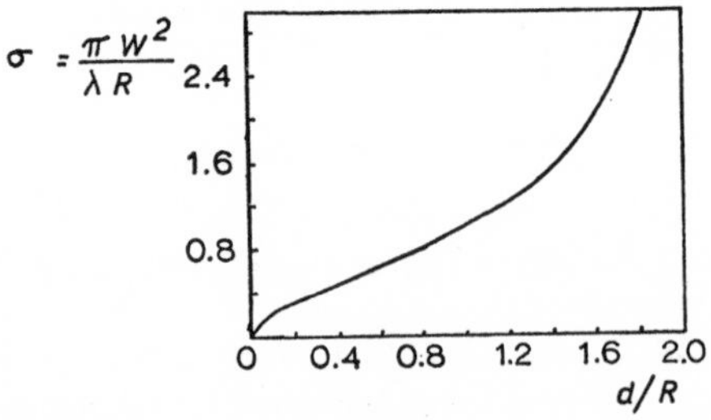

Fig. 7.5. System parameter σ (after Kogelnik [1964]).

Figure 7.6 shows κ as a function of $(\overline{w}/w)^2$ for various values of $\pm\rho$. It may be seen that for mismatch of the phase front curvature ($\rho \neq 0$), the maximum values of κ are obtained for values of $(\overline{w}/w) < 1$.

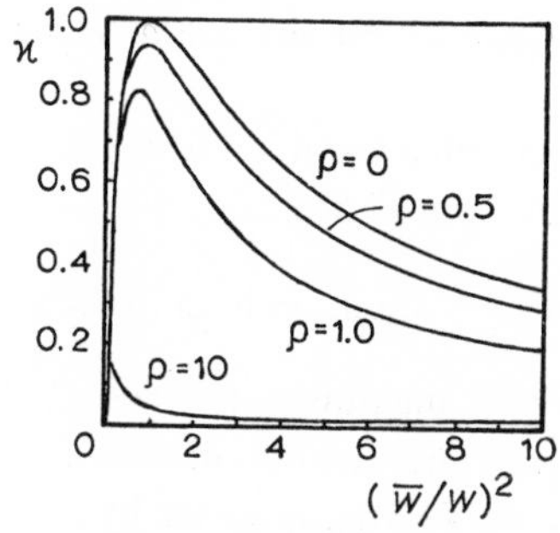

Fig. 7.6. Fractional power κ coupled between fundamental modes (after KOGELNIK [1964]).

E.g. For a cavity consisting of mirrors of radius 100 cm separated by 40 cm, we have $\sigma = 0.5$. If $R = 2\overline{R}$, then $\rho = \sigma$. For a beam radius difference of about 41 % ($\overline{w} = 1.41\, w$), we find that 80 % of the power of the incoming beam is coupled to the fundamental of the cavity.

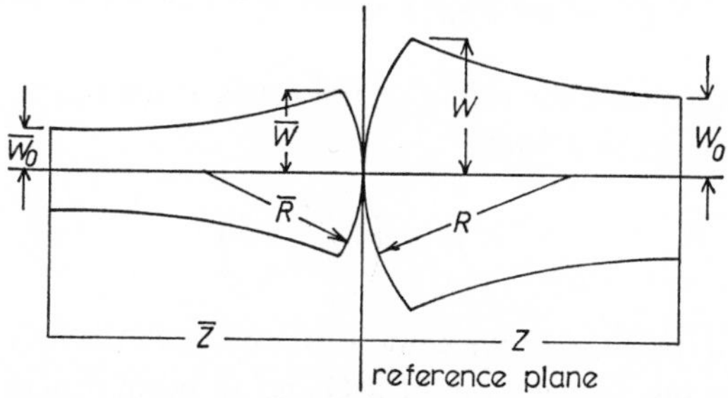

Fig. 7.7. Beam parameters; $\overline{R}$ and R are the radii of curvature of the phase fronts at the reference plane, and $\overline{w}$ and w are the corresponding beam radii; $\overline{w}_0$ and w_0 are the beam radii at the beam waists where the phase fronts are plane (after KOGELNIK [1964]).

It is sometimes convenient to use the distances $\bar{z}$ and z (fig. 7.7) of the two beam waists from the reference plane together with their radii $\overline{w}_0$ and w_0 instead of $\overline{w}$, w, $\overline{R}$, and R. The required relationships have been given in equations (7.18) and (7.19). After some algebra, we obtain

$$q = \frac{1}{\overline{w}_0^2 - \mathrm{i}(\lambda \bar{z}/\pi)} + \frac{1}{w_0^2 - \mathrm{i}(\lambda z/\pi)}$$

References on page 175

and

$$\kappa = \frac{4}{[(w_0/\bar{w}_0)+(\bar{w}_0/w_0)]^2+(\lambda/\pi\bar{w}_0 w_0)^2(\bar{z}+z)^2},$$

which show that κ depends only on the beam waists $\bar{w}_0$ and w_0 and their separation.

The fraction of power coupled from a $\text{TEM}_{\bar{m}\bar{n}}$ mode into a TEM_{mn} mode is $|c_{\bar{m}m}|^2 \cdot |c_{\bar{n}n}|^2$. We obtain from (7.55)

$$|c_{00}|^2 = \kappa^{\frac{1}{2}}, \qquad |c_{02}|^2 = |c_{20}|^2 = \kappa^{\frac{1}{2}}(1-\kappa)/2.$$

The conditions under which maximum power transfer between modes of different order occurs may be readily obtained. E.g. Maximum power conversion from TEM_{02} to TEM_{00} is given by $\kappa = \frac{1}{2}$ for which we have $(|c_{02}|^2|c_{00}|^2)_{\max} = \frac{1}{8}$.

6. Geometrical analogue

We have already shown that

$$\frac{1}{q} = \left(\frac{1}{R} - \frac{2\mathrm{i}}{kw^2}\right) \quad \text{and} \quad q_2 = q_1 + z. \qquad (7.24), (7.25)$$

Suppose that we take the origin of the z-axis at the beam waist, then the second equation can be written,

$$q = \left(\frac{\mathrm{i}kw_0^2}{2} + z\right). \qquad (7.60)$$

Eliminating q from the above equations, we obtain the following equation which describes the propagation of a Gaussian beam in terms of its diameter and radius of curvature,

$$\left(\frac{2}{kw^2} + \frac{\mathrm{i}}{R}\right)\left(\frac{kw_0^2}{2} - \mathrm{i}z\right) = 1. \qquad (7.61)$$

Each factor of this equation is a complex variable, and the equation describes a conformal transformation.

On the basis of this equation, several graphical methods have been proposed for the representation of a Gaussian beam. The first was the circle diagram of COLLINS [1963] which relates beam diameter and radius of curvature of the phase front at any position on the beam to the minimum beam diameter and its location. The Collins chart has two equivalent forms due to Li [1964] and Chu [1966]. Chu derived the counterpart, for the Gaussian beam, of the complex reflection coefficient in transmission line theory, and was thus able to use a Smith chart of complex mismatch coefficients to solve beam matching problems. Various other forms of beam chart have been proposed in the literature and are discussed by Kogelnik [1966].

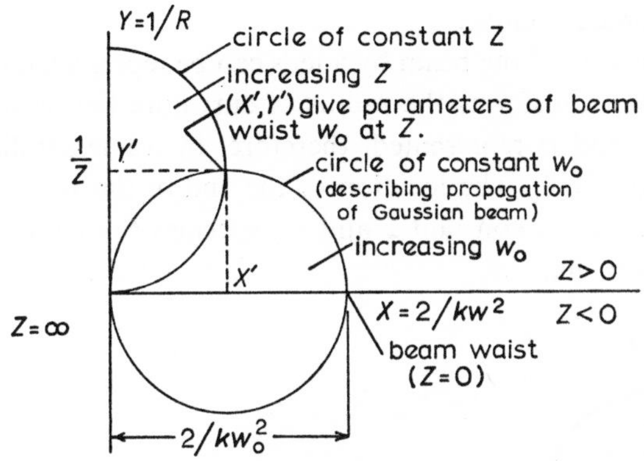

Fig. 7.8. Representation of the propagation of a Gaussian beam by a Collins' circle diagram.

We discuss the Collins' circle diagram in more detail now. We take $X = 2/kw^2$ and $Y = 1/R$ as axes in the complex plane, fig. 7.8. By substituting and then separating the real and imaginary parts of the above equation, we obtain

$$X^2+Y^2-(2/w_0^2k)X = 0; \tag{7.62}$$

$$X^2+Y^2-(1/z)Y = 0. \tag{7.63}$$

A beam is represented in the complex plane by a curve for which w_0 is constant. From (7.62) it can be seen that this curve is a circle which passes through the origin and has the X-axis as its diameter. The diameter is determined by the value of w_0 and is, $2/w_0^2k$. Therefore in the complex XY-plane, the propagation of a Gaussian beam is represented by a circle, the diameter of which is determined by the beam waist.

In order to be able to associate a particular set of values of the beam parameters with a particular point (z) along the direction of propagation of the beam, we examine (7.63). From this equation it can be seen that the curve representing the beam parameters at a fixed value of z, is a circle

References on page 175

passing through the origin with the Y-axis as a diameter. The diameter is determined in this case by the value of z, and is $1/z$.

A chart can be constructed, by plotting in the XY-plane the families of circles associated with constant z and w_0. The propagation of a particular Gaussian beam is represented by the appropriate circle of constant w_0, and the beam parameters at a particular point along the direction of propagation are the co-ordinates of the point of intersection of the beam circle with the appropriate z circle.

The transformation of the beam by a lens can be represented on the circle diagram. The lens transforms the radius of curvature but leaves the beam width unaltered, and is represented, therefore, by a line parallel to the Y-axis and of length $(1/f)$. The location of the end of the vertical line, with respect to the circles of constant z and w_0 specifies the parameters of the transformed beam, fig. 7.9.

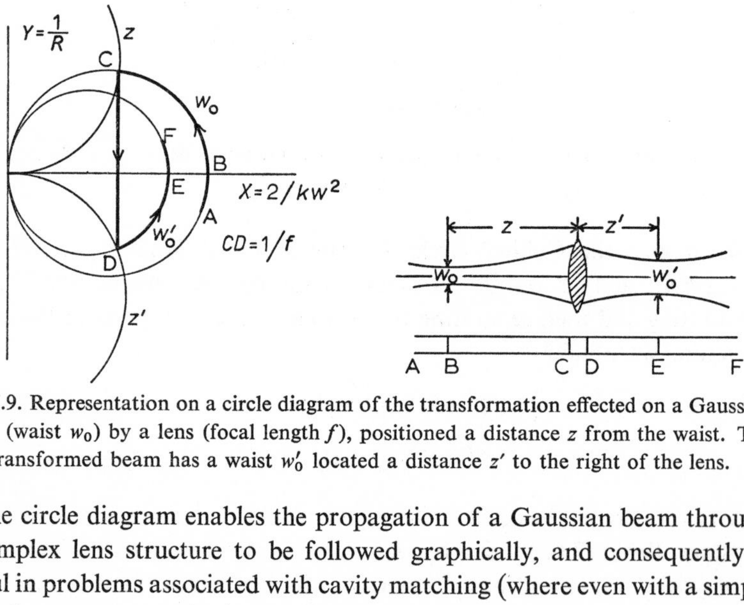

Fig. 7.9. Representation on a circle diagram of the transformation effected on a Gaussian beam (waist w_0) by a lens (focal length f), positioned a distance z from the waist. The transformed beam has a waist w_0' located a distance z' to the right of the lens.

The circle diagram enables the propagation of a Gaussian beam through a complex lens structure to be followed graphically, and consequently is useful in problems associated with cavity matching (where even with a simple optical arrangement the lens action of the cavity mirrors complicates the analysis).

7. Higher order modes

The solution which we have obtained for equation (7.3) and which describes the propagation of a simple Gaussian beam, is, in fact, the lowest order member of a family of solutions, for this equation. If, instead of assuming a solution of the form of (7.4), we had looked for solutions,

$$\chi = \alpha\left(\frac{x}{w}\right)\beta\left(\frac{y}{w}\right)\exp\left[F_2(z)+\frac{x^2+y^2}{F_1(z)}\right],$$

we would have found that the functions, α and β, required were,

$$\alpha\beta = H_m\left(\sqrt{2}\,\frac{x}{w}\right)H_n\left(\sqrt{2}\,\frac{y}{w}\right),$$

where H_m, H_n are Hermite polynomials (appendix H) of order m, n. These solutions form a complete orthogonal set. Each solution is made up of a combination of two Hermite polynomials with a Gaussian function, and is designated by its transverse mode numbers (m and n).

It can be shown that the beam parameter $R(z)$ is the same for all solutions, but that the additional phase shift (over that of a plane wave) depends on m and n,

$$\phi = (m+n+1)\tan^{-1}(2z/kw_0^2).$$

Consequently, the different order transverse modes of a cavity generally have different resonant frequencies as described in ch. 6.

References

CHU, T. S., 1966, Bell Syst. Tech. J. **45**, 287.

COLLINS, S. A., 1963, J. Opt. Soc. Am. **53**, 1339.

ERDELYI, A., et al., 1954, Tables of integral transforms, vols. **1**, **2** (McGraw-Hill Book Co., Inc., New York).

KOGELNIK, H., 1964, Coupling and Conversion Coefficients for Optical Modes, In: Proc. Symp. on Quasi Optics (Polytechnic Press, Brooklyn) pp. 333–347.

KOGELNIK, H., 1965, Bell Syst. Tech. J. **44**, 455.

KOGELNIK, H., 1966, Modes in Optical Resonators, In: Levine, A. K., ed., Lasers, Vol. 1 (Arnold, London) pp. 295–347.

KOGELNIK, H. and T. LI, 1966, Appl. Optics **5**, 1550.

LI, T., 1964, Appl. Optics **3**, 1315.

CHAPTER 8

GAIN AND SATURATION EFFECTS

1. Introduction

The active medium of a laser amplifier is made up of atoms (ions, or molecules), each of which can operate over a frequency range determined by the environmental perturbations of its energy levels and their natural linewidth. The central frequency for each atom also depends upon the velocity of the atom and its environment (e.g. crystal fields, fields due to the charged particles of gas discharges, collisions, and so on). Because of these effects, the frequency range over which the active medium gives gain is very much greater than that of the individual atoms. Likewise, the radiation to be amplified by either an amplifier or an oscillator has a frequency distribution. These three frequency distributions, or lineshapes, i.e. lineshapes of the individual atoms, the active medium, and radiation to be amplified, are most important in studying the interactions between the radiation and the active medium.

2. Gain saturation

The gain of a laser is said to be saturated when the radiation field reaches a steady value. The saturation arises because of the losses due to stimulated emission from the inverted population. This is shown very clearly by Lamb's theory, ch. 9. One of the effects of saturation is that it leads to mode competition. i.e. the reduction of gain available to a given mode by saturation due to some other mode. The behaviour of a laser with respect to saturation depends very much on the type of line broadening which is operative in the system. If the broadening is due to processes which limit the lifetime of the excited state, or if there are fields fluctuating with a period much shorter then the time associated with a transition, the broadening is homogeneous. Certain of these processes result in the energy absorbed from the radiation field in a given transition being distributed to the other particles involved. If the frequency which an atom will emit is not time-dependent on the scale of times involved, as, for example, when the collision period is much longer than the radiative life-time, or when the local fields are static, then the energy is absorbed only by those atoms satisfying the resonance condition and the energy is not shared during the time of an absorption; this is inhomogeneous broadening.

Summarizing, a homogeneously broadened line is one in which all the atoms of the population concerned can contribute to all frequencies of the line; an inhomogeneously broadened line is one in which different parts of the population contribute to different frequencies of the line. The Doppler effect causes inhomogeneous broadening in gases.

2.1 *Theory of gain saturation*

We shall consider gain saturation in a gas discharge which can be described by a 4-level laser system in which levels 3 and 2 are the upper and lower levels of the laser transition. Let the atoms excited to level 3 have a distribution of velocities such that the radiation they emit in undergoing a transition to level 2 has a Doppler profile centred on ν_0 and of half-width at half intensity of $\Delta\nu_D/2$. The actual frequency a given atom emits will itself lie somewhere within the natural line profile of half-width at half intensity of $\Delta\nu_N/2$ and will be centred on ν' which is the particular frequency appropriate to the velocity of the atom. There may be some collision broadening included in this line profile. The frequency of the radiation which causes gain

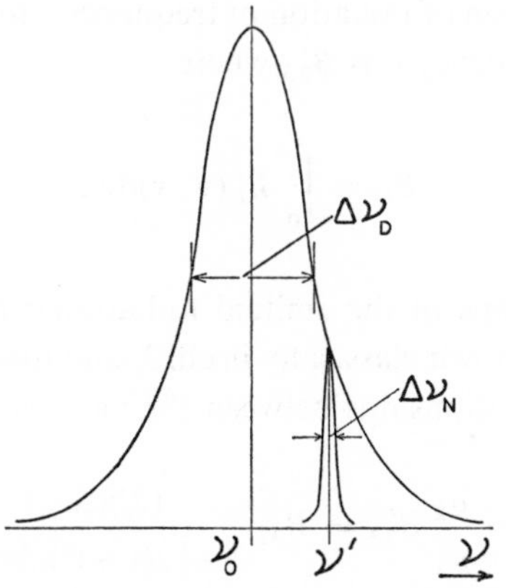

Fig. 8.1. Intensity-frequency profiles for Doppler and natural broadening.

References on page 213

saturation, i.e. the laser frequency, is ν. Figure 8.1 shows the profiles and frequencies to be considered.

We now consider the processes populating and depopulating levels 3 and 2, the upper and lower laser levels and shall use the notation and analysis of GORDON, WHITE and RIGDEN [1963]. Many of the basic principles used have been described by MITCHELL and ZEMANSKY [1961].

The pumping rates to levels 3 and 2 from all processes other than those specified below are $S_{3,2}$. As a result of the pumping process the atom arrives at the excited state 3, 2 with a velocity such that the frequency it emits or absorbs is centred on ν'. This frequency is on the Doppler profile. The terms S_3 and S_2 have a Doppler profile given by

$$S_i = S_{i0} \exp\left[-\{2(\nu'-\nu_0)/\Delta\nu_D\}^2 \ln 2\right]. \tag{8.1}$$

The volume densities of atoms in the levels 3, 2 in unit frequency range at ν' located at plane z along the axis of the discharge are $n_{3,2}$.

The intensity of radiation of frequency ν at plane z is I.

The effective total spontaneous decay rates (Einstein coefficients) are A_3 and A_2. These include the trapped radiation. When radiation to the ground state is trapped, the ground state atoms are re-excited to the level from which the original transition occurred.

The rate of spontaneous emission from level 3 to level 2 is A_{32}.

The Einstein coefficient giving the rate of stimulated emission for atoms with resonance at the Doppler frequency ν' when the stimulating frequency is ν is B'_{32}.

The rate of absorption of radiation of frequency ν for atoms with resonance at the Doppler frequency ν' is B'_{23} where

$$B_{ij} = \int_0^\infty B'_{ij}(\nu', \nu)\mathrm{d}\nu.$$

The natural line-shape of the emitted radiation centred on ν' is Lorentzian. (This has been shown classically in ch. 2, and quantum mechanically in ch. 3.) Hence, the relationships between the various B's are given by

$$B'_{32} = \frac{g_2}{g_3} B'_{23} = B_{32} \frac{(2/\pi\Delta\nu_N)}{1+[2(\nu-\nu')/\Delta\nu_N]^2}, \tag{8.2}$$

where the g's are the statistical weights.

The rate equations for the levels 3 and 2 are given by

$$\dot{n}_3 = S_3 - n_3\left[A_3 + B'_{32}\frac{I}{4\pi}\right] + n_2 B'_{23}\frac{I}{4\pi},$$

$$\dot{n}_2 = S_2 + n_3\left[A_{32} + B'_{32}\frac{I}{4\pi}\right] - n_2\left[A_2 + B'_{23}\frac{I}{4\pi}\right]. \tag{8.3}$$

Consider radiation of frequency $\mathrm{d}\nu$ about ν propagating along the axis of the discharge of cross-sectional area A through atoms located in the volume $A\mathrm{d}z$ at z. Neglecting the effect of spontaneous re-emission since it occurs in all directions, the change in energy content of the radiation is

$$\mathrm{d}I\,\mathrm{d}\nu\,A = B'_{32}n_3 A\,\mathrm{d}z\,h\nu\frac{I}{4\pi}\mathrm{d}\nu - B'_{23}n_2 A\,\mathrm{d}z\,h\nu\frac{I}{4\pi}\mathrm{d}\nu. \tag{8.4}$$

Simplifying and integrating over the profile of ν', we get

$$\frac{\mathrm{d}I}{\mathrm{d}z} = \frac{h\nu I}{4\pi}\int_0^\infty (B'_{32}n_3 - B'_{23}n_2)\mathrm{d}\nu'. \tag{8.5}$$

The values of n_3 and n_2 are not directly measurable so to solve the integral we need to express the integrand in terms of measurable quantities. This we do by solving (8.3) for the steady state $\dot{n}_3 = \dot{n}_2 = 0$ to obtain n_3 and n_2. In assuming that the populations attain steady values, we have neglected the "population pulsations" predicted by Lamb (ch. 9) which apply to the case in which more than one mode is oscillating; the populations of the upper and lower laser level have pulsating components at the beat frequencies of the modes. After some algebra we get,

$$B'_{32}n_3 - B'_{23}n_2 = \frac{B'_{32}\left[\dfrac{S_3}{A_3} - \dfrac{g_3}{g_2}\left(\dfrac{S_3 A_{32} + S_2 A_3}{A_2 A_3}\right)\right]}{1 + \left[\dfrac{g_2}{g_3}\left(\dfrac{A_3 - A_{32}}{A_2 A_3}\right) + \dfrac{1}{A_3}\right]\dfrac{B'_{32}I}{4\pi}}. \tag{8.6}$$

Substituting (8.1), (8.2) and (8.6) into (8.5) and making the substitutions

$$k_0 = h\nu\frac{B_{32}}{4\pi}\left[\frac{S_{30}}{A_3} - \frac{g_3}{g_2}\frac{S_{30}A_{32} + S_{20}A_3}{A_2 A_3}\right] \tag{8.7}$$

and

$$\eta = \left[\frac{g_3}{g_2}\frac{A_3 - A_{32}}{A_2 A_3} + \frac{1}{A_3}\right]\frac{B_{32}}{4\pi}, \tag{8.8}$$

we finally arirve at

$$\frac{1}{I}\frac{\mathrm{d}I}{\mathrm{d}z} = k_0 \int_0^\infty \frac{\dfrac{2}{\pi\Delta\nu_\mathrm{N}} \exp\left[-\left\{\dfrac{2(\nu'-\nu_0)}{\Delta\nu_\mathrm{D}}\right\}^2 \ln 2\right] \mathrm{d}\nu'}{1+\left[\dfrac{2(\nu-\nu')}{\Delta\nu_\mathrm{N}}\right]^2 + \dfrac{2\eta I}{\pi\Delta\nu_\mathrm{N}}}. \tag{8.9}$$

By adjusting the cavity we can ensure that the laser operates at the line centre where the gain is a maximum; thus we may put $\nu = \nu_0$ in (8.9). We may further simplify (8.9) by making the following substitutions;

$$\begin{aligned} x &= 2(\nu'-\nu_0)/\Delta\nu_\mathrm{N}, & \mathrm{d}\nu' &= \Delta\nu_\mathrm{N}\mathrm{d}x/2, \\ \mathscr{I} &= 2\eta I/\pi\Delta\nu_\mathrm{N} = I/I_0, & \mathrm{d}I &= (\pi\Delta\nu_\mathrm{N}/2\eta)\mathrm{d}\mathscr{I}, \\ \varepsilon &= (\ln 2)^{\frac{1}{2}}\Delta\nu_\mathrm{N}/\Delta\nu_\mathrm{D}, \end{aligned} \tag{8.10}$$

which then give

$$\frac{1}{\mathscr{I}}\frac{\mathrm{d}\mathscr{I}}{\mathrm{d}z} = \frac{k_0}{\pi}\int_{-\infty}^{+\infty} \frac{\exp-(\varepsilon x)^2\,\mathrm{d}x}{1+x^2+\mathscr{I}}. \tag{8.11}$$

The quantity I_0 is called the saturation parameter. It is of fundamental importance and is a measure of the radiation intensity required for a given degree of gain saturation. The parameter ε is most important because this contains the information about the type of line broadening. We shall consider two extreme cases, $\varepsilon \to 0$ and $\varepsilon \to \infty$. When $\varepsilon \to 0$, the line is broadened exclusively by the Doppler process; this is the case of inhomogeneous broadening. When $\varepsilon \to \infty$, Lorentzian broadening is very much greater than Doppler; this is the case of homogeneous broadening. Thus, ε is a measure of the degree of inhomogeneous broadening.

To solve the integral of (8.11), we make the substitution $y = x^2$, $\mathrm{d}x = \mathrm{d}y/2y^{\frac{1}{2}}$ and get

$$2\int_0^\infty \frac{\exp-(\varepsilon x)^2\,\mathrm{d}x}{1+\mathscr{I}+x^2} = \int_0^\infty \frac{\exp-(\varepsilon^2 y)\mathrm{d}y}{(1+\mathscr{I}+y)y^{\frac{1}{2}}}.$$

The integral on the right is tabulated in the Bateman Series [1954]. Equation (8.11) now becomes

$$\frac{1}{\mathscr{I}}\frac{\mathrm{d}\mathscr{I}}{\mathrm{d}z} = \frac{k_0 \exp\{(1+\mathscr{I})\varepsilon^2\}}{(1+\mathscr{I})^{\frac{1}{2}}}[1-\mathrm{Erf}\,\{(1+\mathscr{I})^{\frac{1}{2}}\varepsilon\}], \tag{8.12}$$

where Erf is the error function. In the limit $\varepsilon \to 0$ (i.e. for pure inhomogeneous broadening), $\operatorname{Erf}\{(1+\mathscr{I})^{\frac{1}{2}}\varepsilon\} \to 0$ and (8.12) gives,

$$\frac{1}{\mathscr{I}}\frac{\mathrm{d}\mathscr{I}}{\mathrm{d}z} = \frac{k_0}{(1+\mathscr{I})^{\frac{1}{2}}}. \tag{8.13}$$

In the limit $\varepsilon \to \infty$ (i.e. for pure homogeneous broadening),

$$\operatorname{Erf}\{(1+\mathscr{I})^{\frac{1}{2}}\varepsilon\} \to 1-[\exp\{-(1+\mathscr{I})\varepsilon^2\}]/(1+\mathscr{I})^{\frac{1}{2}}\varepsilon\pi^{\frac{1}{2}}.$$

Using this expression in (8.12), we get

$$\frac{1}{\mathscr{I}}\frac{\mathrm{d}\mathscr{I}}{\mathrm{d}z} = \frac{k_0}{\varepsilon\pi^{\frac{1}{2}}}\frac{1}{1+\mathscr{I}}. \tag{8.14}$$

Expressing (8.13) and (8.14) in terms of intensity I by using (8.10) we get

$$\frac{1}{I}\frac{\mathrm{d}I}{\mathrm{d}z} = \frac{k_0}{[1+(I/I_0)]^{\frac{1}{2}}} \tag{8.15}$$

and

$$\frac{1}{I}\frac{\mathrm{d}I}{\mathrm{d}z} = \frac{k_0}{\varepsilon\pi^{\frac{1}{2}}}\frac{1}{[1+(I/I_0)]}, \tag{8.16}$$

where $I_0 = \pi\Delta\nu_{\mathrm{N}}/2\eta$.

The gain coefficient g may be defined by the equation

$$\mathrm{d}I = gI\mathrm{d}z. \tag{8.17}$$

Thus, if the suffixes i and h refer to inhomogeneous and homogeneous broadening, respectively, we can write (8.15) and (8.16) in the form

$$g_i = g_{i0}(1+I/I_0)^{-\frac{1}{2}}; \tag{8.18}$$

$$g_h = g_{h0}(1+I/I_0)^{-1}. \tag{8.19}$$

When the laser starts, the radiation field intensity I is zero and the gain of the system in this case is g_0, the unsaturated gain or the small signal gain. Equation (8.18) shows that when the radiation intensity $I = I_0$, the gain coefficient decreases to $g_{i0}/2$. In (8.19), when $I = I_0$, g_h decreases to $g_{h0}/\sqrt{2}$.

Predictions based on the above theory have been investigated by GORDON, WHITE, and RIGDEN [1963] and good agreement between theory and experiment has been found.

References on page 213

3. Gain narrowing

A laser amplifier consists of a medium in which a population inversion is established between two levels so that stimulated emission can amplify an incoming beam of appropriate frequency. We shall assume that the spontaneous line emitted by the amplifying transition as inhomogeneously broadened (Doppler). The question then arises as to whether the Doppler width of the amplifier line is also the width of the final amplified line emerging from the amplifier. In this context it is important to distinguish between *incremental gain* and *net gain*. Incremental gain is given by $(1/I)(\mathrm{d}I/\mathrm{d}z)$ and refers to an element of length dz of the amplifying medium; net gain is given by $(I_{\mathrm{out}}/I_{\mathrm{in}})$ and describes the increase in intensity of radiation resulting from travelling the full length of the amplifying medium. The Doppler width can always be considered the line width for incremental gain but not necessarily for the net gain. HOTZ [1965] has investigated the variation of gain across the amplifying line of neon at 3.39 μ $(5\mathrm{s}'[\frac{1}{2}]_1 \rightarrow 4\mathrm{p}'[\frac{3}{2}]_2)$ using a fixed frequency laser oscillator and a Zeeman tuned single pass amplifier and has found gain narrowing to agree with the theory given below.

We shall start with the fractional gain per unit length derived by GORDON, WHITE and RIGDEN [1963] and given in (8.9). Simplifying this equation by making the substitutions

$$\begin{aligned} \Omega &= 2(\ln 2)^{\frac{1}{2}}(\nu'-\nu_0)/\Delta\nu_{\mathrm{D}}, \qquad & \omega &= 2(\ln 2)^{\frac{1}{2}}(\nu-\nu_0)/\Delta\nu_{\mathrm{D}}, \\ \nu-\nu' &= \Delta\nu_{\mathrm{D}}(\omega-\Omega)/2(\ln 2)^{\frac{1}{2}}, \qquad & \mathrm{d}\nu' &= \Delta\nu_{\mathrm{D}}\mathrm{d}\Omega/2(\ln 2)^{\frac{1}{2}}, \\ g_0 &= k_0, \end{aligned} \tag{8.20}$$

we get

$$\frac{1}{I}\frac{\mathrm{d}I}{\mathrm{d}z} = g = g_0 \cdot \int_{-\infty}^{+\infty} \frac{\dfrac{2}{\pi\Delta\nu_{\mathrm{N}}} \cdot \dfrac{\Delta\nu_{\mathrm{D}}}{2(\ln 2)^{\frac{1}{2}}} \exp(-\Omega^2)\mathrm{d}\Omega}{\left[\dfrac{2}{\Delta\nu_{\mathrm{N}}} \cdot \dfrac{\Delta\nu_{\mathrm{D}}}{2(\ln 2)^{\frac{1}{2}}}(\omega-\Omega)\right]^2 + 1 + \dfrac{2\eta I}{\pi\Delta\nu_{\mathrm{N}}}}.$$

We can simplify this by making two more substitutions

$$a = (\Delta\nu_{\mathrm{N}}/\Delta\nu_{\mathrm{D}})(\ln 2)^{\frac{1}{2}}; \qquad \alpha = a[1+(2\eta I/\pi\Delta\nu_{\mathrm{N}})]^{\frac{1}{2}}, \tag{8.21}$$

which give

$$g = g_0 \cdot \int_{-\infty}^{+\infty} \frac{\dfrac{1}{\pi a} \exp(-\Omega^2)\mathrm{d}\Omega}{\left(\dfrac{\omega-\Omega}{a}\right)^2 + \left(\dfrac{\alpha}{a}\right)^2},$$

from which we get

$$\frac{\alpha}{a}\frac{g}{g_0} = \frac{\alpha}{\pi}\int_{-\infty}^{+\infty} \frac{\exp(-\Omega^2)\mathrm{d}\Omega}{\alpha^2+(\omega-\Omega)^2}. \tag{8.22}$$

For values of $\alpha \leqslant 10^{-2}$, integration of (8.22) (MITCHELL and ZEMANSKY [1961]) gives

$$\frac{\alpha}{a}\frac{g}{g_0} = \exp(-\omega^2) - \frac{2\alpha}{\pi^{\frac{1}{2}}}[1-2\omega F(\omega)]; \tag{8.23*}$$

$$\frac{g}{g_0} = \frac{\exp(-\omega^2)}{[1+(2\eta I/\pi\Delta\nu_N)]^{\frac{1}{2}}} - \frac{2a}{\pi^{\frac{1}{2}}}[1-2\omega F(\omega)]. \tag{8.24*}$$

If $\Delta\nu_N \ll \Delta\nu_D$ we may neglect the second term on the right and we get

$$g = \frac{1}{I}\frac{\mathrm{d}I}{\mathrm{d}z} = \frac{g_0 \exp(-\omega^2)}{(1+I/I_0)^{\frac{1}{2}}}. \tag{8.25}$$

If we let $I/I_0 = \beta$, $\mathrm{d}\beta = \mathrm{d}I/I_0$ and then integrate over L we get, finally

$$2(1+\beta)^{\frac{1}{2}} + \ln \left.\frac{(1+\beta)^{\frac{1}{2}}-1}{(1+\beta)^{\frac{1}{2}}+1}\right|_{\beta_1}^{\beta_2} = Lg_0 \exp(-\omega^2). \tag{8.26}$$

For small signal gain where $\beta_2-\beta_1 \ll 1$, (8.26) becomes

$$g_0 L \exp(-\omega^2) = \ln(\beta_2/\beta_1) = \ln G_0(\omega), \tag{8.27}$$

where $G_0(\omega)$ is the unsaturated gain defined by

$$G_0(\omega) = I_2/I_1 = \beta_2/\beta_1.$$

At the line centre $\nu-\nu_0 = 0$, so $\omega = 0$ and $G_0(\omega) = G_0(0)$. The value of $G_0(\omega)$ decreases on both sides of the line centre to a value depending on $(\nu-\nu_0)$ at the frequency considered. We wish to determine the frequency at which the value of $G_0(\omega)$ has fallen from $G_0(0)$ to $G_0(0)/2$. Let the value of ω at which this occurs be μ. Then, we have

$$g_0 L = \ln G_0(0); \tag{8.28}$$

$$g_0 L \exp(-\mu^2) = \ln[G_0(0)/2]. \tag{8.29}$$

Dividing (8.28) by (8.29) we get

$$\exp(\mu^2) = \ln G_0(0)/\ln[G_0(0)/2]. \tag{8.30}$$

* $F(\omega) = \mathrm{e}^{-\omega^2}\int_0^{\omega} \mathrm{e}^{x^2}\,\mathrm{d}x = \int_0^{\infty} \mathrm{e}^{-t^2}\sin 2\omega t\,\mathrm{d}t.$

References on page 213

Using (8.20) and assuming a symmetrical line of width at half gain given by

$$\delta = 2(\nu - \nu_0), \qquad \text{then } \mu = (\delta/\Delta\nu_D)(\ln 2)^{\frac{1}{2}},$$

so we obtain

$$\delta = 1.2\Delta\nu_D\,\mu. \tag{8.31}$$

In his experiments, Hotz found that with a gain $G_0(0) = 5000$ and a Doppler width of 340 MHz, the value of the output gain half width, δ, given by (8.30) and (8.31) was 119 MHz. Figure 8.2 shows the variation of δ with input intensity.

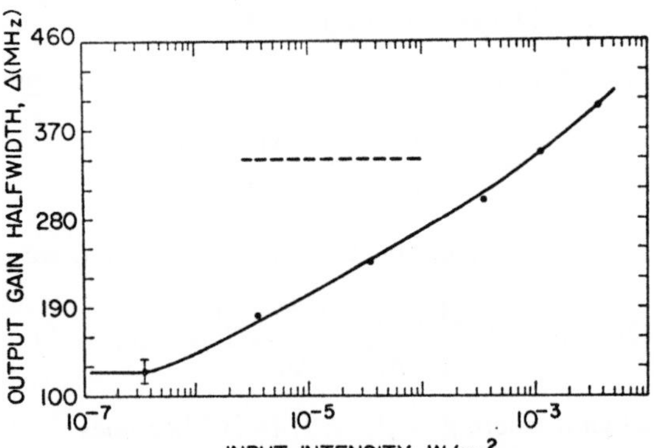

Fig. 8.2. Relationship between the half-width of the output gain and the input intensity. The dashed line is the measured Doppler width (reproduced from D. F. Hotz, Appl. Optics **4**, 527 (1965) with permission).

4. Laser amplifier gain

Rigrod [1963] has used the equation derived in § 2.1 to study the gain characteristics of laser amplifiers at high beam intensities. We shall follow his treatment. Consider a parallel beam of radiation of uniform intensity, which we shall call the signal. Let the signal propagate in the z-direction through an active medium which has an average loss coefficient, a (absorption, scattering, etc.), and homogeneous line broadening. The net gain coefficient is

$$g_0[1+(I/I_0)]^{-1} - a = (1/I)\mathrm{d}I/\mathrm{d}z. \tag{8.32}$$

Let the normalized radiation intensity be $\beta = I/I_0$; $\mathrm{d}I = I_0\,\mathrm{d}\beta$ and transpose (8.32) to obtain

$$g_0\mathrm{d}z = (\mathrm{d}\beta/\beta)(1+\beta)/[1-(a/g_0)(1+\beta)]. \tag{8.33}$$

When the loss parameter $a/g_0 = 0$, we get

$$g_0 \mathrm{d}z = (\mathrm{d}\beta/\beta) + \mathrm{d}\beta, \tag{8.34}$$

which, when integrated over L, the length of the amplifying medium, gives

$$g_0 L = \ln(\beta_2/\beta_1) + \beta_2 - \beta_1, \tag{8.35}$$

where suffixes 1 and 2 refer to input and output, respectively. Figure 8.3 shows a plot of (8.35) for several values of the normalized input power β_1.

When the signal entering the amplifier is small, (8.35) becomes

$$g_0 L \approx \ln(\beta_2/\beta_1). \tag{8.35a}$$

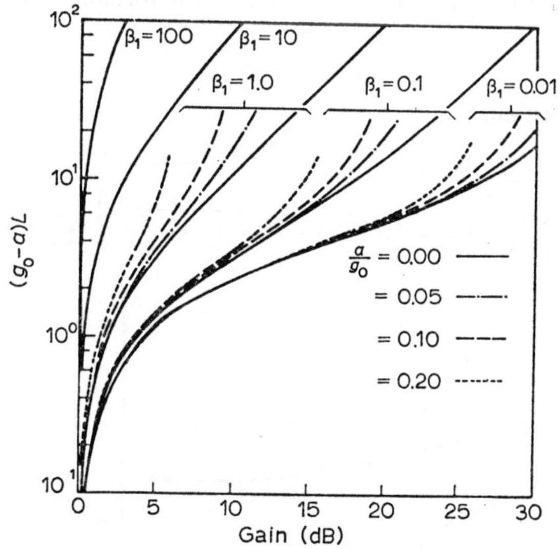

Fig. 8.3. The ordinates indicate the optical path length L needed to obtain a specified over-all gain in a homogeneous-line laser amplifier with unsaturated gain coefficient g_0 for various levels of input power density $\beta_1 = I_1/I_0$ and the distributed loss parameter a/g_0 (reproduced from W. W. Rigrod, J. Appl. Phys. **36**, 2487 (1965) with permission).

5. Saturation in high-gain lasers

Rigrod [1965] has calculated the radiation intensity obtainable from high-gain, high-loss lasers with homogeneous line broadening in which the line shape does not change during gain saturation. As the radiation travels back and forth between the mirrors of the cavity it is amplified by each transit through the active medium of length L and then suffers losses at the

References on page 213

mirrors. We shall neglect losses occurring in the active medium and assume all the losses are at the mirrors. We wish to describe the steady state which is established by the combined effect of amplification and losses. Let the axis of the laser be the z-direction and let the normalized radiation intensities in the $+z$ and $-z$ directions be β_+ and β_- respectively. If we put $I = I_+ + I_-$ in (8.34) we get, finally

$$g_h = g_0/(1+\beta_+ + \beta_-). \tag{8.36}$$

where

$$g_h = (1/\beta_+)(\mathrm{d}\beta_+/\mathrm{dz}) = (-1/\beta_-)(\mathrm{d}\beta_-/\mathrm{dz}), \tag{8.37}$$

because the gain is isotropic. Considering a plane located at an arbitrary point z between 0 and L, by integrating (8.37) and forming their product we get

$$\beta_+\beta_- = c, \tag{8.38}$$

where c is a constant.

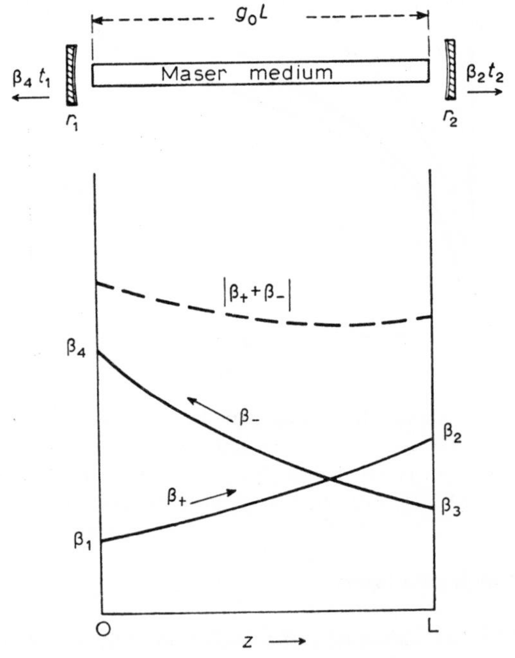

Fig. 8.4. and Fig. 8.5.

Schematic diagram showing normalized levels of flux intensity in both directions in an asymmetric laser oscillator (reproduced from W. W. Rigrod, J. Appl. Phys. **36**, 2487 (1965) with permission).

Figure 8.4 shows the active medium and the location of the mirrors of reflectance r_1 and r_2 respectively. When the radiation meets a mirror a fraction t is transmitted and a fraction a is lost by scattering, absorption, etc.; thus, the reflectances are given by

$$\begin{aligned} r_1 &= 1-a_1-t_1 \quad \text{at} \quad z = 0; \\ r_2 &= 1-a_2-t_2 \quad \text{at} \quad z = L. \end{aligned} \tag{8.39}$$

Figure 8.5 shows the normalized growth of intensity during each transit, the loss occurring when the radiation reaches a mirror, and the combined normalized intensity in the active medium due to the radiation travelling in both directions (+ and −). The case discussed is that for which $r_2 \neq r_1$, i.e. the laser is asymmetric. From fig. 8.5 and the definition of reflectance we get

$$\beta_3 = r_2\beta_2 \quad \text{and} \quad \beta_1 = r_1\beta_4. \tag{8.40}$$

From (8.38) and (8.40) we get

$$\beta_1\beta_4 = \beta_2\beta_3 = c, \quad \text{and} \quad r_4\beta_4^2 = r_2\beta_2^2. \tag{8.41}$$

Therefore, we have

$$\beta_2/\beta_4 = (r_1/r_2)^{\frac{1}{2}}. \tag{8.42}$$

Equations (8.36), (8.37) and (8.38) can be combined to give

$$\frac{1}{\beta_+}\frac{\mathrm{d}\beta_+}{\mathrm{d}z} = \frac{g_0}{1+\beta_+ +(c/\beta_+)}. \tag{8.43}$$

Integrating (8.43) over the length L, we get

$$g_0 L = \ln(\beta_2/\beta_1)+\beta_2-\beta_1-c(1/\beta_2-1/\beta_1). \tag{8.44}$$

In the same way, for gain in the negative direction, we get

$$g_0 L = \ln(\beta_4/\beta_3)+\beta_4-\beta_3-c(1/\beta_4-1/\beta_3). \tag{8.45}$$

Adding (8.44) and (8.45) and using (8.40) and (8.41) we get, finally,

$$\beta_2 = [g_0 L+\ln(r_1 r_2)^{\frac{1}{2}}]r_1^{\frac{1}{2}}/[r_1^{\frac{1}{2}}+r_2^{\frac{1}{2}}][1-(r_1 r_2)^{\frac{1}{2}}]. \tag{8.46}$$

Hence, from (8.42), we get

$$\beta_4 = [g_0 L+\ln(r_1 r_2)^{\frac{1}{2}}]r_2^{\frac{1}{2}}/[r_1^{\frac{1}{2}}+r_2^{\frac{1}{2}}][1-(r_1 r_2)^{\frac{1}{2}}]. \tag{8.47}$$

Laser cavities are very often made from multi-layer dielectric mirrors which do not have zero transmittance, thus radiation is emitted at both ends

References on page 213

of the cavity. The total emitted by both mirrors is given by

$$(I_2/I_0)+(I_4/I_0) = \beta_4 t_1+\beta_2 t_2. \tag{8.48}$$

When the losses at each mirror are equal, $a_1 = a_2 = a$, therefore, from (8.39) we get

$$t_1 = 1-a-r_1, \qquad t_2 = 1-a-r_2. \tag{8.49}$$

Hence, we get

$$(I_2/I_0)+(I_4/I_0) = [g_0 L+\ln (r_1 r_2)^{\frac{1}{2}}][1-a-(r_1 r_2)^{\frac{1}{2}}]/[1-(r_1 r_2)^{\frac{1}{2}}]. \tag{8.50}$$

When one mirror is opaque and lossless ($t_1 = 0$ and $r_1 = 1$), the output intensity is given by the product of (8.46) and t_2

$$(I_2/I_0) = \beta_2 t_2 = [g_0 L+\ln (r_2)^{\frac{1}{2}}]t_2/(a_2+t_2). \tag{8.51}$$

The optimum output coupling (transmittance) of the mirror is that which makes the ratio I_2/I_0 a maximum. Remembering that $r_2 = 1-a_2-t_2$ and differentiating (8.51) with respect to t_2 we obtain an expression for the optimum output coupling, t

$$(t/a) = [g_0 L+\ln r^{\frac{1}{2}}]2(1-a-t)/(a+t), \tag{8.52}$$

where we have dropped the subscript 2. Combining (8.51) and (8.52), we get

$$2I_2/I_0 = t^2/a(1-a-t), \tag{8.53}$$

where t is the optimum transmittance of the putput mirror.

Figure 8.6 shows the normalized maximum output intensity obtainable with optimum output coupling, as given by (8.53) for given values of a and the single pass gain 4.343 $g_0 L$.

When radiation is decoupled from the cavity we may expect this loss and saturation effects in the active medium to cause the radiation intensity within the cavity to vary along the length of the active medium. We shall obtain the ratio of the intensity at the mid-point to the intensity at the mirrors as a function of mirror reflectance for the case of a symmetrical resonator in which $r_1 = r_2 = r$ and $r = 1-a-t$. Since the resonator is symmetric, $\beta_3 = \beta_1$ and $\beta_4 = \beta_2$, and (8.41) gives $\beta_2 = c/\beta_3 = c/\beta_1$. Integrating (8.43) between 0 and $L/2$ we get

$$g_0 L/2 = \ln (\beta_+/\beta_1)+\{\beta_+-(c/\beta_+)\}+(\beta_2-\beta_1).$$

At $L/2$ the flux is the same in both directions so we have

$$\beta_+ = \beta_- = c/\beta_+ \tag{8.54}$$

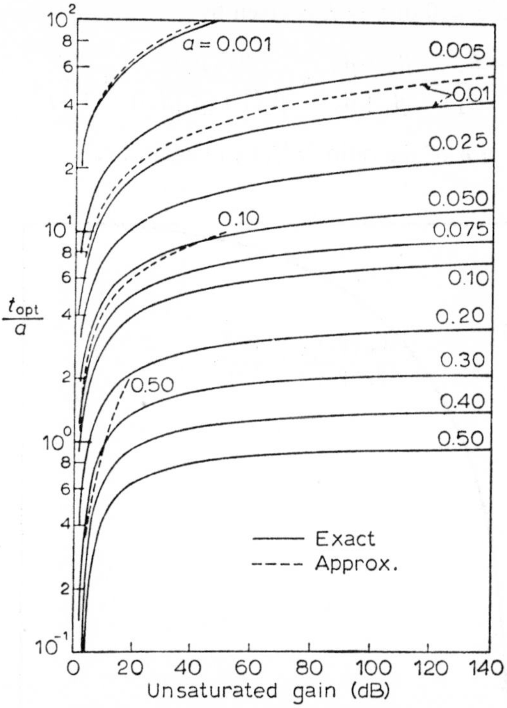

Fig. 8.6. Optimum coupling coefficient (t_{opt}) for maximum oscillator output power, as a function of the loss fraction (a) and the unsaturated gain in dB, 4.343 g_0L. The solid lines represent the exact solution, and the dashed lines the approximate version (reproduced from W. W. RIGROD, J. Appl. Phys. **36**, 2487 (1965) with permission).

which leads to

$$g_0L/2 = \ln(\beta_+/\beta_1) + \beta_2 - \beta_1 . \tag{8.55}$$

For the special case we are considering, (8.44) can be written

$$g_0L = \ln(\beta_2/\beta_1) + 2(\beta_2 - \beta_1). \tag{8.56}$$

Combining (8.55) and (8.56) we get

$$\ln(\beta_+/\beta_1) = (\tfrac{1}{2})\ln(\beta_2/\beta_1) = \ln r. \tag{8.57}$$

Equations (8.44) and (8.57) lead to

$$\beta_- = \beta_+ = (\beta_1\beta_2)^{\frac{1}{2}}.$$

Referring to fig. 8.5 and remembering that we are considering the symmetrical case where $\beta_4 = \beta_2$ and $\beta_3 = \beta_1$ the ratio of the two-way intensity at

$z = L/2$ to that at $z = 0$ or $z = L$ is given by

$$\frac{\beta_+ + \beta_-}{\beta_1 + \beta_4} = \frac{\beta_+ + \beta_-}{\beta_1 + \beta_2} = \frac{2(\beta_1 \beta_2)^{\frac{1}{2}}}{\beta_2(1 + (\beta_1/\beta_2))} = \frac{2r^{\frac{1}{2}}}{1+r}. \tag{8.58}$$

Fig. 8.7 shows a plot of the ratio $2r^{\frac{1}{2}}/(1+r)$ against r.

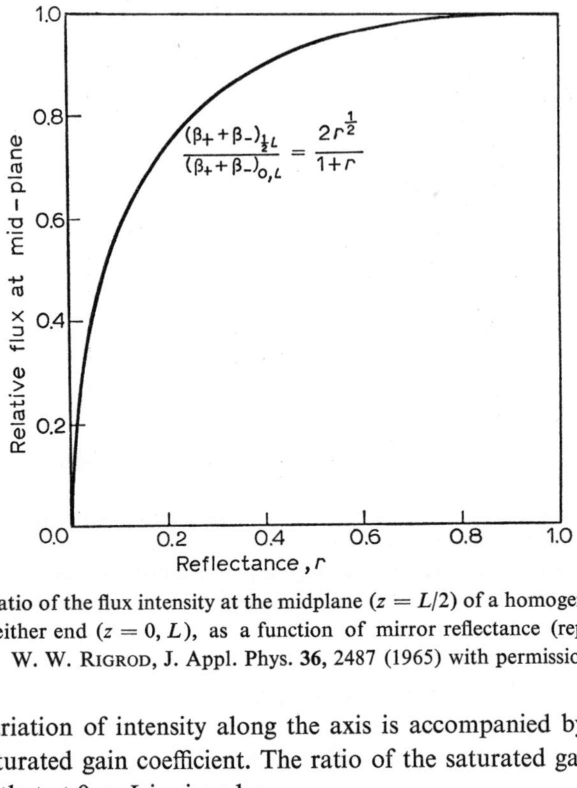

Fig. 8.7. Ratio of the flux intensity at the midplane ($z = L/2$) of a homogeneous-line laser to that at either end ($z = 0, L$), as a function of mirror reflectance (reproduced from W. W. RIGROD, J. Appl. Phys. **36**, 2487 (1965) with permission).

The variation of intensity along the axis is accompanied by a variation of the saturated gain coefficient. The ratio of the saturated gain coefficient at $L/2$ to that at 0 or L is given by

$$\frac{g(L/2)}{g(L)} = \frac{1+\beta_2(1+r)}{1+2\beta_2 r^{\frac{1}{2}}} = \frac{1+\beta_2}{1+\beta_2[2r^{\frac{1}{2}}/(1+r)]}. \tag{8.59}$$

5.1 *Effect of gain saturation on modes*

Most investigations of mode formation in optical cavities have been made with the simplifying assumption that the cavity is passive. In this case, the higher modes of stable cavities made up of concave spherical mirrors have the greater losses. However, when an active medium giving gain is present, the higher order modes do not necessarily have the greatest losses since the maintenance of a mode now depends upon the availability of atoms giving gain. All the modes are competing for atoms in the central region, but since the higher order modes occupy larger volumes of the active medium, they can obtain energy from atoms not available to the lower order modes and so may even become dominant*. Spatial cross relaxation transverse to the laser axis is generally negligible since an atom travels a distance $v\tau(= 10^6 \times 10^{-8}$ cm) before radiating spontaneously. RIGROD [1964] found experimentally that for nearly confocal resonators, the highest order mode allowed by the apertures was normally dominant. The calculations of FOX and LI [1966] also show this behaviour.

Passive cavity theory is applicable to the oscillation threshold condition of a laser. Above the threshold the relative gains and losses of the modes, gain saturation, and variation of refractive index with power become important considerations. If the transition is homogeneously broadened, the dispersion saturates as well as the gain. However, this effect is small since the maximum phase delay for a Lorentzian line is 3.3 degrees per dB of gain (FOX and LI [1966]) and this would be the maximum phase change for complete saturation. If the transition is inhomogeneously broadened, saturation effects produce no first order change in the dielectric constant (refractive index). Because of these considerations, we are concerned only with variations in gain due to non-uniform saturation and their effect on mode formation.

One of the first problems in calculating the mode formation in a cavity with an active medium of saturable gain is that the medium is non-linear. The Fresnel-Kirchhoff formulation of Huygens' principle can be applied to determine the field over the surface of a mirror as in the case of a passive cavity but this is not the total field at the mirror. To obtain the total field we must add the contributions to the radiation field made by each volume element of the active medium between the mirrors. This complete problem is intractable so FOX and LI [1966] have made the same assumption as STATZ and TANG [1965] and considered the active material to be concentrated in two infinitely thin sheets next to the mirror surfaces. The integral equation thus obtained is not separable into x and y components (ch. 6

* In the case of concave mirror cavities where the frequencies of the transverse modes are quite widely spaced, the modes do not compete for the same atoms. For plane-parallel cavities, the transverse modes are closely spaced and compete for the same atoms with the result that the lower order transverse modes dominate because their losses are less.

§§ 1, 2) because, at any point, the gain saturation depends upon the total radiation field which is interacting with the atoms of the medium. The x and y directions are coupled by the atoms. FOX and LI [1966] have separated the integral equation into two independent equations by formulating the problem for circular mirrors with modes having circularly symmetric intensity distributions and using cylindrical coordinates. They used their iterative method of successive approximations to solve the integral equations obtained.

The conclusions of FOX and LI are that for both a homogeneously or an inhomogeneously broadened line, the distribution of field amplitude over the mirror, the diffraction loss, and the phase shift per transit are all essentially unchanged by saturation. The phase distribution over the mirrors showed a small change. Thus for low gain laser oscillators, all the previous calculations made for passive resonators (containing linear media) are valid.

6. Power output

The factors to be considered in calculating the power output from a gas laser are the Doppler linewidth, the pressure broadened "natural" linewidth, gain saturation effects, cavity losses, and output coupling. We shall start with (8.22)

$$\frac{\alpha}{a}\frac{g}{g_0} = \frac{\alpha}{\pi}\int_{-\infty}^{+\infty}\frac{\exp(-\Omega^2)\mathrm{d}\Omega}{(\omega-\Omega)^2+\alpha^2}. \tag{8.60}$$

The integral is given by ABRAMOWITZ and STEGUN [1964] page 302, integral 7.4.13,

$$\int_{-\infty}^{+\infty}\frac{\alpha\exp(-\Omega^2)\mathrm{d}\Omega}{(\omega-\Omega)^2+\alpha^2} = \pi\,\mathscr{Re}\,w(\omega+\mathrm{i}\Omega), \tag{8.61}$$

where $\mathscr{Re}\,w(z)$ is the real part of the error function for complex arguments and $w(z) = [1+\mathrm{erf}(\mathrm{i}z)]\exp(-z^2)$.

From (8.60) and (8.61) and the values given in ch. 8, § 3 we get*

$$\frac{g}{g_0} = \frac{1}{[1+(I/I_0)]^{\frac{1}{2}}}\,\mathscr{Re}\,w\left[\frac{2(\ln 2)^{\frac{1}{2}}}{\Delta\nu_{\mathrm{D}}}(\nu-\nu_0)+\frac{\mathrm{i}(\ln 2)^{\frac{1}{2}}\Delta\nu_{\mathrm{N}}}{\Delta\nu_{\mathrm{D}}}\left(1+\frac{I}{I_0}\right)^{\frac{1}{2}}\right]. \tag{8.62}$$

For a laser of length L, the gain per pass for small gains is gL. The un-

* The value $\Delta\nu_{\mathrm{D}} = (2kT/m\lambda^2)^{\frac{1}{2}}$ was used by SMITH [1966a], see below; the value of $\Delta\nu_{\mathrm{D}}$ used here is $2(\ln 2)^{\frac{1}{2}}(2kT/m\lambda^2)^{\frac{1}{2}}$. Also $\Delta\nu_{\mathrm{N}} = 2\gamma'$ (Smith).

saturated gain is the gain at zero intensity and the maximum, G_m, is at $\nu = \nu_0$. Hence, from (8.62) we get

$$G_m = g_0 L\, \mathscr{R}e\, w[0+i(\ln 2)^{\frac{1}{2}}\Delta\nu_N/\Delta\nu_D], \tag{8.63}$$

which gives the maximum unsaturated gain per pass.

A laser oscillator settles to an intensity I_1 at which the saturated gain is equal to the losses. Following Smith, we now define an excitation parameter η, to be the ratio of the maximum unsaturated gain (8.63) to the total loss per pass. Since the saturated gain is equal to the losses, and assuming that $g(\nu)$ is constant across the laser tube diameter, we can write

$$\eta = \left[1+\frac{I}{I_0}\right]^{\frac{1}{2}} \left[\frac{\mathscr{R}e\, w\{0+i(\ln 2)^{\frac{1}{2}}\Delta\nu_N/\Delta\nu_D\}}{\mathscr{R}e\, w\left\{\dfrac{2(\ln 2)^{\frac{1}{2}}}{\Delta\nu_D}(\nu-\nu_0)+\dfrac{i(\ln 2)^{\frac{1}{2}}\Delta\nu_N}{\Delta\nu_D}\left(1+\dfrac{I_1}{I_0}\right)^{\frac{1}{2}}\right\}}\right]. \tag{8.64}$$

Smith has considered the application of (8.64) to single mode and multimode operation. We shall consider the single mode case.

6.1 *Single mode operation*

At the central frequency of oscillation $\nu = \nu_0$, the radiation travelling in each direction in the laser cavity interacts with the same group of atoms and the total intensity is $2I$. From (8.64) we get

$$\eta = \left[1+\frac{2I_1}{I_0}\right]^{\frac{1}{2}} \left[\frac{\mathscr{R}e\, w\{0+i(\ln 2)^{\frac{1}{2}}\Delta\nu_N/\Delta\nu_D\}}{\mathscr{R}e\, w\left\{0+\dfrac{i(\ln 2)^{\frac{1}{2}}\Delta\nu_N}{\Delta\nu_D}\left(1+\dfrac{2I_1}{I_0}\right)^{\frac{1}{2}}\right\}}\right]. \tag{8.65}$$

Figure 8.8 shows I_1/I_0 at $\nu = \nu_0$ plotted against η for various values of $\Delta\nu_N$, the full width of the collision broadened natural line at half its maximum, It is seen that $\Delta\nu_N$ has a very marked influence on the normalized intensity I_1/I_0. The Doppler width used corresponds to temperatures in the range 300–500 °K (Smith [1966a]) and is given by $\Delta\nu_D = (2kT/m\lambda^2)^{\frac{1}{2}}$ (see footnote § 6). Smith has verified the above theory. He found the value of η from the experimentally measured threshold frequency ν_t by using (8.64) with $I_1 = 0$. Having determined η in this way he measured the intensity I_1 at $\nu = \nu_0$ as a function of η. These intensities were then normalized by dividing by the saturation parameter I_0 (7.5 W cm^{-2} for the results given). Figure 8.9 shows I/I_0 as a function of η. The solid curve is a plot of (8.65) for $\Delta\nu_N = 180$ MHz. The values I_0 and $\Delta\nu_N$ were determined from previous experiments (Smith [1966b]).

References on page 213

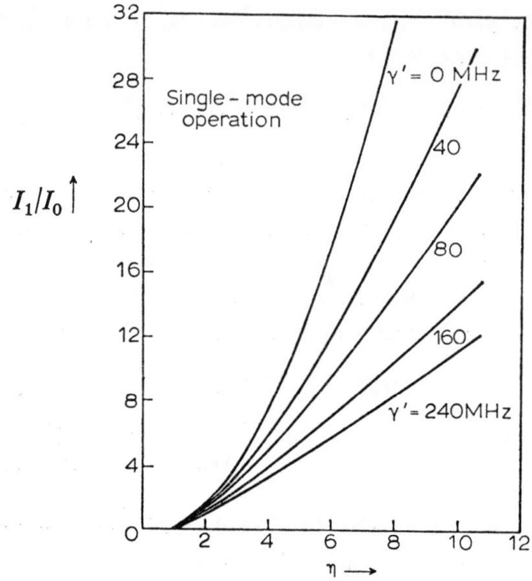

Fig. 8.8. Theoretical plots of normalized single-mode laser intensity I_1/I_0 as a function of the excitation parameter η for various values of the linewidth parameter $\gamma' = \Delta\nu_N/2$ (after SMITH [1966a]).

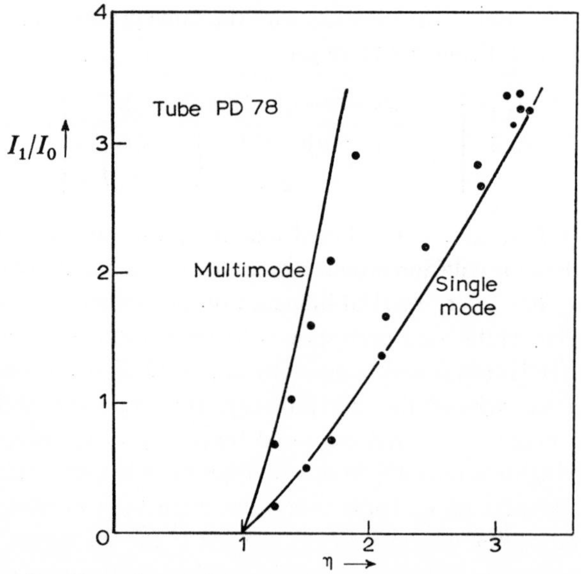

Fig. 8.9. Experimental measurements of laser intensity as a function of the excitation parameter η for a particular tube for both single mode and multimode operation. The solid line is the theoretical for $\Delta\nu_N = 180$MHz (after SMITH [1966a]).

6.2 *Mirror transmission and power optimization*

The gain of a laser depends upon stimulated emission in the active medium of the cavity and since stimulated emission, in turn, depends upon the energy density (or intensity) of the radiation field in the cavity, the gain too is a function of the radiation energy density. Thus, the power extracted from the cavity effects the gain conditions and the question of optimization arises.

An expression for the gain coefficient g for a homogeneously broadened line is (see equation 8.19)

$$g = g_0/(1+I/I_0), \tag{8.66}$$

where g_0 is the unsaturated gain coefficient, and I_0 is a saturation parameter equal to the power density at which the gain falls to $g_0/2$. MENEELY [1967] has considered the optimization problem.

Consider a laser of active length l and mirrors of transmittances and losses t_1, t_2, α_1, α_2 respectively. In addition to the mirror losses there are also scattering losses of fraction β per unit length of the medium. Thus the gain is $g' = g-\beta$. In the steady state, the intensity of the radiation in the cavity is unchanged after one complete round trip in the cavity. If the intensity at the start is I, then after one round trip we have,

$$I \exp[l(g-\beta)][1-t_1-\alpha_1] \exp[l(g-\beta)][1-t_2-\alpha_2] = I.$$

To first order, and with $\alpha = \alpha_1+\alpha_2+2l\beta$ and $t = t_1+t_2$ we get

$$(1-\alpha-t)\exp(2gl) = 1. \tag{8.67}$$

The power output P is proportional to the transmittance, and to the energy density in the cavity, hence

$$P = cIt, \tag{8.68}$$

where c is a constant of proportionality. Combining (8.66) and (8.67) and solving for I, we get

$$I = -I_0[1+2lg_0/\ln(1-t-\alpha)]. \tag{8.69}$$

Putting this into (8.68) and determining the maximum value of P with respect to t we get

$$2lg_0 = -\frac{(1-\alpha-t_m)[\ln(1-\alpha-t_m)]^2}{(1-\alpha-t_m)[\ln(1-\alpha-t_m)]+t_m}. \tag{8.70}$$

Where t_m is the value of t for which the output power is a maximum. Figure

References on page 213

8.10 shows $2lg_0$ as a function of t_m for several values of α and figs. 8.11 and 8.12 show the power P plotted against t_m for typical values of $2lg_0$ and α respectively. The systems to which the above analysis refers most appropriately are the high gain, high power systems such as the CO_2–N_2–He laser. Meneely has discussed results obtained with such systems.

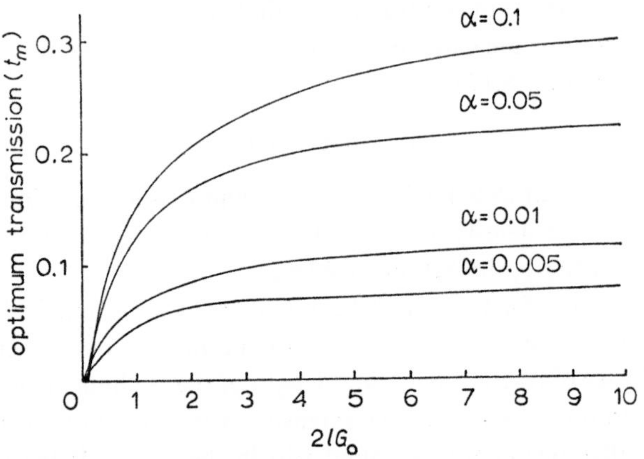

Fig. 8.10. Optimum transmission as a function of $2lg_0$ for several values of the losses, α (after MENEELY [1967]).

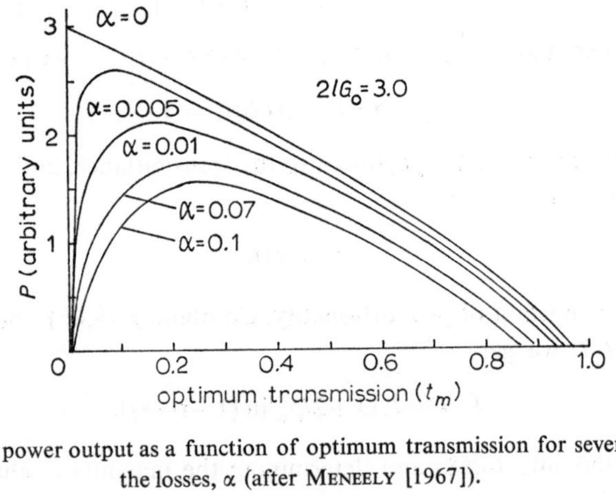

Fig. 8.11. The power output as a function of optimum transmission for several values of the losses, α (after MENEELY [1967]).

7. Hole-burning effects

An inverted population may be described in terms of its spatial distribution, its velocity distribution, the orientation of its members with respect to a given direction (usually the direction of polarization of the radiation field), or in terms of some other characteristic. The term hole-burning is used to describe the selective depletion of the population. Thus a particular hole-burning process may deplete the population of atoms with velocity between v and $v+dv$ (BENNETT [1962a, b]). Another process may remove atoms of a particular energy state within an energy level, and so on. To detect hole-burning it is necessary to be able to identify atoms radiating in a given frequency range. This means that the energy exchange processes operative among them (cross relaxation processes) must occur in times longer than the radiative lifetimes for stimulated emission. If collision rates (say)

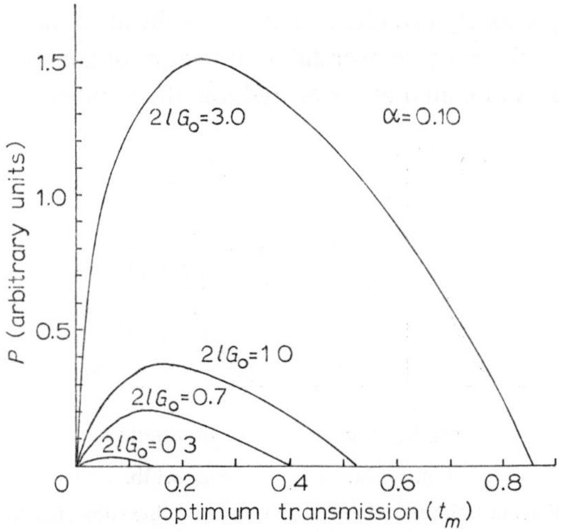

Fig. 8.12. The power output as a function of optimum transmission for several values of $2lg_0$ (after MENEELY [1967]).

are greater than the stimulated emission rate, it is not possible to burn a hole in the frequency distribution of population inversion. Hole-burning occurs in predominantly inhomogeneously broadened lines. In gaseous laser amplifiers or oscillators the cross relaxation within a homogeneous line is slower than the radiation processes. In a solid-state system (e.g. Nd-glass) the fluorescent line is inhomogeneously broadened by interaction with the host fields and the cross relaxation processes are fast.

Hole-burning, due to stimulated emission, occurs at frequencies near to each mode oscillating and the depth and width of a hole depend upon the power in the mode and the natural line width of the transition involved.

References on page 213

Typically, the width appears to be in the range 3–30 MHz. Essentially, it is the requirement that in the steady state, the gain is equal to the losses, i.e. $g(\nu) = \alpha$ which results in hole-burning in inhomogeneously broadened transitions. If the unsaturated gain is g_0, the depth of the hole in the gain curve is $g_0(\nu) - \alpha$. When the line is inhomogeneously broadened, and the natural line width is less than the spacing between cavity modes, the atoms contributing to oscillation in a given cavity mode cannot couple to the radiation field from atoms contributing to another mode. Thus the laser can oscillate simultaneously and independently in a number of modes depending on the gain conditions for each mode. A consequence of hole-burning is that only a fraction of the available energy stored in the system is used. In the case of a homogeneously broadened line (e.g. ruby at room temp.) the same atoms are responsible for gain over different regions of the line, so removal of excited atoms by stimulated emission reduces the number of excited atoms

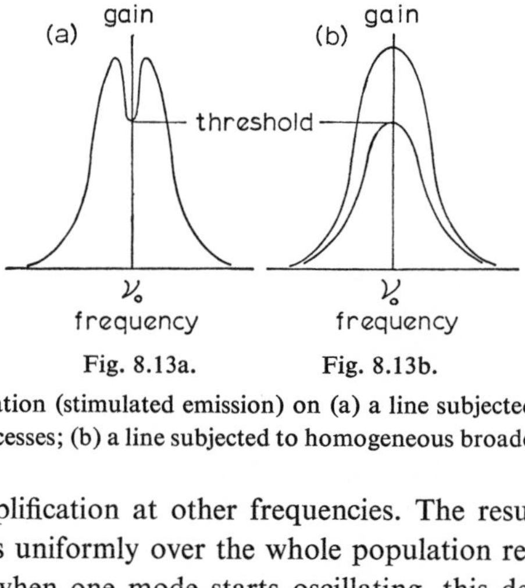

Fig. 8.13a. Fig. 8.13b.

Effect of laser oscillation (stimulated emission) on (a) a line subjected to inhomogeneous broadening processes; (b) a line subjected to homogeneous broadening processes.

available for amplification at other frequencies. The result of this is that the gain saturates uniformly over the whole population responsible for the line profile and when one mode starts oscillating, this depletes the whole population and so prevents oscillation of other modes. Figures 8.13a and 8.13b show the effect of stimulated emission on lines subjected to inhomogeneous and homogeneous broadening processes, respectively.

7.1 *Hole width*

The width of the hole burned in the gain profile of an inhomogeneously broadened line is the width of the spectral region over which the atoms interact with the laser radiation field of intensity I. Since the atoms in this region are those producing gain in the system over the wavelength range concerned, we can obtain the hole width by calculating the gain. We start with (8.4) and the situation on which it is based. We repeat the equation here

$$\mathrm{d}I\,\mathrm{d}\nu\,A = (B'_{32}n_3A\,\mathrm{d}z\,h\nu I\,\mathrm{d}\nu - B'_{23}n_2A\,\mathrm{d}z\,h\nu I\,\mathrm{d}\nu)/4\pi. \tag{8.4}$$

This may be rearranged to give

$$\frac{1}{I}\frac{\mathrm{d}I}{\mathrm{d}z} = \frac{h\nu}{4\pi}[B'_{32}n_3 - B'_{23}n_2]. \tag{8.71}$$

With the algebra, and substitutions described in (8.6), (8.7), (8.8) and (8.17) we get for the gain coefficient

$$\frac{1}{I}\frac{\mathrm{d}I}{\mathrm{d}z} = g = k_0\frac{2}{\pi\Delta\nu_\mathrm{N}}\exp\left[-\left(\frac{2(\nu'-\nu_\mathrm{D})}{\Delta\nu_\mathrm{D}}\right)^2\ln 2\right]\times$$
$$\times\left[1+\left(\frac{2(\nu-\nu')}{\Delta\nu_\mathrm{N}}\right)^2+\frac{2\eta I}{\pi\Delta\nu_\mathrm{N}}\right]^{-1}. \tag{8.72}$$

This expression gives the gain at ν. The maximum gain, $g_{\max}$ occurs at $\nu = \nu'$, where ν' is fixed by the cavity resonance and any mode pulling which may be present. The frequency at which the gain is $g_{\max}/2$ can readily be obtained from (8.72), and if $\Delta\nu_\mathrm{h}$ is the full width at half maximum, we obtain, finally,

$$\Delta\nu_\mathrm{h} = \Delta\nu_\mathrm{N}(1+I/I_0)^{\frac{1}{2}}. \tag{8.73}$$

At the oscillation threshold the hole width is simply the natural line width (there may be some collision broadening). As the laser field intensity increases, the hole width increases. The longitudinal mode separation is given by $\Delta\nu = c/2d$ (4.1). To maximize power output when a number of longitudinal modes are oscillating simultaneously, it is necessary to ensure that the holes overlap; the condition for this is that $\Delta\nu = \Delta\nu_\mathrm{h}$, hence

$$\frac{c}{2d} = \Delta\nu_\mathrm{N}\left(1+\frac{I}{I_0}\right)^{\frac{1}{2}}. \tag{8.74}$$

E.g. If $I/I_0 = 8$ and $\Delta\nu_\mathrm{N} = 10^8$ Hz, the length of laser for which holes overlap is 50 cm.

7.2 *Effect of mode degeneracy on gaseous laser output*

For certain cavity configurations, the resonant frequencies of some modes coincide, that is the modes are degenerate in frequency. A laser with a high gain active medium usually oscillates in many of the modes, longitudinal and transverse, within the profile of the line of the transition associated with laser

References on page 213

action. If the line is Doppler broadened, each oscillating mode saturates and so burns a hole in the profile. If the hole width is greater than the mode spacing, the entire line above threshold is saturated, that is, the entire profile is burned off, and the power output is a maximum because atoms in all parts of the gain profile above threshold contribute to the output. However, if the mode spacing is greater than the hole width, as it may be if many of the modes are degenerate, the parts of the profile between the modes will be unsaturated, that is, the atoms associated with these portions of the gain profile are unable to couple to the radiation field and so the power output for a given number of modes is reduced. HARDING and LI [1964] have studied this aspect of laser behaviour. In addition to reduced output, instabilities in output arise because of competition between the frequency degenerate modes within each group.

Figure 8.14 shows the gain-frequency curve for an inhomogeneously broadened line and the mode spectrum together with the image of each mode reflected in the line centre, (BENNETT [1962a, b]) for the two cases: (a) arbitrary spacing corresponding to maximum output; (b) degenerate spacing corresponding to a reduced power output. To ensure that the entire top of the Doppler gain curve is used, the axial mode spacing ($c/2d$) may be decreased until it becomes equal to a hole width by simply increasing the length of the cavity. In this case the population inversion is reduced over all the oscillating frequencies.

The condition for degeneracy may be obtained by considering some of the longitudinal modes within the lower order transverse modes. Using the resonance condition of BOYD and KOGELNIK [1962],

$$\frac{2d}{\lambda} = q + \frac{1}{\pi}(1+m+n)\cos^{-1}(g_1 g_2)^{\frac{1}{2}}. \tag{8.75}$$

Equation (8.75) applies to the TEM_{mnq} mode. The same value $2d/\lambda$ is obtained for the $\text{TEM}_{m'n'q'}$ mode and so

$$q + \frac{1}{\pi}(1+m+n)\cos^{-1}(g_1 g_2)^{\frac{1}{2}} = q' + \frac{1}{\pi}(1+m'+n')\cos^{-1}(g_1 g_2)^{\frac{1}{2}}.$$

Let $q-q' = n_l$ and $(m'+n')-(m+n) = n_t$ where n_l and n_t are integers, and we get the condition for degeneracy, which is

$$\pi^{-1}\cos^{-1}(g_1 g_2)^{\frac{1}{2}} = n_l/n_t. \tag{8.76}$$

This condition should be avoided if maximum power output is desired.

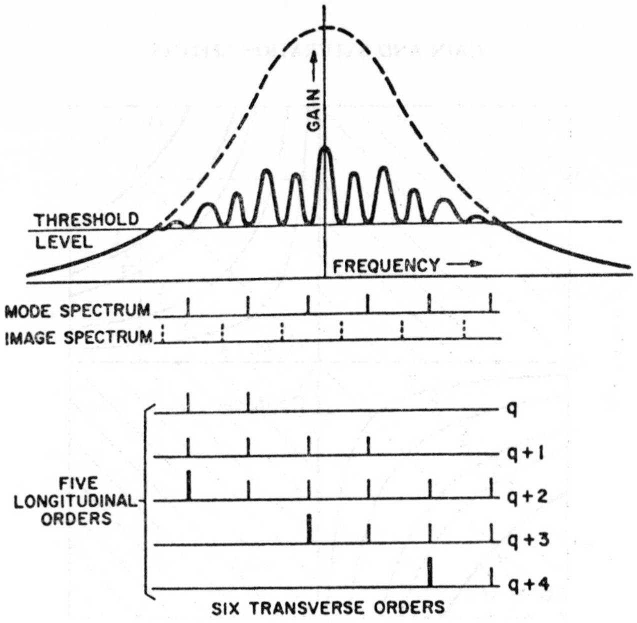

(a)

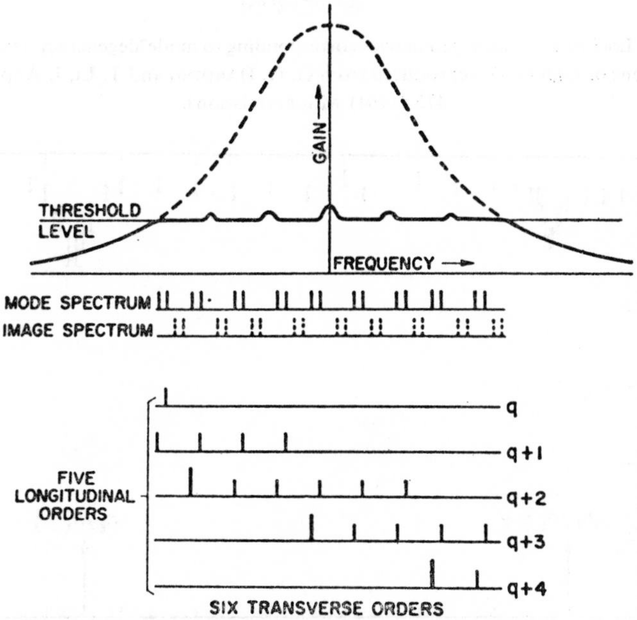

(b)

Fig. 8.14. Gain-frequency curve with mode and image spectra compounded from five longitudinal orders of six transverse modes: (a) degenerate spacing corresponding to a dip; (b) arbitrary spacing corresponding to maximum output (reproduced from G. O. Harding and T. Li, J. Appl. Phys. **35**, 475 (1964) with permission).

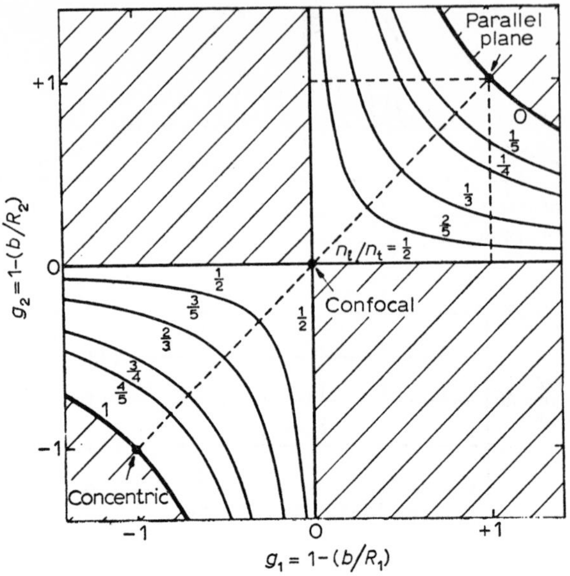

Fig. 8.15. Loci of resonator geometries corresponding to mode degeneracy (shading indicates regions of high loss) (reproduced from G. O. HARDING and T. LI, J. Appl. Phys. **35**, 475 (1964) with permission).

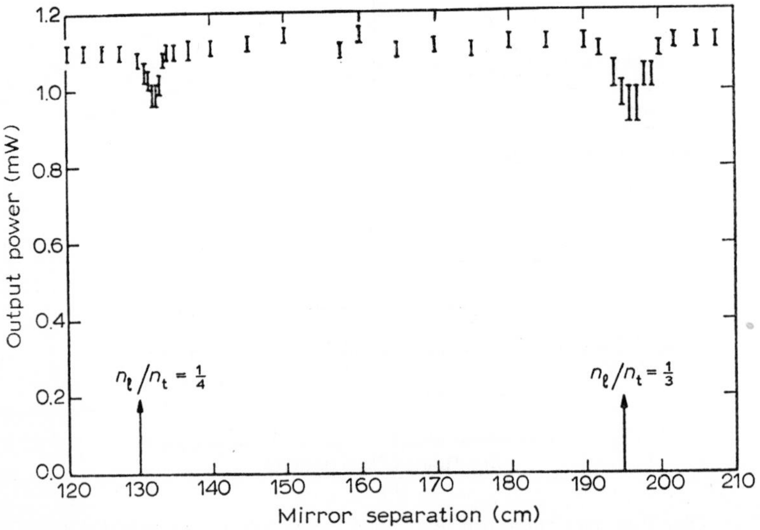

Fig. 8.16. Output power versus mirror separation. The arrows indicate calculated positions of dips (reproduced from G. O. HARDING and T. LI, J. Appl. Phys. **35**, 475 (1964) with permission).

Figure 8.15 is a g_1, g_2 stability diagram showing the loci of resonator geometries corresponding to mode degeneracy. If the separation between two mirrors of equal radius of curvature is varied, this corresponds to moving in the stability diagram along the line $g_1 = g_2$ and dips in the multimode power output of the laser occur at the intersections of this line with the

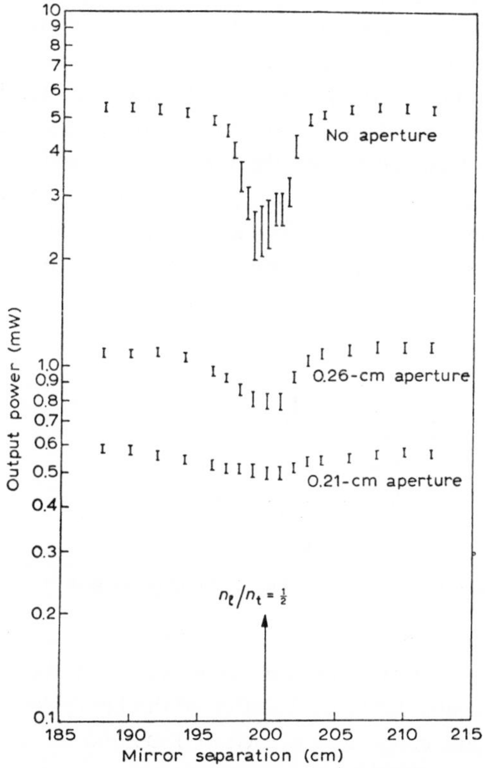

Fig. 8.17. Effect of apertures on the confocal dip for a pair of 2*m* mirrors. The laser tube has a bore of 6 mm. The arrow indicates the calculated position of dip (reproduced from G. O. Harding and T. Li, J. Appl. Phys. **35**, 475 (1964) with permission).

hyperbolas. The results of such an experiment by Harding and Li [1964] are given in fig. 8.16 for mirrors which were flat and 260 cm radius of curvature. The theory was also confirmed by using apertures to control the number of oscillating transverse modes, fig. 8.17.

7.3 *Hole burning and Lamb dip*

In a gas discharge laser, broadening is by the Doppler effect and is in-

References on page 213

homogeneous, and only those atoms whose velocities are such that the radiation they can be stimulated to emit corresponds to a cavity resonance contribute to the oscillation in the cavity. Since the oscillation consists of waves travelling in both directions along the laser axis, atoms with both $+v$ and $-v$ velocity components can contribute to radiation at a given frequency. Thus, each frequency which oscillates, depletes the populations of two groups of atoms, one group with velocity $+v$, and the other with velocity, $-v$, and, provided the mirrors are of equal reflectivity, the holes burned in the gain curve are symmetrical about the centre, fig. 8.18. There are two holes for one resonance at all resonant frequencies above threshold unless the frequency coincides with the line centre, in which case, of course, there is only one. The power output is proportional to the total population contributing to the radiation field. The area of the hole in the gain curve is a measure of the contributing population, and the width is approximately

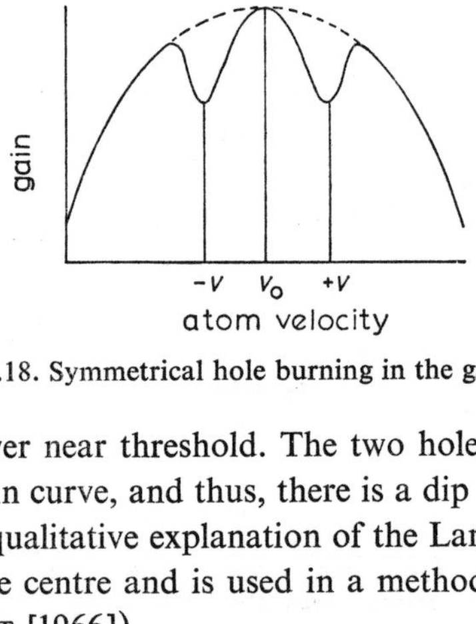

Fig. 8.18. Symmetrical hole burning in the gain curve.

independent of power near threshold. The two holes start to overlap near the centre of the gain curve, and thus, there is a dip in the power output of the laser. This is a qualitative explanation of the Lamb dip (ch. 9). The dip is located at the line centre and is used in a method of mode stabilization (BLOOM and WRIGHT [1966]).

8. Gain saturation in amplifiers with hole burning and cross relaxation

As described in the introduction to this chapter, the three lineshapes to be considered are those of the radiation to be amplified (the signal), the active medium (amplifier), and the atoms of the amplifier. A thorough study of gain saturation in amplifiers with hole burning and cross relaxation has been made by CABEZAS and TREAT [1966] whose analysis we now follow.

Consider an amplifier with an inhomogeneously broadened medium and with cylindrical symmetry about axis z and uniform gain in a radial direction.

The number of photons per second per unit area per unit frequency interval in the amplifier at plane z is $\mathscr{I}(\nu)$. Let the population inversion in the amplifier be $n(\nu)$ per unit volume per unit frequency interval. When there is no feed-back and we neglect losses due to scattering and absorption the gain conditions are described by

$$\frac{\partial \mathscr{I}(\nu)}{\partial t} + c\frac{\partial \mathscr{I}(\nu)}{\partial z} = c\mathscr{I}(\nu)\int_0^\infty \sigma(\nu, \nu')n(\nu')\mathrm{d}\nu', \tag{8.77}$$

where c is the velocity of light in the amplifier, $\sigma(\nu, \nu')$ is the stimulated emission cross section of the atomic line centred at ν', and ν is the stimulating frequency. The bandwidth of the atomic line is narrow compared with the inhomogeneous linewidth of the medium and is assumed to be the minimum frequency interval for which cross relaxation processes are infinitely fast. (This is the hole width described in § 7.1.)

Consider a 4-level system in which the terminal level empties instantaneously. The rate of change of the inverted population is given by

$$\frac{\partial n(\nu)}{\partial t} = -n(\nu)\int_0^\infty \sigma(\nu, \nu')\mathscr{I}(\nu')\mathrm{d}\nu' - An(\nu) + W[N_0 g(\nu, \nu_a) - n(\nu)] +$$

$$+F\left[g(\nu, \nu_a)\int_0^\infty n(\nu')\mathrm{d}\nu' - n(\nu)\right]. \tag{8.78}$$

The first term on the right hand side is the loss due to stimulated emission and the second is the loss due to spontaneous emission. In the third term, N_0 is the total density of particles available for pumping, and the distribution $g(\nu, \nu_a)$ centred at ν_a and normalized to unity is the fraction able to be pumped giving $N_0 g(\nu, \nu_a) - n(\nu)$ as the actual number pumped. The pump rate W is of the same dimensions as the Einstein A, as is also F, in the last term, which gives the cross-relaxation rate. The total number of inverted atoms per unit volume is given by

$$N = \int_0^\infty n(\nu)\mathrm{d}\nu.$$

In the steady state (8.77) becomes

$$\partial \mathscr{I}(\nu)/\partial z = \mathscr{I}(\nu)\int_0^\infty \sigma(\nu, \nu')n(\nu')\mathrm{d}\nu'. \tag{8.79}$$

Integrating over all frequencies, we get

$$d\mathscr{I}/\mathrm{d}z = \int_0^\infty \mathscr{I}(\nu)\left[\int_0^\infty \sigma(\nu, \nu')n(\nu')\mathrm{d}\nu'\right]\mathrm{d}\nu, \tag{8.80}$$

References on page 213

where $\mathscr{I} = \int_0^\infty \mathscr{I}(\nu)d\nu$ is the flux at any point z along the axis of the amplifier. Remembering that the contents of the square brackets refer to stimulated emission occurring within the atomic line profile, and that $\mathscr{I}(\nu)$ is the signal being amplified, we now consider the case in which $\Delta\nu_s \ll \Delta\nu_i$ (fig. 8.19).

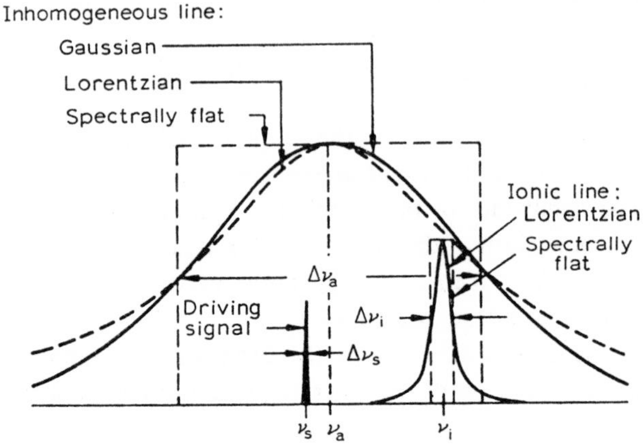

Fig. 8.19. Spectra of the inhomogeneous line, the ionic line, and the driving signal (reproduced from A. Y. CABEZAS and R. P. TREAT, J. Appl. Phys. **37**, 3556 (1966) with permission).

In this case the quantity in square brackets varies very much more slowly with ν than does $\mathscr{I}(\nu)$, and so may be approximated as a constant and taken outside the integral to give

$$d\mathscr{I}/dz = \mathscr{I}\int_0^\infty \sigma(\nu_s, \nu')n(\nu')d\nu', \tag{8.81}$$

where ν_s is the centre frequency of the signal. Similarly, we have

$$\int_0^\infty \sigma(\nu, \nu')\mathscr{I}(\nu')d\nu' = \sigma(\nu, \nu_s)\mathscr{I}, \tag{8.82}$$

so that, in the steady state, integrating (8.78) over ν' gives

$$0 = -n(\nu)\sigma(\nu, \nu_s)\mathscr{I} - (A+W)n(\nu) + WN_0 g(\nu, \nu_a) + \\ + F[g(\nu, \nu_a)N - n(\nu)]. \tag{8.83}$$

Integrating over all ν gives

$$0 = -\int_0^\infty n(\nu)\sigma(\nu, \nu_s)\mathscr{I}d\nu - \int_0^\infty (A+W)n(\nu)d\nu + \\ + WN_0\int_0^\infty g(\nu, \nu_a)d\nu + F\int_0^\infty [g(\nu, \nu_a)N - n(\nu)]d\nu.$$

Remembering that $\int_0^\infty g(\nu, \nu_a)\mathrm{d}\nu = 1$ and using (8.81) we obtain, finally

$$N = (WN_0 - \mathrm{d}\mathscr{I}/\mathrm{d}z)/A', \tag{8.84}$$

where $A' = A+W$.

Substituting this value for N into (8.83) gives

$$n(\nu) = \frac{\left[\dfrac{WN_0}{A'} - \dfrac{F}{A'(A'+F)}\dfrac{\mathrm{d}\mathscr{I}}{\mathrm{d}z}\right] g(\nu, \nu_a)}{1 + \dfrac{\sigma(\nu, \nu_s)\mathscr{I}}{A'+F}}. \tag{8.85}$$

Substituting this into (8.81) gives

$$\frac{\mathrm{d}\mathscr{I}}{\mathrm{d}z} = \mathscr{I}\left[\frac{WN_0}{A'} - \frac{F}{A'(A'+F)}\frac{\mathrm{d}\mathscr{I}}{\mathrm{d}z}\right]\int_0^\infty \frac{g(\nu, \nu_a)\sigma(\nu, \nu_s)\mathrm{d}\nu}{1+[\sigma(\nu, \nu_s)\mathscr{I}/(A'+F)]}. \tag{8.86}$$

We are now in a position to study gain saturation for systems with various profiles for $g(\nu, \nu_a)$ and $\sigma(\nu, \nu_s)$. The inhomogeneous lines $g(\nu, \nu_a)$ considered by CABEZAS and TREAT [1966] are the following with the values of k such that $g(\nu, \nu_a)$ is normalized to unity.

Gaussian: $$g(\nu, \nu_a) = \frac{k}{\Delta\nu_a} \exp\left[-4 \ln 2 \left(\frac{\nu-\nu_a}{\Delta\nu_a}\right)^2\right], \tag{8.87}$$

where $k = (4\pi \ln 2)^{\frac{1}{2}}/\pi$.

Lorentzian: $$g(\nu, \nu_a) = \frac{k}{\Delta\nu_a} \frac{1}{1+4[(\nu-\nu_a)/\Delta\nu_a]^2}, \tag{8.88}$$

where $k = 2/\pi$.

Spectrally Flat:

$$g(\nu, \nu_a) = \frac{k}{\Delta\nu_a}, \quad \text{for} \quad \left(\nu_a - \frac{\Delta\nu_a}{2}\right) < \nu < \left(\nu_a + \frac{\Delta\nu_a}{2}\right), \tag{8.89}$$

where $k = 1$.

The following distributions are considered for the individual atoms.

Lorentzian: $$\sigma(\nu, \nu_i) = \sigma_i/[1+4\{(\nu-\nu_i)/\Delta\nu_i\}^2]. \tag{8.90}$$

Spectrally flat: $$\sigma(\nu, \nu_i) = \sigma_i, \quad \text{for} \quad (\nu_i - \Delta\nu_i/2) < \nu < (\nu_i + \Delta\nu_i/2). \tag{8.91}$$

Results obtained with spectrally flat profiles provide a useful extreme case to set limits.

References on page 213

When the atomic line is Lorentzian, the integral in (8.86) is

$$g(\nu_a, \nu_a)(\pi/2)\sigma_i \Delta\nu_i/[1+\{\sigma_i \mathscr{I}/(A'+F)\}]^{\frac{1}{2}} \tag{8.92}$$

for the case where $\nu_s \approx \nu_a$ and $\Delta\nu_s \ll \Delta\nu_a$. It may be seen by considering (8.87), (8.88) and (8.89) that $g(\nu, \nu_a)$ in (8.86) is given by

$$g(\nu_a, \nu_a) = k/\Delta\nu_a. \tag{8.93}$$

Using (8.93) and the value of the integral, (8.92), (8.86) becomes

$$\frac{1}{\mathscr{I}}\frac{d\mathscr{I}}{dz} = \left[\frac{WN_0}{A'} - \frac{F}{A'(A'+F)}\frac{d\mathscr{I}}{dz}\right]\frac{\pi}{2}k\frac{\Delta\nu_i}{\Delta\nu_a}\sigma_i\left[1+\frac{\sigma_i\mathscr{I}}{A'+F}\right]^{-\frac{1}{2}}. \tag{8.94}$$

We now define a stimulated emission cross section for the amplifier, σ_a, thus

$$\sigma_a \equiv (\pi/2)k\sigma_i(\Delta\nu_i/\Delta\nu_a). \tag{8.95}$$

For weak signals ($\mathscr{I} \to 0$) and σ_a defined as in (8.95), eq. (8.94) becomes

$$(1/\mathscr{I})(d\mathscr{I}/dz) = WN_0\sigma_a/A'. \tag{8.96}$$

This equation can now be integrated over the length of the amplifier to get

$$\ln \mathscr{I}(\text{output}) - \ln \mathscr{I}(\text{input}) = WN_0\sigma_a l/A'. \tag{8.97}$$

We may also define the steady state weak signal gain G_w thus,

$$\ln G_w \equiv WN_0\sigma_a l/A'. \tag{8.98}$$

Two other useful definitions can be made, namely

$$\bar{\mathscr{I}} \equiv \frac{\sigma_i\mathscr{I}}{A'+F} \quad \text{and} \quad f \equiv \frac{F}{A'}\frac{\Delta\nu_i}{\Delta\nu_a}, \tag{8.99}$$

where f is called the narrow band cross relaxation coefficient. With these various substitutions, (8.94) may now be written

$$\left[\frac{(1+\bar{\mathscr{I}})^{\frac{1}{2}}}{\bar{\mathscr{I}}} + \frac{\pi}{2}kf\right]\frac{d\bar{\mathscr{I}}}{dz} = \ln\left(\frac{G_w}{l}\right). \tag{8.100}$$

The equation governing gain saturation of a Lorentzian line is obtained by integrating (8.100) over the amplifier length which gives, finally

$$(1+\bar{\mathscr{I}}_0 G)^{\frac{1}{2}} - 2(1+\bar{\mathscr{I}}_0)^{\frac{1}{2}} + \ln\left[\frac{(1+\bar{\mathscr{I}}_0 G)^{\frac{1}{2}}-1}{(1+\bar{\mathscr{I}}_0 G)^{\frac{1}{2}}+1}\right] + $$

$$+\ln\left[\frac{(1+\bar{\mathscr{I}}_0)^{\frac{1}{2}}+1}{(1+\bar{\mathscr{I}}_0)^{\frac{1}{2}}-1}\right] + \frac{\pi}{2}kf(G-1)\bar{\mathscr{I}}_0 = \ln G_w, \tag{8.101}$$

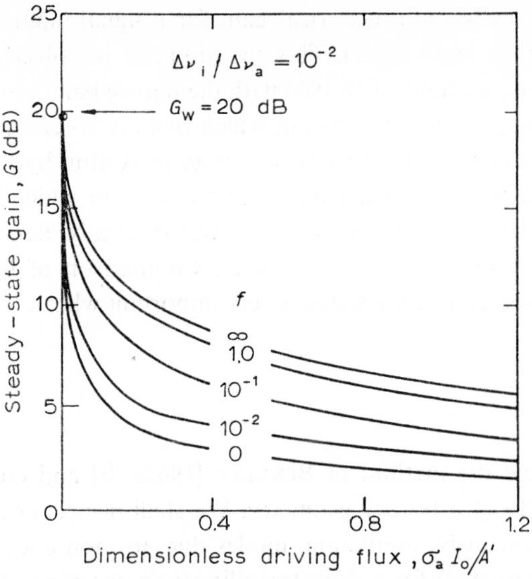

Fig. 8.20. Solutions of equation (8.101) for gain saturation with the narrow band cross-relaxation coefficient f as parameter. Lorentzian ionic, and Gaussian inhomogeneous line (reproduced from A. Y. Cabezas and R. P. Treat, J. Appl. Phys. **37**, 3556 (1966) with permission).

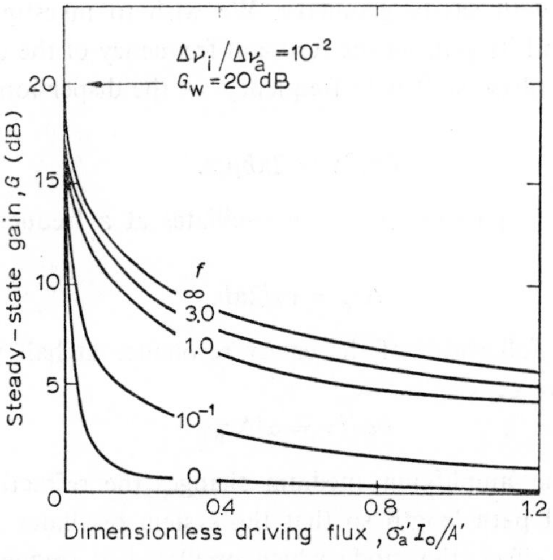

Fig. 8.21. Solutions for the case in which both the inhomogeneous line and the atomic lines are spectrally flat. The cross-relaxation coefficient f is the parameter $\sigma_a' = k\sigma_i\Delta\nu_i/\Delta\nu_a$ (reproduced from A. Y. Cabezas and R. P. Treat, J. Appl. Phys. **37**, 3556 (1966) with permission).

References on page 213

where G is the steady-state numerical gain for a signal input $\mathscr{I}_0$, and $\bar{\mathscr{I}}_0$ is the dimensionless input signal. This equation can be solved numerically. Figure 8.20 shows solutions of (8.101) with the narrowband cross relaxation coefficient f as parameter. The limit in which there is no cross relaxation is given by $f \to 0$, and that of infinitely fast cross relaxation by $f \to \infty$.

Equations can be developed readily for the case in which both the inhomogeneous line and the atomic lines are spectrally flat. Figure 8.21 shows their solutions with f as parameter. Comparison of figs. 8.20 and 8.21 shows that the atomic lineshape is very important when cross relaxation is slow ($f < 1$).

9. Mode pulling

We shall follow the method of BENNETT [1962a, b] and consider mode pulling in a cavity of arbitrary geometry. We shall neglect coupling effects between simultaneously oscillating modes due to non-linearities of the medium. The phase shift for a wave travelling from one mirror to the other is

$$\phi = \omega t = 2\pi \nu b \mu / c, \tag{8.102}$$

where μ is the refractive index of the medium in the cavity, b is the cavity length, ν is the frequency of the cavity resonance, and c is the phase velocity which can vary with cavity geometry. We wish to investigate the effect of the medium and its gain on the resonant frequency of the cavity.

The change of phase shift with frequency, or the dispersion of the cavity is given by

$$\partial\phi/\partial\nu = 2\pi b\mu/c. \tag{8.103}$$

When the cavity is passive, $\mu = 1$, it oscillates at a frequency ν_p. From (4.15) we get

$$\Delta\nu_p = c\alpha/2\pi b. \tag{8.104}$$

where $\Delta\nu_p$ is the full width of the cavity resonance at half maximum intensity. Hence, we get

$$\partial\phi/\partial\nu = \alpha/\Delta\nu_p. \tag{8.105}$$

Introducing the amplifying medium changes the refractive index and hence the optical path length so that the system oscillates at a different frequency ν_a. In effect, the mode which oscillated at frequency ν_p in the passive case has changed its frequency to ν_a for the active case. Part of the change in refractive index is due to the unexcited particles of the medium, and part is due to the excited particles.

At frequencies close to that of the laser transition, or any other transition, there is anomalous dispersion. For the other transitions, the medium absorbs, but for the laser transition, the medium has gain (negative absorption). The absorption and gain profiles together with their associated anomalous dispersion curves are shown in fig. 8.22. We are only concerned with frequencies near to the laser transition frequency ν_D, where it may be seen that a cavity resonance of frequency less than ν_D has its frequency increased by the amplifying medium while a cavity resonance of frequency greater than ν_D has its frequency decreased. The net result is that the cavity resonance is pulled towards the centre of the line of the laser transition.

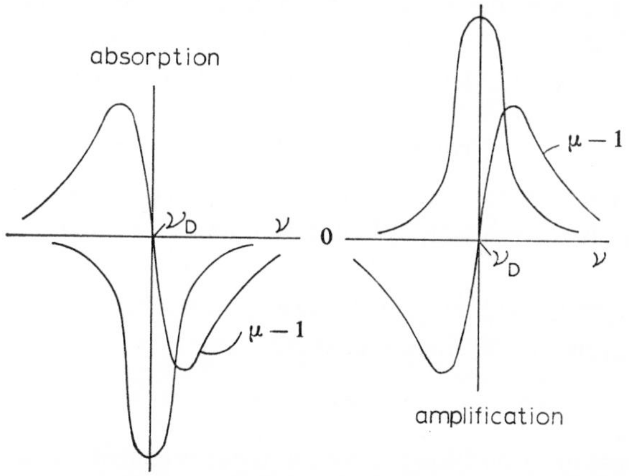

Fig. 8.22. Absorption and gain profiles together with their associated anomalous dispersion curves.

The total change of phase caused by introducing the active medium is given by

$$(\partial\phi/\partial\nu)(\nu_p-\nu_a) = \Delta\Phi_m(\nu_a), \tag{8.106}$$

where $\Delta\Phi_m(\nu_a)$ is the total change in single pass phase shift at the frequency of oscillation which is caused by inserting the active medium and in the presence of the oscillation. As noted above, the refractive index of the medium inserted may be considered to be made up of two parts; the first (μ_e) is due to medium introduced and includes the contributions from all excited states other than those of the laser transition; the other is due to the laser transition, thus, the total phase shift is given by

$$\begin{aligned}\Delta\Phi_m(\nu_a) &= (2\pi b/c)[(\mu_e-1)+(\mu-1)]\nu_a = \\ &= (2\pi b/c)(\mu_e-1)\nu_a+\Delta\phi_m(\nu_a).\end{aligned} \tag{8.107}$$

References on page 213

From (8.104), (8.105), (8.106) and (8.107) we get

$$\frac{\nu_p}{\mu_e} - \frac{\Delta\nu_p}{\alpha\mu_e}\Delta\phi_m(\nu_a) = \nu_a\,.$$

This may be simplified by defining

$$\nu_p/\mu_e \equiv \nu_c \quad \text{and} \quad \Delta\nu_p/\mu_e \equiv \Delta\nu_c\,.$$

Thus, we get

$$\nu_a = \nu_c - \Delta\nu_c \Delta\phi_m(\nu_a)/\alpha. \tag{8.108}$$

The change in phase shift $\Delta\phi(\nu_a)$ depends upon the shape of the amplifying transition of the active medium. The gain per pass may be related to the phase shift per pass by means of the Kramers-Kronig relations as shown in ch. 2 § 4.10.2. The result for a Lorentzian line is

$$\Delta\phi_m(\nu_a) = -g(\nu_a)(\nu_m - \nu_a)/\Delta\nu_m\,, \tag{8.109}$$

where ν_m is the frequency at the line centre (i.e. the spontaneous emission line) and $\Delta\nu_m$ is the full width of the line at half maximum fractional energy gain.

When a laser is oscillating in the steady state, the fractional energy gain per pass saturates at the fractional energy loss, thus

$$g(\nu_a) = \alpha. \tag{8.110}$$

For homogeneous broadening, since all particles contribute to the gain at all frequencies over the line profile, proportional decreases in gain to satisfy (8.110) occur over all the frequencies and the oscillation frequency is always its value at threshold and there is no power dependent frequency pulling.

Combining (8.108), (8.109) and (8.110), we get, for a homogeneously broadened Lorentzian line,

$$\nu_a = \frac{\nu_c \Delta\nu_m + \nu_m \Delta\nu_c}{\Delta\nu_m + \Delta\nu_c}\,. \tag{8.111}$$

For $\Delta\nu_c \ll \Delta\nu_m$, (8.111) gives

$$\nu_a = \nu_c + (\nu_m - \nu_c)(\Delta\nu_c/\Delta\nu_m). \tag{8.112}$$

The beat frequency between two modes ν_a' and ν_a is

$$\nu_a' - \nu_a = (\nu_c' - \nu_c)\left(1 - \frac{\Delta\nu_c}{\Delta\nu_m}\right). \tag{8.113}$$

In a system characterized by inhomogeneous broadening, hole burning occurs when it operates above threshold. BENNETT [1962a, b, 1964] gives a thorough discussion of mode pulling and hole burning. For a Gaussian line in threshold conditions, Bennett shows that when $\Delta\nu_c \ll \Delta\nu_m$ for frequencies near the line centre, the frequency ν_a is given by $\nu_a \approx \nu_c + (\nu_m - \nu_c)(0.94\Delta\nu_c/\Delta\nu_m)$ which shows a 6 % reduction in the pulling factor of a Lorentzian (8.112).

BENNETT [1962a] measured beat frequencies in a He-Ne laser and found that the beat separation was less than $c/2b$ by about 1 part in 800. The conditions of the experiment were such that $\Delta\nu_c \approx 1$ MHz and the Doppler width $\Delta\nu_m \approx 800$ MHz, which shows that (8.113) accounted for the discrepancy observed.

References

ABRAMOWITZ, M. and I. A. STEGUN, 1964, Handbook of Mathematical Functions, Third Printing, 1965, National Bureau of Standards, Applied Mathematics Series, 55, Integral 7.4.13, p. 302.

BATEMAN, H. 1954, Tables of Integral Transforms, Vol. I (McGraw-Hill, New York) p. 136, No. 4.2.25.

BENNETT, Jr., W. R., 1962a, Phys. Rev. **126**, 580.

BENNETT, Jr., W. R., 1962b, Appl. Optics, Supplement on Optical Masers, pp. 24–61.

BENNETT, Jr., W. R., 1964, Relaxation Mechanisms, Dissociative Excitation Transfer, and Mode Pulling Effects in Gas Lasers, In: Grivet, P. and N. Bloembergen, eds., Proceedings of the Third International Congress on Quantum Electronics (Columbia University Press, New York) pp. 441–458.

BLOOM, A. L. and D. L. WRIGHT, 1966, Proc. I.E.E.E. **54**, 1290.

BOYD, G. D. and H. KOGELNIK, 1962, Bell Syst. Tech. J. **41**, 1347.

CABEZAS, A. Y. and R. P. TREAT, 1966, J. Appl. Phys. **37**, 3556.

FOX, A. G. and T. LI, 1966, I.E.E.E. J. Quantum Electronics, **QE–2**, 774.

GORDON, E. I., A. D. WHITE, and J. D. RIGDEN, 1963, Gain Saturation at 3.39 Microns in the He–Ne Maser. In: Proceedings of the Symposium on Optical Masers, New York, 1963 (Polytechnic Press, Brooklyn, New York) pp. 309–318.

HARDING, G. O. and T. LI, 1964, J. Appl. Phys. **35**, 475.

HOTZ, D. F., 1965, Appl. Optics, **4** 527.

MENEELY, C. T., 1967, Appl. Optics, **6**, 1434.

MITCHELL, A. C. G., and M. W. ZEMANSKY, 1961, Resonance Radiation and Excited Atoms, Second Impression (Cambridge University Press, London) pp. 94–96.

RIGROD, W. W., 1963, J. Appl. Phys. **34**, 2602.

RIGROD, W. W., 1964, Isolation of Axi-Symmetrical Optical-Resonator Modes, In: Grivet, P. and N. Bloembergen, eds., Proceedings of the Third International Congress on Quantum Electronics (Columbia University Press, New York) pp. 1285–1290.

RIGROD, W. W., 1965, J. Appl. Phys. **36** 2487.

SMITH, P. W., 1966(a), I.E.E.E. J. Quantum Electronics, **QE–2**, 62.

SMITH, P. W., 1966(b), J. Appl. Phys. **37**, 2089.

STATZ, H. and C. L. TANG, 1965, J. Appl. Phys. **36**, 1816.

CHAPTER 9

LAMB'S THEORY OF THE LASER

1. Introduction

LAMB [1964a] has proposed a theoretical description for the operation of a multimode laser oscillator, based on a classical description of the electromagnetic field in terms of high-Q modes, and a quantum mechanical description of the active medium in terms of the density matrix. The electromagnetic field is coupled to the medium through a macroscopic, electric polarization term acting as the field source in accordance with Maxwell's equations. The model has proved extremely useful in the description of gaseous lasers, particularly when the effect of some characteristic of the active medium on the radiation field is to be investigated. Nonlinear polarization terms are included so that phenomena such as saturation, frequency pulling and pushing, mode competition and frequency locking are considered.

However, the theory has its limitations which we now summarize. Firstly, since the field is described in terms of classical modes of well defined amplitude and phase, effects which introduce statistical fluctuations (e.g. spontaneous emission) are excluded. The theory does not therefore predict the ultimate theoretical linewidth of a laser mode, nor does it describe the coherence properties of the field. Secondly, since the coupling between the field and the medium is dealt with by low order perturbation methods, the theory only applies to the weak signal case. Although saturation effects are considered, these do not appear until the third order terms in the perturbation theory, and it is at this point that the calculations stop.

Lamb's original paper considers the case in which the emission line is Doppler broadened to an extent large compared to the natural linewidth, in which the natural linewidth is radiatively determined, in which only two atomic states contribute to the laser oscillation, and in which the electromagnetic field is scalar (polarization effects neglected).

A generalization of Lamb's theory with regard to these latter two aspects has attracted a considerable amount of investigation, since in the presence of static magnetic fields the upper and lower laser levels are split into their magnetic sublevels, and the different sublevels interact to differing extents with the laser radiation field depending on its polarization state. The most detailed and comprehensive treatment is that due to SARGENT et al. [1967a, b], although earlier investigators including HEER and GRAFT [1965], FORK and SARGENT [1966], D'YAKONOV and FRIDRIKHOV [1966], DURAND [1966], VAN HAERINGEN [1967] have made significant contributions to various aspects of the problem.

Collisional broadening of the natural linewidth has also been considered by many investigators, usually in attempts to extend Lamb's theory to cover particular experimental situations, and this work is discussed shortly. More recently, however GYORFFY et al. [1968] have given a comprehensive theoretical treatment.

UEHARA and SHIMODA [1965] and CULSHAW [1967] have extended the theory from third, to fifth and higher order terms. UCHIDA [1967] has employed Lamb's theory to investigate the response of gas lasers to resonator loss modulation and to small external signal injection (due to spontaneous emission, for example), with particular regard to mode locking and the intrinsic noise of the laser.

As well as his principal paper in "Physical Review", Lamb has published several expositions which are helpful in developing the background necessary for its understanding. One of these, on quantum mechanical amplifiers (LAMB [1960]) preceeds the publication of the "Physical Review" paper, and outlines the classical and semiclassical theories of emission and absorption of radiation, coherent excitation, weak and strong signal theory, etc. The other two expositions were both published at about the same time as the Physical Review paper. One (LAMB [1964b]) is concerned solely with developing the theory, in a more simple-minded fashion, while the other (LAMB [1965]) has a more general content as well.

As regards work concerned with the experimental investigation, either directly or indirectly, of the predictions of Lamb's theory, the following resume, while by no means exhaustive, includes most of the principal papers: SZOKE and JAVAN [1963] investigated the saturation behaviour of the power gain of the 1.15 μ transition in Ne, and compared their experimental results with the predictions of Lamb's theory (9.157); MCFARLANE et al. [1963] investigated the 'Lamb dip' (§ 7.4) in the 1.15 μ He-Ne laser with both naturally occurring and isotopically enriched Ne; SZOKE and JAVAN

References on page 259

[1966] extended their earlier work to a full investigation of the influence of collisions on the saturation behaviour of the 1.15 μ Ne transition and on the frequency and shape of the "Lamb dip"; SMITH [1966] investigated the saturation behaviour of the 6328 Å transition in the He-Ne laser, and included a consideration of collision effects (earlier work of a similar kind was carried out on this transition by ALPERT and WHITE [1963], but their findings were not at the time related to Lamb's theory).

Whereas the work summarized above was carried out with single mode oscillation, FORK and POLLACK [1965] investigated mode intensities and beat frequencies for simultaneous oscillation on two axial modes, and related their experimental findings to a modified form of Lamb's original theory, making allowance for collisional effects. BOLWIJN [1964, 1965, 1966] has studied the single mode output power modulation of the 1.15 μ Ne transition that is brought about either by modulation of the cavity losses or the excitation processes, and has related such effects to the small signal gain and saturation behaviour of the transition using the approach of Lamb's theory. BOERSCH et al. [1967] using the 6328 Å Ne transition and HAISMA and BOUWHUIS [1964] using the 1.15 μ Ne transition have performed beat frequency measurements on the 'combination tones' predicted by third order Lamb theory (§ 8).

Predictions from Lamb theory for the case when weak, axial magnetic fields are present have been explored experimentally by TOMLINSON and FORK [1967] and SETTLES and HEER [1968], but mention should also be made of earlier work by CULSHAW and KANNELAUD [1966].

2. Representation of field in an active cavity

We begin our discussion of Lamb's theory by deriving the form of Maxwell's wave equation appropriate to a description of the field within an active cavity (rationalized MKS units used throughout this chapter). The wave equation for propagation in free space has the well-known form

$$\text{curl curl}\, \boldsymbol{E} + \mu_0 \varepsilon_0 \frac{\partial^2 \boldsymbol{E}}{\partial t^2} = 0. \tag{9.1}$$

In dealing with the usual types of laser cavity, for low order transverse modes, there are no rapid variations in the electric field in the xy-plane perpendicular to the direction of propagation (z) of the radiation ($\partial^2/\partial x^2 \approx 0$, etc.). If we consider only radiation polarized in a direction perpendicular to the propagation direction, then we can simplify the above vector equation to the corresponding scalar equation

$$-\frac{\partial^2 E}{\partial z^2} + \mu_0 \varepsilon_0 \frac{\partial^2 E}{\partial t^2} = 0. \tag{9.2}$$

When (9.2) is subject to the boundary conditions of the cavity, the solutions are the lossless, normal modes of the cavity. The longitudinal modes are described by the eigenfunctions (standing waves),

$$E_n = E_{0n} \sin(K_n z) \cos(\Omega_n t), \tag{9.3}$$

where for a cavity of length L,

$$K_n = n\pi/L \tag{9.4}$$

and n is a large integer specifying the mode. (This is the longitudinal mode number designated by q in ch. 4 § 6.) Substitution of (9.3) and (9.4) into (9.2) gives Ω_n, the angular frequency of the nth mode

$$\Omega_n = \pi n c/L. \tag{9.5}$$

Any arbitrary field configuration can be expanded in terms of these normal longitudinal modes, since they form a complete set of orthogonal functions (Appendix I), thus we have

$$E = \sum_n E_{0n} \sin(K_n z) \cos(\Omega_n t). \tag{9.6}$$

So far, we have neglected losses from the radiation field (for example, due to diffraction or output coupling). Lamb's theory deals with this damping of the field by introducing an Ohmic conductivity, σ, into the medium in the cavity. This method of describing losses in terms of a volume effect avoids a complicated boundary value problem, but we lose information about the spatial variation of the field due to the losses being predominantly through the mirrors at each end of the cavity (ch. 4 § 5.2, ch. 8 § 5). Introducing the conductivity term into (9.2), we obtain

$$-\frac{\partial^2 E}{\partial z^2} + \mu_0 \sigma \frac{\partial E}{\partial t} + \mu_0 \varepsilon_0 \frac{\partial^2 E}{\partial t^2} = 0. \tag{9.7}$$

The damping is assumed to be small, and we therefore retain the form of the solution in terms of the normal modes as given by (9.6), but allow the amplitudes (E_{0n}) to become slowly-varying functions of time, thus

$$E = \sum_n E_{0n}(t) \sin(K_n z) \cos(\Omega_n t). \tag{9.8}$$

Substitution of (9.8) into (9.7) and remembering that the modes are independent, leads to

$$\dot{E}_{0n}(t) = -\sigma E_{0n}(t)/2\varepsilon_0 , \tag{9.9}$$

where terms such as $\ddot{E}_{0n}(t)$ and $\sigma\dot{E}_{0n}(t)$ have been neglected. The approximate solution to (9.7) is therefore

$$E = \sum_n E_{0n} \exp(-\sigma t/2\varepsilon_0) \sin(K_n z) \cos(\Omega_n t). \tag{9.10}$$

With the damping we can associate a quality factor, Q, defined by (see ch. 4 § 5)

$$Q = \varepsilon_0 \Omega_n/\sigma. \tag{9.11}$$

In general the damping varies from mode to mode, and so a more general solution would be

$$E = \sum_n E_{0n} \exp(-\Omega_n t/2Q_n) \sin(K_n z) \cos(\Omega_n t). \tag{9.12}$$

We now consider the manner in which the electric field is driven by the active medium in the cavity. As we have already pointed out (§ 1), Lamb does this by introducing a macroscopic polarization term, $P(z, t)$, into the wave equation, which now becomes (see appendix L)

$$-\frac{\partial^2 E}{\partial z^2} + \mu_0 \sigma \frac{\partial E}{\partial t} + \mu_0 \varepsilon_0 \frac{\partial^2 E}{\partial t^2} = -\mu_0 \frac{\partial^2 P}{\partial t^2} . \tag{9.13)*$$

As was the case with the conductivity term, we assume the polarization term to be small compared with $\varepsilon_0 E$, so that we can retain the form of the solution as that of the sum of the normal mode eigenfunctions associated with the passive cavity. The polarization driving term is significant over a frequency range corresponding to the linewidth of the appropriate transition in the active medium, reaching a maximum at the line centre. In general, therefore, the effective frequency of this driving term differs from the frequencies of the passive modes (Ω_n), and so we allow the frequencies of the active modes (ω_n) to differ slightly from those in the passive case. We further allow additional, slowly varying, time dependent phase terms into the eigenfunctions, as well as the slowly varying amplitude terms already introduced.

We are therefore looking for a solution of the form

$$E = \sum_n E_{0n}(t) \sin(K_n z) \cos[\omega_n t + \phi_n(t)]. \tag{9.14}$$

* The polarization term is due to (a) permanent dipole moments which may be associated with individual states, and (b) dipole moments associated with transitions between states.

Such a solution can be said to describe the quasi-stationary forced oscillations of the system; stationary in the sense that it involves an expansion in terms of the normal (undriven) modes of the passive cavity, and quasi-stationary in that these normal modes are allowed to be perturbed in amplitude, phase and resonance frequency by the driving force.

If we substitute (9.14) into (9.13), multiply throughout by sin $(K_m z)$ and then integrate over z, (L is the cavity length) we can make use of the orthogonal properties of the normal modes to obtain

$$K_m^2 E_m + \mu_0 \sigma_m \frac{\partial E_m}{\partial t} + \mu_0 \varepsilon_0 \frac{\partial^2 E_m}{\partial t^2} = -\mu_0 \frac{\partial^2 P_m}{\partial t^2}, \tag{9.15}$$

where

$$E_m(t) = E_{0m}(t) \cos [\omega_m t + \phi_m(t)]; \tag{9.16}$$

$$P_m(t) = \frac{2}{L} \int_0^L P(z, t) \sin (K_m z) \mathrm{d}z. \tag{9.17}$$

The macroscopic polarization, $P(z, t)$, is produced by all the modes that are oscillating, and equation (9.17) gives the component, $P_m(t)$, of $P(z, t)$ that influences a particular (the mth) mode. Equation (9.17) shows that if $P(z, t)$ is expanded into its spatial Fourier components, then the particular component that influences a particular mode is the one having the same spatial form as that mode.

Since the frequency spectrum of the polarization is essentially determined by the spectral line shape of the exciting transition, it is approximately monochromatic (at optical frequencies) in the sense that if we associate a frequency, ω, with the polarization, then its second time derivative in (9.15) can be approximated by $-\omega^2 P_m(t)$, for all the modes. We also define the Q's of the various modes in terms of ω (ch. 4 § 5):

$$Q_m = \varepsilon_0 \omega / \sigma_m, \tag{9.18}$$

so that eq. (9.15) becomes

$$\frac{\partial^2 E_m}{\partial t^2} + \left(\frac{\omega}{Q_m}\right) \frac{\partial E_m}{\partial t} + \Omega_m^2 E_m = \left(\frac{\omega^2}{\varepsilon_0}\right) P_m. \tag{9.19}$$

We now assume that $P_m(t)$ is made up of two components; one in phase with the cavity field, the other in quadrature (we do this since we cannot assume that the induced polarization remains in phase with the field). Thus we obtain

$$P_m(t) = C_m(t) \cos [\omega_m t + \phi_m(t)] + S_m(t) \sin [\omega_m t + \phi_m(t)]. \tag{9.20}$$

References on page 259

In assuming this form for $P_m(t)$ we have essentially neglected frequency components of the polarization that are far from the cavity resonance. Since in any usual laser configuration, the mode separation ($\sim$ 150 MHz) greatly exceeds the cavity mode bandwidth ($\sim$ 1 MHz), this is a reasonable approximation (ch. 4 § 5.2). On substituting (9.20) and (9.16) into (9.19), and separating sine and cosine terms we obtain two equations which relate the amplitude and phase terms associated with the field to the amplitudes of the two components of the polarization.

The following approximations are made in deriving these equations: firstly that terms of the order $(\omega \dot{E}_{0m}/Q_m)$, $(\dot{\phi}_m \dot{E}_{0m})$ $(\omega \dot{\phi}_m E_{0m}/Q_m)$, and second derivatives of time can be neglected, secondly that (ω_m/ω) is approximately unity, and thirdly that the effective frequency of the active cavity mode, $(\omega_m + \dot{\phi}_m)$, lies very close to the frequency of the passive cavity mode, Ω_m so that $(\omega_m + \dot{\phi}_m)^2 - \Omega_m \approx 2\omega(\omega_m + \dot{\phi}_m - \Omega_m)$. The equations we now obtain, are called the 'self consistency' equations

$$\dot{E}_{0m} + \tfrac{1}{2}(\omega/Q_m)E_{0m} = -\tfrac{1}{2}(\omega/\varepsilon_0)S_m(t); \tag{9.21}$$

$$(\omega_m + \dot{\phi}_m - \Omega_m)E_{0m} = -\tfrac{1}{2}(\omega/\varepsilon_0)C_m(t). \tag{9.22}$$

The first equation describes the effect of the damping and the active medium on the mode amplitude. If the in-quadrature component of the polarization is zero, the amplitude is exponentially damped out, as for a passive, lossy cavity (9.9). The in-quadrature component of the polarization, represents the gain introduced by the active medium, which overcomes the cavity losses, and allows oscillations to occur. The second equation describes the part the in-phase component of the polarization plays in altering the frequency of the field from that associated with the passive cavity; and therefore describes frequency pushing and pulling effects, etc.

To illustrate this it is instructive to consider the case when the polarization is a linear function of the electric field, thus

$$C_m = \varepsilon_0 E_{0m} \chi'; \tag{9.23}$$

$$S_m = \varepsilon_0 E_{0m} \chi''. \tag{9.24}$$

On substitution of the above into equations (9.21), and (9.22) we obtain

$$(\omega_m + \dot{\phi}_m - \Omega_m) = -\tfrac{1}{2}\omega\chi'_m; \tag{9.25}$$

$$\dot{E}_{0m} = -\tfrac{1}{2}\omega[\chi''_m + (1/Q_m)]E_{0m}. \tag{9.26}$$

The first equation indicates that the oscillation frequency differs from the eigenvalue for the mode, Ω_m, by a "pulling term" $(-\omega\chi'_m)/2$ due to the presence of the dielectric. From the second equation it can be seen that if $\chi''_m > 0$, the dielectric adds to the damping already present. On the other hand, if $\chi''_m < (-Q_m)^{-1}$, the equation describes an exponential build-up of the oscillation, and it is at this point that higher order terms (since in general the polarization is not a linear function of field) need to be considered to take into account the saturation behaviour of the medium.

3. The macroscopic polarization

We have so far developed a description of the cavity field in terms of the macroscopic polarization associated with the active medium. It is now necessary to relate this macroscopic polarization, as a function of the electric field, to the atomic properties of the active medium. Once this is done, equations (9.21) and (9.22) are used as self-consistency relationships for the system, and enable the amplitude and phase behaviour of the field to be derived.

We assume that the active medium is composed of a collection of two level atomic systems, so that the wave function for a particular system is of the form (3.60)

$$\Psi = a(t)u_a + b(t)u_b, \tag{9.27}$$

We have already derived the equations of motion for the coefficients, a and b (3.61) in the case when the two level system is interacting with monochromatic radiation and when the finite lifetimes of the levels, due to decay processes to other levels not explicitly considered, are taken into consideration by the introduction of decay constants. These equations are

$$\begin{aligned} \hbar i\dot{b} &= aV + bE_b - i\gamma_b \hbar b/2, \\ \hbar i\dot{a} &= bV + aE_a - i\gamma_a \hbar a/2. \end{aligned} \tag{9.28}$$

We shall see shortly that the quantities required are aa^*, bb^*, a^*b, ab^* (aa^* and bb^* are the probabilities of the system being in the upper and lower energy levels respectively, while a^*b and ab^* are involved in the polarization). We also require to take various averages over all the systems in the active medium, since they are excited at different times and with different velocity components. It is therefore convenient to adopt the density matrix formalism (ch. 3 § 4.1), and to do this we define the following matrix

$$\boldsymbol{\rho} = \begin{bmatrix} aa^* & b^*a \\ a^*b & b^*b \end{bmatrix} = \begin{bmatrix} \rho_{aa} & \rho_{ab} \\ \rho_{ba} & \rho_{bb} \end{bmatrix}. \tag{9.29}$$

Equations (9.28) are then substituted into the elements of (9.29) to derive its equation of motion

$$\mathrm{i}\frac{\mathrm{d}}{\mathrm{d}t}\boldsymbol{\rho} = \boldsymbol{H}\boldsymbol{\rho} - \boldsymbol{\rho}\boldsymbol{H} - \frac{\mathrm{i}}{2}(\boldsymbol{\Gamma}\boldsymbol{\rho} + \boldsymbol{\rho}\boldsymbol{\Gamma}), \tag{9.30}$$

where

$$\boldsymbol{H} = \hbar^{-1}\begin{bmatrix} E_a & V \\ V & E_b \end{bmatrix} \quad \text{and} \quad \boldsymbol{\Gamma} = \begin{bmatrix} \gamma_a & 0 \\ 0 & \gamma_b \end{bmatrix}.$$

It should be remembered that although we have introduced the density matrix formalism, no averaging is, as yet, involved in (9.30), and the equation describes the behaviour of only a single atomic system (and not the active medium as a whole). The similarity between the equation of motion of the density matrix (ch. 3 § 4.1) and eq. (9.30) should be noted. The latter equation has an additional term to take into account the decrease in the probability of finding the system in one or other of the two energy levels that are considered explicitly, whereas in the derivation of the equation of motion of the density matrix (3.153) we assumed that all possible states of the system had been considered. In other words, because we are restricting our attention to only two of the states, we are forced to work with a density matrix that is not normalized, and the additional term on the right hand side of (9.30) takes this into account.

The average dipole moment (scalar) of the two level atom is given by

$$P = e\langle r\rangle = e\int \Psi^* r \Psi \,\mathrm{d}q. \tag{9.31}$$

If we assume that the atom has no permanent dipole moment associated with either of its two states, then from (9.27) the average dipole moment can be written (3.93)

$$P = e\langle r\rangle = \mathscr{P}(\rho_{ab} + \rho_{ba}), \tag{9.32}$$

where

$$\mathscr{P} = e\int u_a^* r u_b \,\mathrm{d}q. \tag{9.33}$$

(In chs. 2 and 3, $\mathscr{P}$ is designated D_{mn}.)

The dipole moment can also be expressed in matrix form, using (9.29)

$$e\langle r\rangle = \mathrm{Tr}\,(\mathscr{P}\boldsymbol{\rho}). \tag{9.34}$$

4. Theory for stationary atoms

4.1 *Equation of motion of density matrix with excitation*

We have described a two level atomic system that is both interacting with a monochromatic radiation field and also relaxing through processes described by damping constants, and have shown how the equations for the time development of such a system can be conveniently written in matrix form. This enabled us to derive the polarization term that appears in the equations describing the radiation field (9.21) and (9.22), for the case when only a single atomic system is interacting with this field.

We now extend the description to cover the interaction of the radiation field with the whole of the active medium, and so need to consider a macroscopic system consisting of all the two level atomic systems that comprise the active medium. Since we do not possess all possible knowledge of the active medium, in that we do not choose to specify, for a particular atomic system, the time and position at which it was excited, nor its associated velocity components, etc., we take averages over these properties. As a consequence, we are unable to associate a definite wave function with the macroscopic system, and the matrix, ρ, takes on the character of a density matrix.

The different averages that may need to be carried out for a particular active medium are considered below:

(i) Averages over the times and positions at which the atomic systems are excited into one or the other of their two energy levels.

(ii) Averages over the velocity components associated with the atomic systems. The radiation field "seen" by a moving atomic system differs from that seen by a stationary atomic system, and as a consequence its interaction with the radiation field is modified.

(iii) Averages over the different environments in which the atomic systems are located.

Averages of the type (ii) and (iii) are associated with inhomogeneous broadening, and we shall consider the former case, that of Doppler broadening, later (§ 5.1). In this section we restrict our attention to the averaging associated with the excitation of the atomic systems, and assume that they are all stationary. As a consequence they all "see" the same radiation field and have the same interaction term in (9.28), so allowing the rotating wave approximation to be made (ch. 3 § 3.4).

Suppose that $\lambda_a(t_0)\mathrm{d}t_0$ is the average number of atoms, per unit volume, excited to state a in the time interval $(t_0, t_0+\mathrm{d}t_0)$ and that $\rho(a, t_0, t)$ is the matrix describing one of these atoms at a later time, t at which, of course, the atom may not be purely in state a. The equation of motion of $\rho(a, t_0, t)$ is still (9.30).

The excitation rate $\lambda_a(t_0)$ may be a function of position within the active medium, but for the present we do not denote this explicitly. If we assume that there is only excitation to the state a, the density matrix for the active medium is found by summing over all time intervals previous to the observation time, so that

$$\rho(a, t) = \int_{-\infty}^{t} \lambda_a(t_0)\rho(a, t_0, t)\mathrm{d}t_0. \tag{9.35}$$

The above equation corresponds to (3.138) which we used previously to define the density matrix. In this case, however, our definition is in terms of a macroscopic time average over a collection of atomic systems created at different times, rather than in terms of an ensemble average as previously.

Differentiation of (9.35) gives

$$\frac{\partial}{\partial t}\rho(a, t) = \lambda_a(t)\rho(a, t_0, t_0) + \int_{-\infty}^{t} \lambda_a(t_0)\frac{\partial}{\partial t}[\rho(a, t_0, t)]\mathrm{d}t_0, \tag{9.36}$$

where

$$\rho(a, t_0, t_0) = \begin{bmatrix} 1 & 0 \\ 0 & 0 \end{bmatrix} = \rho(a). \tag{9.37}$$

If (9.30) is multiplied throughout by $\lambda_a(t_0)$ and integrated with respect to t_0 between the limits, $-\infty$ and t, and if we substitute† from (9.36), we obtain the following as the equation of motion of the density matrix, $\rho(a, t)$ on the assumption that $\boldsymbol{H}$ is independent of t_0,

$$\mathrm{i}\frac{\partial}{\partial t}\rho(a, t) = \mathrm{i}\lambda_a\rho(a) + \boldsymbol{H}\rho(a, t) - \rho(a, t)\boldsymbol{H} - \frac{\mathrm{i}}{2}[\boldsymbol{\Gamma}\rho(a, t) + \rho(a, t)\boldsymbol{\Gamma}]. \tag{9.38}$$

We see that (9.38) differs from (9.30) in that it includes an additional term which describes the rate of excitation to the state a.

If we also allow excitation to take place to state b as well as to state a, then the density matrix becomes

$$\rho(t) = \rho(a, t) + \rho(b, t) \tag{9.39}$$

† We use partial differentials, since $\rho(a, t_0, t)$ is now a function of several variables. Since the atoms in the system are not moving, it is permissible to make such a substitution in going from (9.30) to (9.38).

and $\lambda_a(t)$ is replaced by the matrix

$$\boldsymbol{\lambda}(t) = \begin{bmatrix} \lambda_a & 0 \\ 0 & \lambda_b \end{bmatrix}. \tag{9.40}$$

The cross terms do not appear because coherent excitation (ch. 3 § 4.2) has been neglected. Equation (9.38) now becomes

$$\mathrm{i}\frac{\partial}{\partial t}\boldsymbol{\rho}(t) = \mathrm{i}\boldsymbol{\lambda} + \boldsymbol{H}\boldsymbol{\rho}(t) + \boldsymbol{\rho}(t)\boldsymbol{H} - \frac{\mathrm{i}}{2}[\boldsymbol{\Gamma}\boldsymbol{\rho}(t) + \boldsymbol{\rho}(t)\boldsymbol{\Gamma}]. \tag{9.41}$$

This equation is written out fully in appendix E.

It should be pointed out that the designation of $\boldsymbol{\rho}(t)$ as a "density matrix" is not, perhaps, a satisfactory one. Whilst $\boldsymbol{\rho}(t)$ embodies the main character of the density matrix as described in (ch. 3 § 4.1) in that it represents a macroscopic averaging (albeit over time) carried out on a quantum mechanical system, it is not normalized, nor is its trace constant. However, it is convenient to maintain the terminology, provided it is remembered that not all the properties previously attributed to the density matrix apply here.

We now investigate the solutions to (9.21), (9.22) and (9.41) under different operating conditions (single, multimode, etc.) and with different approximations. Our first approximation is to assume that the population difference between the two levels is time independent and we solve the simplified equations for single mode operation.

4.2 *First order theory (stationary atoms)*

If we write out the equation for the diagonal element, of the density matrix $\boldsymbol{\rho}(t)$ using equation (9.41) we have (see appendix E)

$$\dot{\rho}_{ab} = +\frac{\mathrm{i}V}{\hbar}(\rho_{aa}-\rho_{bb}) - (\mathrm{i}\omega_{ab}+\gamma)\rho_{ab}, \tag{9.42}$$

where

$$\begin{aligned} \omega_{ab} &= (E_a - E_b)/\hbar, \\ \gamma &= (\gamma_a + \gamma_b)/2. \end{aligned} \tag{9.43}$$

In order to find a first order solution to the above, we assume that the population difference between the two states is constant in time, so that the first term can be written

$$(\rho_{aa}-\rho_{bb}) = N(z), \tag{9.44}$$

References on page 259

i.e. the population inversion is a function only of position in the active medium. Equation (9.42) becomes

$$\dot{\rho}_{ab} = +\frac{\mathrm{i}V}{\hbar}N(z)-(\mathrm{i}\omega_{ab}+\gamma)\rho_{ab}. \tag{9.45}$$

From (3.74) we can substitute for V

$$V = \mathrm{i}\omega_{ab}A\mathscr{P}. \tag{9.46}$$

In § 2 we derived equations describing the electromagnetic field configuration within the cavity in terms of the scalar electric field, and so we now write (9.46) in terms of the electric field as

$$V = -\mathscr{P}\cdot E. \tag{9.47}$$

We consider single mode operation when (9.14) takes the following form for the mth mode

$$E_m(t) = E_{0m}(t)\sin(K_m z)\cos(\omega_m t+\phi_m), \tag{9.48}$$

where $E_{0m}(t)$ and $\phi_m(t)$ satisfy (9.21) and (9.22) respectively. The interaction term (9.47) therefore becomes

$$V = -\mathscr{P}E_{0m}(t)\sin(K_m z)\cos(\omega_m t+\phi_m). \tag{9.49}$$

Equation (9.45) describes a simple harmonic oscillator with damping, γ, which is being driven (first term) at a frequency of approximately ω_m, close to its natural frequency, ω_{ab}. If we neglect the driving term, we obtain a solution of the form,

$$\rho_{ab} = \rho_{ab}^{(0)}\exp[-(\mathrm{i}\omega_{ab}+\gamma)t]. \tag{9.50}$$

In the presence of the driving term we keep the same solution and allow $\rho_{ab}^{(0)}$ to be a function of time, when substitution of (9.50) into (9.45) gives

$$\dot{\rho}_{ab}^{(0)} = -\frac{1}{\hbar}E_{0m}(t)\mathscr{P}\sin(K_m z)\exp[+(\mathrm{i}\omega_{ab}+\gamma)t]N(z)\times$$
$$\times\frac{\mathrm{i}}{2}[\exp(\mathrm{i}\omega_m t+\mathrm{i}\phi_m)-\exp(-\mathrm{i}\omega_m t-\mathrm{i}\phi_m)]. \tag{9.51}$$

We neglect the time variation of $E_{0m}(t)$ and $\phi_m(t)$ in comparison with the other factors in the exponentials (i.e. we assume quasi-monochromaticity), and we also neglect the anti-resonance term and transients in the above, so that the solution for ρ_{ab} becomes:

$$\rho_{ab} = \frac{1}{2\hbar} \cdot \frac{E_{0m}(t)\mathscr{P} \sin(K_m z)}{(\omega_m - \omega_{ab}) - \mathrm{i}\gamma} N(z) \exp(-\mathrm{i}\omega_m t - \mathrm{i}\phi_m). \tag{9.52}$$

We now use equations (9.33) and (9.52) to determine the macroscopic polarization

$$P = -\frac{\mathscr{P}^2}{2\hbar} \cdot \frac{N(z) \cdot E_{0m}(t) \cdot \sin(K_m z)}{\{(\omega_m - \omega_{ab})^2 + \gamma^2\}} \times$$
$$\times \{2(\omega_{ab} - \omega_m) \cos(\omega_m t + \phi_m) + 2\gamma \sin(\omega_m t + \phi_m)\}. \tag{9.53}$$

Since we are considering the *m*th mode, we must take the spatial Fourier component of $P(z, t)$ corresponding to this mode, which according to equation (9.17) is given by

$$P_m(z, t) = -\frac{\mathscr{P}^2}{2\hbar} E_{0m}(t) \left\{\frac{2}{L}\int_0^L N(z) \sin^2(K_m z) \mathrm{d}z\right\} \times$$
$$\times \left\{\frac{2(\omega_{ab} - \omega_m) \cos(\omega_m t + \phi_m) + 2\gamma \sin(\omega_m t + \phi_m)}{(\omega_m - \omega_{ab})^2 + \gamma^2}\right\}. \tag{9.54}$$

Comparing the above equation with (9.20) we therefore have

$$C_m = -\frac{\mathscr{P}^2}{\hbar} \cdot E_{0m}(t) \cdot \bar{N} \cdot (\omega_{ab} - \omega_m)/\{(\omega_m - \omega_{ab})^2 + \gamma^2\}; \tag{9.55}$$

$$S_m = -\frac{\mathscr{P}^2}{\hbar} E_{0m}(t) \cdot \bar{N} \cdot \gamma/\{(\omega_m - \omega_{ab})^2 + \gamma^2\}, \tag{9.56}$$

where

$$\bar{N} = \frac{2}{L}\int_0^L N(z) \cdot \sin^2(K_m z) \mathrm{d}z. \tag{9.57}$$

It will be noted that under the approximation of a constant population inversion, the macroscopic polarization is a linear function of the electric field. We have already discussed this case in § 2. Substitution of (9.56) into the self-consistency equation (9.21) gives the following equation describing the manner in which the amplitude of the field $E_{0m}(t)$, varies with time

$$\dot{E}_{0m} = \left[-\omega/2Q_m + \frac{\omega \mathscr{P}^2 \bar{N} \gamma}{2\varepsilon_0 \hbar\{(\omega_n - \omega_{ab})^2 + \gamma^2\}}\right] E_{0m}. \tag{9.58}$$

If the amplitude is to increase with time, rather than being exponen-

References on page 259

tially damped by the losses of the cavity, then (9.58) shows that

$$\frac{\mathscr{P}^2 \cdot \bar{N} \cdot \gamma}{\varepsilon_0 \hbar \{(\omega_m - \omega_{ab})^2 + \gamma^2\}} > \frac{1}{Q_m} . \tag{9.59}$$

The threshold condition for laser oscillation to occur is defined by the above expression with an equality sign, and in particular if the cavity is tuned to resonance ($\omega_m = \omega_{ab}$), the threshold population inversion for the mth mode, $\bar{N}_\mathrm{T}$, is given by

$$\frac{\mathscr{P}^2 \bar{N}_\mathrm{T}}{\varepsilon_0 \hbar \gamma} = \frac{1}{Q_m} . \tag{9.60}$$

From the above expression it can be seen that, in order to achieve a low value for the threshold population inversion associated with a particular active medium, the transition probability (which is proportional to $\mathscr{P}^2$) for the laser transition must be large, while the damping constant, γ, must be small (i.e. the homogeneous line width of the laser transition must be narrow).

For a population inversion in excess of the threshold inversion, (9.58) predicts that the amplitude of the mode builds up exponentially, and without limit in this approximation. In practice, of course, this does not happen, for as the field builds up in amplitude, the population of the upper level decreases (due to stimulated emission), while that of the lower level increases, and it is just such effects that we have neglected in the first-order theory. If we solve the equations without assuming a constant population inversion, then the macroscopic polarization becomes a non-linear function of electric field, and it is these higher order terms that predict a saturation value for the amplitude of the electric field. We consider the non-linear behaviour of the macroscopic polarization, and saturation effects in § 4.3. We have, therefore, seen that as regards the behaviour of the amplitude of the electric field, our first-order theory is only capable of making predictions about the threshold behaviour of the laser, and not about the manner in which the system adjusts itself to steady state operation above threshold.

We now consider the solution of the second self-consistency equation (9.22) which describes the influence of the active medium on the oscillation frequency of the mode. Substitution from (9.55) into (9.22), and neglecting $\dot{\phi}_m$ gives

$$(\omega_m - \Omega_m) \approx \frac{1}{2} \left(\frac{\omega}{\varepsilon_0} \right) \frac{\mathscr{P}^2}{\hbar} \frac{\bar{N}(\omega_{ab} - \omega_m)}{[(\omega_m - \omega_{ab})^2 + \gamma^2]} . \tag{9.61}$$

If we consider threshold operation, then we can write (9.59) with an equality sign, when substitution into the above gives

$$(\omega_m - \Omega_m) \approx \omega(\omega_{ab} - \omega_m)/2Q_m\gamma. \tag{9.62}$$

We now define a "stabilization factor", σ, which is the ratio between the cavity bandwidth, $\omega/2Q_m$ and the natural line width, γ, of the lasing transition,

$$\sigma = \omega/2Q_m\gamma, \tag{9.63}$$

so that the above expression becomes

$$\omega_m = (\Omega_m + \sigma\omega_{ab})/(1+\sigma). \tag{9.64}$$

For a typical gas laser, the value of σ lies in the range 0.1 to 0.01.

The oscillation frequency of the mode is determined by a "centre-of-mass" type of expression involving the oscillation frequency of the passive cavity (Ω_m) and the central frequency of the atomic transition (ω_{ab}), the weighting factors being the inverses of the respective line widths. The oscillation frequency is pulled towards the centre of the natural line width by an amount proportional to the detuning of the cavity, $(\Omega_m - \omega_{ab})$, from this central frequency.

Examination of (9.55) and (9.56) for the in-phase and in-quadrature components respectively of the macroscopic polarization, shows that they are related by the dispersion relations, that we have discussed in ch. 2 § 4.10 (and in particular the dispersion relation for a Lorentzian profile derived in ch. 2 § 4.10.1). This is, of course, to be expected, since the in-quadrature component describes how the amplitude of the field changes with propagation through the medium (9.21), while the in-phase component describes the corresponding change of phase with propagation (9.22).

If we extend the first-order theory to several modes above threshold, we find that the modes behave independently in this approximation; each mode being described by equations similar to (9.55) and (9.56).

To show this we note that the perturbation, V, for multimode operation becomes

$$V = -\mathrm{i}\mathscr{P}\sum_n E_{0n}(t)\sin(K_n z)\cos(\omega_n t + \phi_n), \tag{9.65}$$

where the sum extends over all modes above threshold. We then obtain on substitution of the above into (9.45)

$$\rho_{ab} = \frac{N(z)}{2\hbar}\mathscr{P}\sum_n \left[\frac{E_{0n}(t)\sin(K_n z)}{(\omega_n - \omega_{ab}) - \mathrm{i}\gamma}\exp\{-\mathrm{i}(\omega_n t + \phi_n)\}\right], \tag{9.66}$$

References on page 259

which yields for the macroscopic polarization

$$P = -\frac{\mathscr{P}^2}{2\hbar} N(z) \times$$

$$\times \sum_n \left[\frac{E_{0n}(t) \sin(K_n z)\{2(\omega_{ab}-\omega_n)\cos(\omega_n t+\phi_n)+2\gamma \sin(\omega_n t+\phi_n)\}}{(\omega_n-\omega_{ab})^2+\gamma^2} \right]. \tag{9.67}$$

We now evaluate the spatial Fourier component of the above polarization that corresponds to the particular mode in which we are interested, (the mth mode, say) when from the orthogonal properties of the sine function we obtain

$$P_m = -\frac{\mathscr{P}^2}{2\hbar} \cdot \frac{2}{L} \cdot \int_0^L N(z) \sin^2(K_m z) \mathrm{d}z \times$$

$$\times \frac{E_{0m}(t)[2(\omega_{ab}-\omega_m) \cos(\omega_m t+\phi_m)+2\gamma \sin(\omega_m t+\phi_m)}{(\omega_m-\omega_{ab})^2+\gamma^2}. \tag{9.68}$$

Since the above expression only involves the mth mode itself, we see that in this approximation the modes behave independently.

In conclusion, therefore, this section on the first-order approximation to Lamb's theory has demonstrated the manner in which we set about solving the various equations describing the cavity field and the active medium. The predictions of the theory in this approximation, however, are only applicable to threshold oscillation, and so we now consider higher-order approximations in order to describe above-threshold operation and saturation effects.

4.3 *Non-linear theory (stationary atoms)*

In order to describe higher order effects in the interaction of the radiation with the active medium, we must take into consideration the influence of the stimulated emission on the populations of the laser levels. If we extract the diagonal terms from (9.41), we obtain the following pair of equations that describe the populations of the upper and lower laser levels respectively

$$\begin{aligned} \dot{\rho}_{aa} &= -\gamma_a \rho_{aa} + \lambda_a + \frac{\mathrm{i}V}{\hbar}(\rho_{ab}-\rho_{ba}), \\ \dot{\rho}_{bb} &= -\gamma_b \rho_{bb} + \lambda_b + \frac{\mathrm{i}V}{\hbar}(\rho_{ba}-\rho_{ab}). \end{aligned} \tag{9.69}$$

In each of these equations, the first term on the right hand side describes the decay of the populations due to the phenomenological damping, the second term describes the pumping of the populations through the excitation mechanism, while the third describes the influence of the radiation field on the populations. The procedure adopted for the solution of (9.69) is to substitute for the non-diagonal elements those solutions (9.66) which were obtained under the approximation that $(\rho_{aa}-\rho_{bb})$ was independent of time. In this case, however, $(\rho_{aa}-\rho_{bb})$ is not put equal to $N(z)$ since we now require to find its new steady state value in the presence of the radiation. If this is carried out, then (9.69) take on the form of rate equations which depend only on the diagonal components (i.e. only on the populations)

$$\begin{aligned} \dot{\rho}_{aa} &= -\gamma_a \rho_{aa} + \lambda_a + R(\rho_{bb}-\rho_{aa}), \\ \dot{\rho}_{bb} &= -\gamma_b \rho_{bb} + \lambda_b + R(\rho_{aa}-\rho_{bb}), \end{aligned} \tag{9.70}$$

where

$$R = \frac{\gamma \mathscr{P}^2}{2\hbar^2} \cdot \frac{E_{0m}^2 \sin^2 K_m z}{\gamma^2 + (\omega_{ab}-\omega_m)^2}. \tag{9.71}$$

We have restricted our attention to the case of single mode operation only. The third term on the right hand side of (9.70) describes a net pumping process by the radiation from the more populated to the less populated level. This process is dependent on the intensity of the radiation, on the homogeneous line width of the transition (γ), on the distance of the mode frequency (ω_m) from the line centre (ω_{ab}), and on the transition probability between the two levels (proportional to $\mathscr{P}^2$). The steady state solution for the population difference between the two states in the presence of the radiation field, is readily obtained from (9.70) as

$$(\rho_{aa}-\rho_{bb}) = \frac{\lambda_a/\gamma_a - \lambda_b/\gamma_b}{1+2\gamma R/\gamma_a \gamma_b} = \frac{N(z)}{1+R/R_s}, \tag{9.72}$$

where

$$R_s = \gamma_a \gamma_b / 2\gamma. \tag{9.73}$$

The quantity R_s is a measure of the rate at which the system saturates i.e. of the rate at which the population difference adjusts to the influence of the radiation field.

In order to calculate the effect that this change in the population difference has on the radiation field, the value of $(\rho_{aa}-\rho_{bb})$ given by (9.72) is now substituted into the expression for ρ_{ab} (9.42), so enabling the polarization to be calculated to higher (third) order

$$\rho_{ab} = \frac{\mathscr{P}}{2\hbar} \cdot \frac{E_{0m} \sin K_m z}{(\omega_m - \omega_{ab}) - i\gamma} \cdot \frac{N(z)}{(1+R/R_s)} \cdot \exp(i\omega_m t + i\phi_m) \tag{9.74}$$

References on page 259

from which the spatial Fourier component of the polarization corresponding to the single mode can be obtained by comparison with (9.17) as

$$P_m(z,t) = -\frac{\mathscr{P}^2}{2\hbar}\cdot E_{0m}\cdot\left\{\frac{2}{L}\int_0^L \frac{N(z)\sin^2 K_m z}{(1+R/R_s)}\cdot \mathrm{d}z\right\}\cdot$$

$$\cdot\frac{2(\omega_{ab}-\omega_m)\cos(\omega_m t+\phi_m)+2\gamma\sin(\omega_m t+\phi_m)}{(\omega_m-\omega_{ab})^2+\gamma^2}. \tag{9.75}$$

In order to continue further with this solution, we assume that the system only saturates weakly in the sense that $R \ll R_s$. This allows us to expand out the denominator of the integral in (9.75), so that on substitution from (9.71) for the value of R we finally obtain

$$C_m = -\frac{\mathscr{P}^2 E_{0m}\overline{N}\cdot(\omega_{ab}-\omega_m)}{\hbar\{(\omega_{ab}-\omega_m)^2+\gamma^2\}}[1-\tfrac{3}{4}I_m\gamma^2\mathscr{L}(\omega_m-\omega_{ab})]; \tag{9.76}$$

$$S_m = -\frac{\mathscr{P}^2 E_{0m}\overline{N}\gamma}{\hbar\{(\omega_{ab}-\omega_m)^2+\gamma^2\}}\cdot[1-\tfrac{3}{4}I_m\gamma^2\mathscr{L}(\omega_m-\omega_{ab})], \tag{9.77}$$

where

$$I_m = \frac{\mathscr{P}^2\cdot E_{0m}^2}{\hbar^2\gamma_a\gamma_b} \tag{9.78}$$

which is a dimensionless parameter describing the effectiveness of the radiation field in altering the level populations, and

$$\mathscr{L}(\omega_m-\omega_{ab}) = [\gamma^2+(\omega_m-\omega_{ab})^2]^{-1}. \tag{9.79}$$

Under the assumption of weak saturation (already made) these expressions can be written in a slightly different form as

$$C_m = -\frac{\mathscr{P}^2\cdot E_{0m}\cdot\overline{N}}{\hbar}\cdot(\omega_{ab}-\omega_m)\cdot[(\omega_{ab}-\omega_m)^2+\gamma^2(1+\tfrac{3}{4}I_m)]^{-1}; \tag{9.80}$$

$$S_m = -\frac{\mathscr{P}^2\cdot E_{0m}\cdot\overline{N}}{\hbar}\cdot\gamma\cdot[(\omega_{ab}-\omega_m)^2+\gamma^2(1+\tfrac{3}{4}I_m)]^{-1}. \tag{9.81}$$

We now substitute the new expressions for the polarization into the self-consistency equations for the radiation field (9.21) and (9.22). In the case of the equation describing the amplitude of the field (9.21), we obtain the following expression which describes the steady state ($\dot{E}_{0m} = 0$) intensity of oscillation

$$E_{0m}^2 = \frac{4\hbar^2\gamma_a\gamma_b}{3\mathscr{P}^2}\cdot\left\{\frac{\overline{N}}{\overline{N}_\mathrm{T}} - 1 - \frac{(\omega_{ab}-\omega_m)^2}{\gamma^2}\right\}, \tag{9.82}$$

where $\overline{N}_\mathrm{T}$ has already been defined (9.60) as the population inversion at threshold. When the mode coincides with the line centre, (9.82) reduces to

$$E_{0m}^2 = \frac{4\hbar^2\gamma_a\gamma_b}{3\mathscr{P}^2}\left\{\frac{\overline{N}}{\overline{N}_\mathrm{T}} - 1\right\}. \tag{9.83}$$

We note that for a given value of the population inversion above threshold, the intensity at which the radiation field saturates is inversely proportional to the transition probability associated with the laser transition ($\mathscr{P}^2$), but is proportional to the decay rate of the upper laser level. From this it follows that, other factors being equal, those transitions for which the branching ratio is small saturate at higher powers for a given population inversion in excess of threshold. Of course if the branching ratio is small because of a low transition probability associated with the laser transition, the population inversion for threshold, $\overline{N}_\mathrm{T}$, given by (9.60), is correspondingly increased.

We summarize below the important expressions that have been derived so far with regard to the intensity of the radiation. The expressions apply only when the cavity resonance coincides with the line centre.

Population inversion for threshold (9.60)

$$\overline{N}_\mathrm{T} = \varepsilon_0\hbar\gamma/\mathscr{P}^2 Q_m.$$

Population inversion in absence of radiation field

$$\overline{N} = (\lambda_a/\gamma_a - \lambda_b/\gamma_b). \tag{9.84}$$

Intensity of oscillation at saturation to third order for single mode (9.83)

$$E_{0m}^2 = \frac{4\hbar^2\gamma_a\gamma_b}{3\mathscr{P}^2}\left[\frac{\overline{N}}{\overline{N}_\mathrm{T}} - 1\right].$$

If we substitute for the various quantities in (9.83) using (9.60) and (9.84), then on the assumption that $\gamma_b \gg \gamma_a$ and $\lambda_b \leqslant \lambda_a$ we obtain the following for the intensity of the radiation inside the cavity

$$E_{0m}^2 \approx \frac{4\hbar^2}{3\varepsilon_0}\left(\frac{2Q_m\lambda_a}{\hbar} - \frac{\varepsilon_0\gamma_a\gamma_b}{\mathscr{P}^2}\right). \tag{9.85}$$

This demonstrates that, to third order, the intensity of oscillation is directly

References on page 259

proportional to the rate of pumping of the upper laser level whereas the effectiveness of the second term in the bracket in reducing the intensity of oscillation is inversely proportional to the branching ratio for the transition. It should be stressed, however, that the expressions for intensity of oscillation only apply just above threshold (when $\overline{N}/\overline{N}_{\mathrm{T}} \ll 1$).

If we consider the self-consistency equation for the phase behaviour of the radiation (9.22), then we find that the same expression as was derived in terms of the linear theory applies (9.62), namely

$$(\omega_m - \Omega_m) \approx \omega(\omega_{ab} - \omega_m)/2Q_m\gamma. \tag{9.62}$$

The reason for this, is that in both cases the value of the in-quadrature component of the polarization is determined by the cavity-Q. Since the in-phase component is related to the in-quadrature component by the dispersion relations, the steady state solution to (9.25) remains the same.

5. Theory for moving atoms

5.1 *Density matrix and macroscopic polarization in the presence of atomic motion*

In our discussion of Lamb's theory so far, we have restricted our attention to the case of stationary atoms, so that the only linewidth effects considered have been those due to the homogeneous broadening associated with each individual atom. We now consider the case when the atoms are in motion, so that, with the collection of atoms as a whole, there is associated a Doppler (inhomogeneous) linewidth.

As a consequence of its motion an atom "sees" a modified electric field, which differs from the electric field in the laboratory reference frame (9.14) and which depends on the velocity components of the atomic motion. Since the interaction between a particular atom and the cavity field in now a function of the atom's velocity, the summation procedure described in § 4.1 must be modified to take into consideration the excitation of the atoms with a distribution of velocities.

It will be shown in § 6 that, to first order, the modification to the interaction between an atom and the cavity field can be regarded as a Doppler shifting of the atom's resonance frequency; there thus being a distribution of apparent resonance frequencies amongst the collection of atoms making up the active medium. This picture enables some of the results of first order theory to be interpreted in simple physical terms. It is not, however, applicable to higher order theory (§ 7).

We return to the equation of motion for the density matrix (9.30) derived in § 3 for a single atom:

$$\dot{\rho} = -\mathrm{i}[\boldsymbol{H}, \rho] - \tfrac{1}{2}(\Gamma\rho + \rho\Gamma). \tag{9.30}$$

If the atom is excited at the point $\boldsymbol{r}_0$ to the state α ($\alpha = a, b$) at time t_0 and has a velocity $\boldsymbol{v}$, then at a general time, t ($t \geqslant t_0$), we write its density matrix $\rho(\alpha, \boldsymbol{r}_0, t_0, \boldsymbol{v}, t)$. This density matrix obeys the equation of motion (9.30), but because of the motion of the atom, the function $V(t)$ in (9.30) now has a complicated time dependence. The electric field "seen" by the atom at some general time t is now $E\{\boldsymbol{r}_0 + \boldsymbol{v}(t-t_0), t\}$, i.e. the electric field at the space point $\boldsymbol{r}_0 + \boldsymbol{v}(t-t_0)$ at time t. This electric field replaces the electric field previously described by (9.14), in the equation for $V(t)$ (9.49). If at time t_0 and space point $\boldsymbol{r}_0$, the excitation rate, per unit volume per unit time, to the state α is $\lambda_\alpha(\boldsymbol{r}_0, t_0, \boldsymbol{v})$ for atoms with velocity $\boldsymbol{v}$, then the macroscopic polarization at point $\boldsymbol{r}$ at time t due to the atoms excited in the time interval t_0 to $(t_0 + \mathrm{d}t_0)$ with velocities in the range $\boldsymbol{v}$ to $(\boldsymbol{v} + \mathrm{d}\boldsymbol{v})$ is given by analogy with (9.32) as

$$\mathrm{d}P(\boldsymbol{r}, t) = \mathscr{P} \sum_{\alpha=a,b} [\{\rho_{ab}(\alpha, \boldsymbol{r}_0, t_0, \boldsymbol{v}, t) + \rho_{ba}(\alpha, \boldsymbol{r}_0, t_0, \boldsymbol{v}, t)\} \cdot \\ \cdot \delta\{\boldsymbol{r} - \boldsymbol{r}_0 - \boldsymbol{v}(t-t_0)\} \cdot \lambda_\alpha(\boldsymbol{r}_0, t_0, \boldsymbol{v}) \cdot \mathrm{d}\boldsymbol{r}_0 \cdot \mathrm{d}t_0 \cdot \mathrm{d}\boldsymbol{v}]. \tag{9.86}$$

The delta function in the above ensures that only those atoms whose velocity ($\boldsymbol{v}$), point of creation ($\boldsymbol{r}_0$), and time of creation (t_0) are so related that at time t they are at the point $\boldsymbol{r}$, and so contribute to the macroscopic polarization at that point. The average macroscopic polarization at the point $\boldsymbol{r}$ at time t due to all the atoms is therefore given by appropriate integration of (9.86) as

$$P(\boldsymbol{r}, t) = \mathscr{P} \cdot \sum_{\alpha=a,b} \left[\int_{-\infty}^{t} \mathrm{d}t_0 \int \mathrm{d}\boldsymbol{r}_0 \int \mathrm{d}\boldsymbol{v} \cdot \lambda_\alpha(\boldsymbol{r}_0, t_0, \boldsymbol{v}) \cdot \right. \\ \left. \cdot \{\rho_{ab}(\alpha, \boldsymbol{r}_0, t_0, \boldsymbol{v}, t) + \rho_{ba}(\alpha, \boldsymbol{r}_0, t_0, \boldsymbol{v}, t)\} \cdot \delta\{\boldsymbol{r} - \boldsymbol{r}_0 - \boldsymbol{v}(t-t_0)\}\right]. \tag{9.87}$$

If it is assumed that $\lambda_\alpha(\boldsymbol{r}_0, t_0, \boldsymbol{v})$ is a slowly varying function of position in the sense that it does not vary greatly over distances comparable to the average path length travelled by an atom before it decays through damping, then it can be replaced by $\lambda_\alpha(\boldsymbol{r}, t_0, \boldsymbol{v})$ in (9.87) and so be taken as constant for integration over $\boldsymbol{r}_0$. The integration over $\boldsymbol{r}_0$ in (9.87) may be performed so that we obtain

References on page 259

$$P(\boldsymbol{r}, t) = \mathscr{P} \cdot \sum_{\alpha=a,b} \left[\int_{-\infty}^{t} \mathrm{d}t_0 \int \mathrm{d}\boldsymbol{v} \cdot \lambda_\alpha(\boldsymbol{r}, t_0, \boldsymbol{v}) \cdot \right.$$
$$\left. \cdot \{\rho_{ab}(\alpha, \boldsymbol{r}-\boldsymbol{v}t+\boldsymbol{v}t_0, t_0, \boldsymbol{v}, t)+c.\,c.\} \right]. \qquad (9.88)$$

5.2 *Summary of equations describing radiation field and active medium*

We now possess all the equations that are required for a full description of the radiation field, the active medium and the interaction between them. For convenience these equations are listed again below, with a brief statement of their purpose.

(1) Radiation field in cavity described in terms of normal modes

$$E(z, t) = \sum_n E_{0n}(t) \sin(K_n z) \cos\{\omega_n t + \phi_n(t)\}, \qquad (9.14)$$

where

$$K_n = n\pi/L. \qquad (9.4)$$

(2) Relations between the time-dependent amplitude and phase terms in (9.14) and the macroscopic polarization associated with the active medium (self-consistency equations)

$$\dot{E}_{0m} + \tfrac{1}{2}(\omega/Q_m)E_{0m} = -\tfrac{1}{2}(\omega/\varepsilon_0)S_m(t); \qquad (9.21)$$

$$(\omega_m + \dot{\phi}_m - \Omega_m)E_{0m} = -\tfrac{1}{2}(\omega/\varepsilon_0)C_m(t), \qquad (9.22)$$

where

$$\Omega_m = \pi mc/L, \qquad (9.5)$$

C_m is the component of the macroscopic polarization associated with the mth mode which is in phase with the electric field, and S_m is the component in quadrature with the electric field.

(3) Relation between the macroscopic polarization and the density matrix description of the individual atoms in the active medium

$$P(\boldsymbol{r}, t) = \mathscr{P} \cdot \sum_{\alpha=a,b} \left[\int_{-\infty}^{t} \mathrm{d}t_0 \int \mathrm{d}\boldsymbol{v} \cdot \lambda_\alpha(\boldsymbol{r}, t_0, \boldsymbol{v}) \cdot \right.$$
$$\left. \cdot \{\rho_{ab}(\alpha, \boldsymbol{r}-\boldsymbol{v}t+\boldsymbol{v}t_0, t_0, \boldsymbol{v}, t)+c.\,c.\} \right]. \qquad (9.88)$$

(4) Equation of motion of the density matrix of an individual atom of the active medium in the presence of the radiation field and damping

$$\dot{\boldsymbol{\rho}} = -\mathrm{i}[\boldsymbol{H}, \boldsymbol{\rho}] - \tfrac{1}{2}(\boldsymbol{\Gamma}\boldsymbol{\rho} + \boldsymbol{\rho}\boldsymbol{\Gamma}), \qquad (9.30)$$

where

$$\mathsf{H} = \hbar^{-1}\begin{bmatrix} E_a & V(t) \\ V(t) & E_b \end{bmatrix}; \qquad \mathsf{\Gamma} = \begin{bmatrix} \gamma_a & 0 \\ 0 & \gamma_b \end{bmatrix}.$$

The matrix equation may be written out in element form as

$$\dot{\rho}_{ab} = -\mathrm{i}\omega_{ab}\rho_{ab} - \gamma\rho_{ab} + \frac{\mathrm{i}V(t)}{\hbar}(\rho_{aa} - \rho_{bb}); \tag{9.89}$$

$$\dot{\rho}_{aa} = -\gamma_a\rho_{aa} + \frac{\mathrm{i}V(t)}{\hbar}(\rho_{ab} - \rho_{ba}); \tag{9.90}$$

$$\dot{\rho}_{bb} = -\gamma_b\rho_{bb} - \frac{\mathrm{i}V(t)}{\hbar}(\rho_{ab} - \rho_{ba}); \tag{9.91}$$

$$\rho_{ba} = \rho_{ab}^{*}.$$

(5) For an atom described by the density matrix $\boldsymbol{\rho}(\alpha, \boldsymbol{r}_0, t_0, \boldsymbol{v}, t)$ the interaction term $V(t)$ in (9.30) involves the electric field $E\{\boldsymbol{r}_0 + \boldsymbol{v}(t-t_0), t\}$ and is

$$V(t) = -\mathscr{P}E\{\boldsymbol{r}_0 + \boldsymbol{v}(t-t_0), t\}. \tag{9.92}$$

5.3 *Iterative solution of the Lamb equations*

The equations summarized in the preceeding section may be solved by iterative methods. To zero order the influence of the radiation field on the populations of the levels (ρ_{aa}, ρ_{bb}) is neglected, so that the matrix elements ρ_{aa} and ρ_{bb} for atoms excited at time t_0 are derived from (9.90) as

$$\begin{aligned} \rho_{aa}^{(0)}(a, \boldsymbol{r}_0, \boldsymbol{v}, t) &= \exp\{-\gamma_a(t-t_0)\}, \\ \rho_{bb}^{(0)}(b, \boldsymbol{r}_0, \boldsymbol{v}, t) &= \exp\{-\gamma_b(t-t_0)\}. \end{aligned} \tag{9.93}$$

Using these solutions, (9.89) may be used to calculate ρ_{ab} and ρ_{ba} to first order

$$\dot{\rho}_{ab}^{(1)}(a, \boldsymbol{r}_0, t_0, \boldsymbol{v}, t) = -(\mathrm{i}\omega_{ab} + \gamma)\rho_{ab}^{(1)}(a, \boldsymbol{r}_0, t_0, \boldsymbol{v}, t) + \\ + \frac{\mathrm{i}V(t)}{\hbar}\exp\{-\gamma_a(t-t_0)\} \tag{9.94}$$

and similarly for $\rho_{ab}^{(1)}(b, \boldsymbol{r}_0, t_0, \boldsymbol{v}, t)$ etc. Integration of (9.94) gives

$$\rho_{ab}^{(1)}(a, \boldsymbol{r}_0, t_0, \boldsymbol{v}, t) = \frac{\mathrm{i}}{\hbar}\int_{t_0}^{t} \mathrm{d}t'\, V(t') \exp\{(\gamma + \mathrm{i}\omega_{ab})(t'-t) + \gamma_a(t_0 - t')\}. \tag{9.95}$$

The above solution may then be substituted back into (9.90) to enable

$\rho_{aa}(a, \boldsymbol{r}_0, t_0, \boldsymbol{v}, t)$ to be calculated to second order

$$\dot{\rho}_{aa}^{(2)}(a, \boldsymbol{r}_0, t_0, \boldsymbol{v}, t'') = -\gamma_a \rho_{aa}^{(2)}(a, \boldsymbol{r}_0, t_0, \boldsymbol{v}, t'') - \\ -\frac{V(t'')}{\hbar^2}\left[\int_{t_0}^{t''} \mathrm{d}t''' V(t''') \exp\{(\gamma + \mathrm{i}\omega_{ab})(t''' - t'') + \gamma_a(t_0 - t''')\} + c.\,c.\right], \tag{9.96}$$

which on integration gives

$$\rho_{aa}^{(2)}(a, \boldsymbol{r}_0, t_0, \boldsymbol{v}, t') = -\hbar^{-2}\int_{t_0}^{t'} \mathrm{d}t'' \int_{t_0}^{t''} \mathrm{d}t''' \cdot V(t'') \cdot V(t''') \cdot \\ \cdot [\exp\{(\gamma + \mathrm{i}\omega_{ab})(t''' - t'') + \gamma_a(t_0 - t''') + \gamma_a(t'' - t')\} + c.\,c.]. \tag{9.97}$$

In a similar fashion the second order solution for $\rho_{bb}(a, \boldsymbol{r}_0, t_0, \boldsymbol{v}, t)$ may be obtained as

$$\rho_{bb}^{(2)}(a, \boldsymbol{r}_0, t_0, \boldsymbol{v}, t') = \hbar^{-2}\int_{t_0}^{t'} \mathrm{d}t'' \int_{t_0}^{t''} \mathrm{d}t''' V(t') V(t''') \cdot \\ \cdot [\exp\{\gamma_b(t'' - t') + (\gamma + \mathrm{i}\omega_{ab})(t''' - t'') + \gamma_a(t_0 - t''')\} + c.\,c.]. \tag{9.98}$$

The second order solutions for ρ_{aa} and ρ_{bb} are then substituted back into (9.89) to enable ρ_{ab} to be calculated to third order as

$$\rho_{ab}^{(3)}(a, \boldsymbol{r}_0, t_0, \boldsymbol{v}, t) = \mathrm{i}\hbar^{-1}\int_{t_0}^{t} \mathrm{d}t' V(t') \exp\{(\gamma + \mathrm{i}\omega_{ab})(t' - t)\} \cdot \\ \cdot \{\rho_{aa}^{(2)}(a, \boldsymbol{r}_0, t_0, \boldsymbol{v}, t') - \rho_{bb}^{(2)}(a, \boldsymbol{r}_0, t_0, \boldsymbol{v}, t')\}. \tag{9.99}$$

The iterative solutions developed in this section refer to the behaviour of a single atom, and provide a method for solving (9.30) to second order in ρ_{aa} and ρ_{bb} and to third order in ρ_{ab}, ρ_{ba}. These solutions must now be summed over all the atoms making up the active medium, allowing for the velocity distribution amongst the atoms and for their different times of excitation (t_0). In order to derive the macroscopic polarization associated with the active medium, this summation is carried out by using (9.88) developed in § 5.2.

In the next section we consider the first order solutions. In this approximation the population terms (ρ_{aa}, ρ_{bb}) are taken to zero order, i.e. it is assumed that the radiation field does not alter their values, and the off-diagonal terms are then calculated to first order. In this way the macroscopic polarization is derived to first order.

6. First order theory

The contribution to the off-diagonal element, $\rho_{ab}^{(1)}$, of the density matrix at time t and space point $\boldsymbol{r}$, of all the atoms excited to the state a with velocity $\boldsymbol{v}$, regardless of their time or point of excitation is given by

$$\rho_{ab}^{(1)}(a, \boldsymbol{r}, \boldsymbol{v}, t) = \int_{-\infty}^{t} \mathrm{d}t_0 \cdot \lambda_a(\boldsymbol{r}_0 = \boldsymbol{r}-\boldsymbol{v}t+\boldsymbol{v}t_0, \boldsymbol{v}, t_0) \cdot \cdot \rho_{ab}^{(1)}(a, \boldsymbol{r}_0 = \boldsymbol{r}-\boldsymbol{v}t+\boldsymbol{v}t_0, t_0, \boldsymbol{v}, t). \tag{9.100}$$

If $\lambda_a(\boldsymbol{r}_0, \boldsymbol{v}, t_0)$ is a slowly varying function of time (as well as of position) in the sense that it does not change appreciably in the time taken for an atom to travel from its point of excitation ($\boldsymbol{r}_0$) to the point where the density matrix is being evaluated ($\boldsymbol{r}$), then it may be replaced by $\lambda_a(\boldsymbol{r}, \boldsymbol{v}, t)$ and so be taken outside the integral in (9.100). The interaction at time t' between an atom of velocity $\boldsymbol{v}$ excited at space point $\boldsymbol{r}_0$ at time t_0 is

$$V(t') = -\mathscr{P}E\{\boldsymbol{r}_0+\boldsymbol{v}(t'-t_0), t'\}. \tag{9.101}$$

In (9.101) the atoms are characterized by a point of excitation given by $\boldsymbol{r}_0 = \boldsymbol{r}-\boldsymbol{v}(t-t_0)$, so that the interaction at time t' for such atoms may be written

$$V(t') = -\mathscr{P}E\{\boldsymbol{r}-\boldsymbol{v}(t-t'), t'\}. \tag{9.102}$$

Using (9.102), we can therefore write (9.95) as

$$\rho_{ab}^{(1)}(a, \boldsymbol{r}_0 = \boldsymbol{r}-\boldsymbol{v}t+\boldsymbol{v}t', t_0, \boldsymbol{v}, t) = = -\mathrm{i}\hbar^{-1}\int_{t_0}^{t} \mathrm{d}t' \mathscr{P} \cdot E\{\boldsymbol{r}-\boldsymbol{v}(t-t'), t'\} \exp\{(\gamma+\mathrm{i}\omega_{ab})(t'-t)+\gamma_a(t_0-t')\}. \tag{9.103}$$

On substitution of (9.103) into (9.100) we obtain

$$\rho_{ab}^{(1)}(a, \boldsymbol{r}, \boldsymbol{v}, t) = -\mathrm{i}\hbar^{-1}\lambda_a(\boldsymbol{r}, \boldsymbol{v}, t) \cdot \cdot \int_{-\infty}^{t} \mathrm{d}t_0 \int_{t_0}^{t} \mathrm{d}t' \cdot \mathscr{P} \cdot E\{\boldsymbol{r}-\boldsymbol{v}(t-t'), t'\} \exp\{(\gamma+\mathrm{i}\omega_{ab})(t'-t)+\gamma_a(t_0-t')\}. \tag{9.104}$$

The order of integration in (9.104) may be interchanged as follows

$$\int_{-\infty}^{t} \mathrm{d}t_0 \int_{t_0}^{t} \mathrm{d}t' \cdot F(t, t', t_0) = \int_{-\infty}^{t} \mathrm{d}t' \int_{-\infty}^{t'} \mathrm{d}t_0 F(t, t', t_0) \tag{9.105}$$

References on page 259

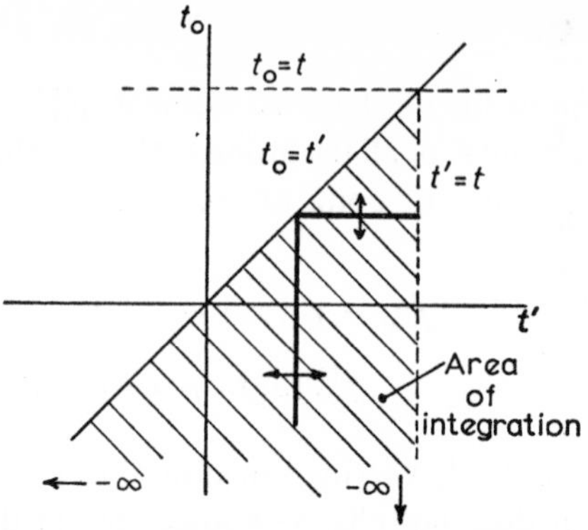

Fig. 9.1. Area of integration (shaded) in the $t't_0$ plane of the double integrals (9.105).

(reference to fig. 9.1 will show that both the double integrals in (9.105) are over the same area in the $t_0 t'$-plane) in which case the integration over t_0 may be performed to give

$$\rho_{ab}^{(1)}(a, \boldsymbol{r}, \boldsymbol{v}, t) = -(\mathrm{i}/\hbar\gamma_a) \cdot \lambda_a(\boldsymbol{r}, \boldsymbol{v}, t) \cdot \\ \cdot \int_{-\infty}^{t} \mathrm{d}t' \mathscr{P} \cdot E\{\boldsymbol{r} - \boldsymbol{v}(t-t'), t'\} \cdot \exp\{(\gamma + \mathrm{i}\omega_{ab})(t'-t)\}. \tag{9.106}$$

In the description of the electric field in terms of modes derived in § 2, only a one-dimensional spatial dependence was considered. Accordingly, from now on, we restrict our attention to the one-dimensional case entirely, and assume that the variables connected with the active medium are constant in the xy-plane. In this case, substitution for the electric field in (9.106) in terms of the mode description gives (dropping subscript 0 from E_{0m} for convenience)

$$\rho_{ab}^{(1)}(a, z, v, t) = -(\mathrm{i}/\hbar\gamma_a) \cdot \lambda_a(z, v, t) \cdot \\ \cdot \sum_m \int_{-\infty}^{t} \mathrm{d}t' \cdot \mathscr{P} \cdot \exp\{(\gamma + \mathrm{i}\omega_{ab})(t'-t)\} \cdot \\ \cdot E_m(t') \cdot \sin[K_m\{z - v(t-t')\}] \cos\{\omega_m t' + \phi_m(t')\}, \tag{9.107}$$

where v is the component of $\boldsymbol{v}$ along the z-axis.

We now make the rotating wave approximation (ch. 3 § 3.4), and further assume that $E_m(t')$ and $\phi_m(t')$ are slowly varying functions of time, in the sense that they do not change much over times comparable to γ^{-1}, so that they may effectively be evaluated at t in (9.107). With these approximations and putting $\tau' = (t-t')$, (9.107) becomes

$$\rho_{ab}^{(1)}(a, z, v, t) = -(\mathrm{i}\mathscr{P}/2\hbar\gamma_a) \cdot \lambda_a(z, v, t) \sum_m E_m(t) \cdot \exp\{-\mathrm{i}\omega_m t - \mathrm{i}\phi_m(t)\} \cdot$$
$$\cdot \int_0^\infty \mathrm{d}\tau' \sin\{K_m(z-v\tau')\} \cdot \exp\{-\gamma\tau' - \mathrm{i}(\omega_{ab}-\omega_m)\tau'\}. \tag{9.108}$$

Substitution of (9.108) into (9.86) enables the macroscopic polarization due to the atoms with velocity component v and initially excited to the state a, to be derived

$$P^{(1)}(a, z, v, t) = -(\mathrm{i}\mathscr{P}^2/2\hbar\gamma_a) \cdot \lambda_a(z, v, t) \sum_m \Big[E_m(t) \cdot \exp\{-\mathrm{i}\omega_m t - \mathrm{i}\phi_m(t)\} \cdot$$
$$\cdot \int_0^\infty \mathrm{d}\tau' \sin\{K_m(z-v\tau')\} \cdot \exp\{-\gamma\tau' - \mathrm{i}(\omega_{ab}-\omega_m)\tau'\} + c.\,c. \Big]. \tag{9.109}$$

We must next consider the contribution to the polarization of atoms excited to the state b. Since this state is below the state a, it may also be populated by spontaneous decay from the state a, so that its excitation function should be written as $\lambda_b(z, v, t) + f \cdot \gamma_a \cdot \rho_{aa}(z, v, t)$ where f is the branching ratio giving the fraction of the population in the state a that decays spontaneously to the state b. However, for this treatment we neglect the latter term, and leave a consideration of its influence until later (§ 7.4). Under this approximation the contribution of atoms excited to state b is exactly similar to (9.109) but with b replacing a, and with a sign change.

We further assume that the velocity distribution of atoms excited to the state b is similar to that for atoms excited to the state a, so that the excitation functions may be written in the form

$$\lambda_\alpha(z, v, t) = \Lambda_\alpha(z, t) \cdot W(v) \qquad \text{for} \quad \alpha = a, b. \tag{9.110}$$

The total macroscopic polarization at a space point z and a time t due to all the atoms at that space point and at that time regardless of their velocity component (v) or the state to which they are initially excited is readily seen to be

$$P^{(1)}(z, t) = -(\mathrm{i}\mathscr{P}^2/2\hbar) \cdot \left\{\frac{\Lambda_a(z, t)}{\gamma_a} - \frac{\Lambda_b(z, t)}{\gamma_b}\right\} \cdot$$
$$\cdot \sum_m \Big[E_m(t) \cdot \exp\{-\mathrm{i}\omega_m t - \mathrm{i}\phi_m(t)\} \cdot$$
$$\cdot \int_{-\infty}^{+\infty} \mathrm{d}v \int_0^\infty \mathrm{d}\tau' \cdot W(v) \sin\{K_m(z-v\tau')\} \cdot \exp\{-\gamma\tau' - \mathrm{i}(\omega_{ab}-\omega_m)\tau'\} + c.\,c. \Big]. \tag{9.111}$$

References on page 259

The first term in the brackets on the right hand side describes the population inversion between the state a and the state b, in the absence of the cavity radiation field, and will be designated $N(z, t)$. (Lamb refers to $N(z, t)$ as the excitation density.) In order to find the influence of the macroscopic polarization on a particular mode (the nth say) then the spatial Fourier component of (9.111) corresponding to that mode is required in accordance with (9.17)

$$P_n^{(1)}(t) = (2/L) \cdot \int_0^L \mathrm{d}z \, P^{(1)}(z, t) \cdot \sin(K_n z). \tag{9.17}$$

On substituting (9.111) into (9.17), terms of the form

$$\sin(K_n z) \sin\{K_m(z - v\tau')\}$$

arise in the integral, and may be re-arranged using standard trigonometrical expressions as follows

$$\sin(K_n z) \sin\{K_m(z - v\tau')\} =$$
$$= \tfrac{1}{2} \cos\{(K_n - K_m)z + K_m v\tau'\} - \tfrac{1}{2} \cos\{(K_n + K_m)z - K_m v\tau'\}. \tag{9.112}$$

As z varies, the last term on the right hand side of the above oscillates rapidly, since its wavelength corresponds to optical frequencies. Since $N(z, t)$ changes only little over an optical wavelength, the contribution of this last term to the integral averages out closely to zero, and will therefore be neglected. If the first term on the right hand side is further expanded as follows

$$\tfrac{1}{2} \cos\{(K_n - K_m)z + K_m v\tau'\} =$$
$$= \tfrac{1}{2} \cos\{(K_n - K_m)z\} \cos(K_m v\tau') - \tfrac{1}{2} \sin\{(K_n - K_m)z\} \sin(K_m v\tau'), \tag{9.113}$$

then since the velocity distribution function, $W(v)$, is normally an even function of v, only that part of the above that is even in v will contribute to the integral over v. We may therefore replace

$$\sin(K_n z) \sin\{K_m(z - v\tau')\}$$

in (9.112) by

$$\tfrac{1}{2} \cos\{(K_n - K_m)z\} \cos(Kv\tau').$$

The subscript has been dropped from the wave-number in the last part of this expression, since all the modes considered have approximately the same frequency ($K = \omega/c$). We therefore have

$$P_n^{(1)}(z,t) = -(\mathrm{i}\mathscr{P}^2/2L\hbar)\cdot\sum_m \Big[E_m(t)\cdot\exp\{-\mathrm{i}\omega_m t-\mathrm{i}\phi_m(t)\}\cdot$$
$$\cdot\int_0^L \mathrm{d}z\int_{-\infty}^{+\infty}\mathrm{d}v\int_0^{\infty}\mathrm{d}\tau'\cdot N(z,t)\cdot W(v)\cdot\cos\{(K_n-K_m)z\}\cdot$$
$$\cdot\cos(Kv\tau')\cdot\exp\{-\gamma\tau'-\mathrm{i}(\omega_{ab}-\omega_m)\tau'\}+c.\,c.\Big]. \tag{9.114}$$

The integration over τ' may now be performed to give

$$P_n^{(1)}(z,t) = -(\mathrm{i}\mathscr{P}^2/2\hbar)\cdot\sum_m \Big[E_m(t)\exp\{-\mathrm{i}\omega_m t-\mathrm{i}\phi_m(t)\}\cdot$$
$$\cdot N_{n-m}(t)\int_{-\infty}^{+\infty}\mathrm{d}v\cdot W(v)\cdot\mathscr{D}(\omega_{ab}-\omega_m+Kv)+c.\,c.\Big], \tag{9.115}$$

where

$$\mathscr{D}(\omega) = (\gamma+\mathrm{i}\omega)^{-1} \tag{9.116}$$

and

$$N_{n-m}(t) = \frac{1}{L}\int_0^L \mathrm{d}z\cdot N(z,t)\cdot\cos\{(K_n-K_m)z\}. \tag{9.117}$$

We now restrict our attention to the case when the velocity distribution function describes a Maxwellian distribution of velocities, and is hence of the form

$$W(v) = (u\pi^{\frac{1}{2}})^{-1}\exp(-v^2/u^2), \tag{9.118}$$

where $u = (2kT/m)^{\frac{1}{2}}$, T being the atom temperature, and m the atomic mass.

For this case it is preferable to perform the integration over v before that over τ'. We therefore return to (9.114), and after substituting from (9.118) for $W(v)$ obtain the following integral over v

$$\int_{-\infty}^{+\infty}\mathrm{d}v\cdot\exp(-v^2/u^2)\cos(Kv\tau') =$$
$$= \int_{-\infty}^{+\infty}\mathrm{d}v\cdot\exp\{-(v^2/u^2)+\mathrm{i}Kv\tau'\}. \tag{9.119}$$

This integral is just the Fourier transform of the velocity distribution function. Since in this case the velocity distribution function is Gaussian, its Fourier transform is also Gaussian, and may readily be evaluated to give for the above $(u\pi^{\frac{1}{2}})\exp(-\frac{1}{4}\cdot K^2u^2\tau'^2)$, so that (9.114) now becomes

$$P_n^{(1)}(t) = -(\mathscr{P}^2/2\hbar Ku)\cdot\sum_m [E_m(t)\cdot\exp\{-\mathrm{i}\omega_m t-\mathrm{i}\phi_m(t)\}\cdot$$
$$\cdot N_{n-m}(t)\cdot Z(\omega_m-\omega_{ab})+c.\,c.], \tag{9.120}$$

References on page 259

where (dropping the prime on τ),

$$Z(\omega_m-\omega_{ab}) = Z(\omega_m-\omega_{ab}, \gamma, Ku) =$$
$$= \mathrm{i}Ku\int_0^\infty \mathrm{d}\tau\cdot\exp\{\mathrm{i}(\omega_m-\omega_{ab})\tau-\gamma\tau-\tfrac{1}{4}K^2u^2\tau^2\}. \tag{9.121}$$

This latter function is a well known complex function in the theory of Doppler broadening, its real, $Z_r(\omega_m-\omega_{ab})$ and imaginary, $Z_i(\omega_m-\omega_{ab})$, parts have been extensively tabulated (see appendix N). In the case when the excitation density, $N(z, t)$, is independent of position within the cavity, and so can be written $\overline{N}(t)$, then we have

$$N_{n-m}(t) = \frac{\overline{N}(t)}{L}\int_0^L \mathrm{d}z\cdot\cos\{(K_n-K_m)z\} = \overline{N}(t)\cdot\delta_{nm}, \tag{9.122}$$

so that the component of macroscopic polarization becomes

$$P_n^{(1)}(t) = -(\mathscr{P}^2/2\hbar Ku)\cdot E_n(t)\cdot\overline{N}(t)\cdot$$
$$\cdot\exp\{-\mathrm{i}\omega_n t-\mathrm{i}\phi_n(t)\}\cdot Z(\omega_n-\omega_{ab})+c.\,c. \tag{9.123}$$

In this approximation, therefore, the component of macroscopic polarization driving a particular mode of the cavity is independent of the other modes that may also be above threshold i.e. the modes act independently of one another. Substitution of (9.123) into the self-consistency equations (9.21) and (9.22) enables the influence of the active medium on the amplitude and phase behaviour of the mode to be determined

$$\dot{E}_n = \{\tfrac{1}{2}(\omega/\varepsilon_0)\cdot(\mathscr{P}^2/\hbar Ku)\cdot\overline{N}\cdot Z_i(\omega_n-\omega_{ab})-\tfrac{1}{2}(\omega/Q_n)\}E_n; \tag{9.124}$$
$$\omega_n+\dot{\phi}_n-\Omega_n = \tfrac{1}{2}(\omega/\varepsilon_0)\cdot(\mathscr{P}^2/\hbar Ku)\cdot\overline{N}\cdot Z_r(\omega_n-\omega_{ab}). \tag{9.125}$$

For a steady state solution to the amplitude equation, then we have

$$(\mathscr{P}^2/\varepsilon_0\hbar Ku)\cdot\overline{N}\cdot Z_i(\omega_n-\omega_{ab}) = Q_n^{-1}. \tag{9.126}$$

For the case when $\gamma \ll Ku$, the plasma dispersion function, $Z_i(\omega_n-\omega_{ab})$ may be expanded (see appendix N) so that (9.126) becomes

$$2\pi^{\frac{1}{2}}(e^2/4\pi\varepsilon_0\hbar c)\cdot(c/u)\cdot(\mathscr{P}/e)^2\cdot\lambda\cdot\overline{N}\cdot\exp\{-(\omega_n-\omega_{ab})^2/(Ku)^2\} = Q_n^{-1}, \tag{9.127}$$

where $\lambda = 2\pi/K$ and $(e^2/4\pi\varepsilon_0\hbar c)$ is the fine structure constant (which is 1/137). The above expression is in fact a threshold condition, since no account has been taken of the influence of the radiation field on $\overline{N}$, and as such is analogous to that derived for stationary atoms in § 4. In this case, however, where the total line shape associated with the active medium is essentially Doppler (i.e. $\gamma \ll Ku$), the threshold dependence on the distance of the mode from the line centre (ω_{ab}) follows a Gaussian function with a half-width determined by the Doppler width of the line (Ku).

In this limit of first order theory, the threshold condition is capable of a simple, physical interpretation.

The atoms in the active medium that can couple to a particular mode of frequency ω_m are such that their apparent resonance frequencies, in the lab. frame of reference, lie within the homogeneous linewidth (γ) of the mode frequency. The number of atoms whose apparent resonance frequencies lie within the range ω to $(\omega + \mathrm{d}\omega)$ as a consequence of their Doppler motion is

$$\text{const. } \bar{N} \cdot (Ku)^{-1} \cdot \exp\{-(\omega - \omega_{ab})^2/(Ku)^2\} \cdot \mathrm{d}\omega, \tag{9.128}$$

and therefore the number of atoms that can couple to a mode is approximately (if $\gamma \ll Ku$)

$$\text{const. } \bar{N} \cdot \gamma \cdot (Ku)^{-1} \cdot \exp\{-(\omega - \omega_{ab})^2/(Ku)^2\}. \tag{9.129}$$

If the above is substituted into the threshold condition for stationary atoms (9.60) in place of the total population inversion, the threshold condition for moving atoms (9.127) is obtained. It is interesting to note that this latter no longer depends on the homogeneous linewidth (γ).

From the phase equation (9.125), the frequency of the mode in the presence of the active medium may be obtained, by putting $\dot{\phi}_n = 0$ without any loss of generality, as

$$\omega_n = \Omega_n + \tfrac{1}{2}(\omega/\varepsilon_0) \cdot (\mathscr{P}^2/\hbar Ku) \cdot \bar{N} \cdot Z_r(\omega_n - \omega_{ab}). \tag{9.130}$$

If the previously derived threshold condition (9.127) is used, $\bar{N}$ may be substituted in terms of Q_n, so that (9.130) becomes

$$\omega_n = \Omega_n + \tfrac{1}{2}(\omega/Q_n) \cdot Z_r(\omega_n - \omega_{ab})/Z_i(\omega_n - \omega_{ab}). \tag{9.131}$$

With the approximation $\gamma \ll Ku$, as previously, then from appendix N it may be seen that (9.131) becomes

$$\omega_n = \Omega_n - (\omega/\pi^{\frac{1}{2}} Q_n) \cdot \int_0^{\xi_n} \mathrm{d}x \cdot \mathrm{e}^{x^2}, \tag{9.132}$$

where

$$\xi_n = (\omega_n - \omega_{ab})/Ku. \tag{9.133}$$

The integral in (9.132) may be expanded to different orders in $(\omega_n - \omega_{ab})/Ku$.

References on page 259

To first order we obtain the following expression which describes linear pulling

$$\frac{(\omega_n-\Omega_n)}{(\omega_{ab}-\omega_n)} = \omega(\pi^{\frac{1}{2}}Q_n Ku)^{-1} = \sigma. \tag{9.134}$$

By linear we mean that the amount by which the active mode frequency (ω_n) is "pulled" from the passive mode frequency (Ω_n) is linearly proportional to the distance of the mode from the line centre (ω_{ab}). The above expression indicates that the mode is always pulled towards the line centre. It should be contrasted with the expression derived for the case of stationary atoms (9.62). In this former case the stabilization factor, σ, was found to be the ratio of the cavity bandwidth to the natural (homogeneous) linewidth, whereas in the present case it is the ratio of the cavity bandwidth to the Doppler (inhomogeneous) linewidth. If the integral in (9.132) is expanded to second order in $(\omega_n-\omega_{ab})/Ku$, then in this case it is found that there is also a non-linear contribution to the pulling which is

$$\frac{(\omega_n-\Omega_n)}{(\omega_{ab}-\omega_n)} = \{1+\tfrac{1}{3}(\Omega_n-\omega_{ab})^2/(Ku)^2\}\sigma. \tag{9.135}$$

This expression is still only valid if $\gamma \ll Ku$, and shows that the frequency of the active mode is pulled closer to the line centre than is implied by the linear approximation.

7. Non-linear theory

7.1 *Third order theory*

The third order contribution to the off-diagonal elements of the density matrix, $\rho_{ab}^{(3)}(z, v, t)$, is given by analogy with (9.100) as

$$\rho_{ab}^{(3)}(z, v, t) = \sum_{\alpha=a,b} \lambda_\alpha(z, v, t)\cdot\int_{-\infty}^{t} \mathrm{d}t_0 \cdot \rho_{ab}^{(3)}(\alpha, z_0 = z-vt+vt_0, t_0, v, t). \tag{9.136}$$

Substitution from (9.99) and (9.97) into (9.136) leads to integrals of the form

$$\lambda_a(z, v, t)\cdot \hbar^{-3}\int_{-\infty}^{t}\mathrm{d}t_0\int_{t_0}^{t}\mathrm{d}t'\int_{t_0}^{t'}\mathrm{d}t''\int_{t_0}^{t''}\mathrm{d}t'''\cdot V(t')\cdot V(t'')\cdot V(t''')\cdot$$
$$\cdot\exp\{(\gamma+\mathrm{i}\omega_{ab})(t'-t)+(\gamma+\mathrm{i}\omega_{ab})(t'''-t'')+\gamma_a(t_0-t''')+\gamma_a(t''-t')\}. \tag{9.137}$$

By analogy with the procedure discussed in relation to the double integral in (9.104), repeated interchange of the orders of integration may be carried out in the above. This allows the integration over t_0 to be performed, so that the integrals now become of the form

$$\lambda_a(z, v, t) \cdot \hbar^{-3} \cdot \gamma_a^{-1} \int_{-\infty}^{t} \mathrm{d}t' \int_{-\infty}^{t'} \mathrm{d}t'' \int_{-\infty}^{t''} \mathrm{d}t''' V(t') \cdot V(t'') \cdot V(t''') \cdot$$
$$\cdot \exp\{(\gamma + \mathrm{i}\omega_{ab})(t' - t + t''' - t'') + \gamma_a(t'' - t')\}. \tag{9.138}$$

We substitute for each interaction term, $V(t)$, in the above in accordance with (9.92), and further substitute for the electric field in terms of the mode description. On doing this we obtain the following expression for $\rho_{ab}^{(3)}(z, v, t)$:

$$\rho_{ab}^{(3)}(z, v, t) = \mathrm{i}\mathscr{P}^3/(8\hbar^3)\, N(z, t) \cdot \sum_\mu \sum_\rho \sum_\sigma E_\mu E_\rho E_\sigma \cdot$$
$$\cdot \Big[\exp\{-\mathrm{i}\omega_\mu t + \mathrm{i}\omega_\rho t - \mathrm{i}\omega_\sigma t - \mathrm{i}\phi_\mu(t) + \mathrm{i}\phi_\rho(t) - \mathrm{i}\phi_\sigma(t)\} \cdot$$
$$\cdot \int_0^\infty \mathrm{d}\tau' \int_0^\infty \mathrm{d}\tau'' \int_0^\infty \mathrm{d}\tau''' \cdot \sin\{K_\mu(z - v\tau')\} \cdot$$
$$\cdot \sin\{K_\rho(z - v\tau' - v\tau'')\} \sin\{K_\sigma(z - v\tau' - v\tau'' - v\tau''')\} \cdot$$
$$\cdot \exp\{-(\gamma - \mathrm{i}\omega_\mu + \mathrm{i}\omega_\rho - \mathrm{i}\omega_\sigma + \mathrm{i}\omega_{ab})\tau' -$$
$$-(\gamma_a + \mathrm{i}\omega_\rho - \mathrm{i}\omega_\sigma)\tau'' - (\gamma + \mathrm{i}\omega_{ab} - \mathrm{i}\omega_\sigma)\tau'''\} +$$
$$+ \exp\{-\mathrm{i}\omega_\mu t - \mathrm{i}\omega_\rho t + \mathrm{i}\omega_\sigma t - \mathrm{i}\phi_\mu(t) - \mathrm{i}\phi_\rho(t) + \mathrm{i}\phi_\sigma(t)\} \cdot$$
$$\cdot \int_0^\infty \mathrm{d}\tau' \int_0^\infty \mathrm{d}\tau'' \int_0^\infty \mathrm{d}\tau''' \cdot \sin\{K_\mu(z - v\tau')\} \cdot$$
$$\cdot \sin\{K_\rho(z - v\tau' - v\tau'')\} \sin\{K_\sigma(z - v\tau' - v\tau'' - v\tau''')\} \cdot$$
$$\cdot \exp\{-(\gamma - \mathrm{i}\omega_\mu - \mathrm{i}\omega_\rho + \mathrm{i}\omega_\sigma + \mathrm{i}\omega_{ab})\tau' -$$
$$-(\gamma_a - \mathrm{i}\omega_\rho + \mathrm{i}\omega_\sigma)\tau'' - (\gamma - \mathrm{i}\omega_{ab} + \mathrm{i}\omega_\sigma)\tau'''\}\Big] +$$
$$+ \text{same with } \gamma_b \text{ replacing } \gamma_a, \tag{9.139}$$

where anti-resonance terms have been neglected as in the previous cases.

We now restrict our attention to the case of single mode operation, but shall return to a consideration of multimode operation in a later section (§ 8).

7.2 *Single mode operation*

In the case of single mode operation, (9.139) becomes greatly simplified. However, as we have seen previously, in order to calculate the macroscopic

References on page 259

polarization driving the mode, (9.139) must be integrated over the velocity distribution, $W(v)$, and further, the spatial Fourier component corresponding to the mode must be evaluated according to (9.17). In the evaluation of the Fourier component, it is apparent that integrals of the following form will arise

$$\frac{2}{L}\int_0^L \mathrm{d}z \cdot N(z,t) \cdot \sin(K_n z) \sin\{K_n(z-v\tau')\} \cdot \\ \cdot \sin\{K_n(z-v\tau'-v\tau'')\} \cdot \sin\{K_n(z-v\tau'-v\tau''-v\tau''')\}. \tag{9.140}$$

We have already discussed, in the case of first order theory (§ 6), the manner of dealing with such integrals. In this case we have

$$\sin(K_n z)\cdot \sin\{K_n(z-v\tau')\} \approx \tfrac{1}{2}\cos(K_n v\tau'), \\ \sin\{K_n(z-v\tau'-v\tau'')\}\cdot \sin\{K_n(z-v\tau-v\tau''-v\tau''')\} \approx \tfrac{1}{2}\cos(K_n v\tau'''), \tag{9.141}$$

where, as before, spatially dependent terms with wavelengths corresponding to optical frequencies have been ignored. The integral (9.140) therefore becomes

$$\tfrac{1}{2}\bar{N}(t)\cos(K_n v\tau')\cos(K_n v\tau'''), \tag{9.142}$$

where

$$\bar{N}(t) = \frac{1}{L}\int_0^L \mathrm{d}z \cdot N(z,t). \tag{9.143}$$

We therefore have

$$P_n^{(3)}(t) = (\mathrm{i}\mathscr{P}^4/16\hbar^3)\cdot \bar{N}(t)\cdot E_n^3 \cdot \exp\{-\mathrm{i}\omega_n t - \mathrm{i}\phi_n(t)\} \cdot \\ \cdot \int_{-\infty}^{+\infty}\mathrm{d}v \int_0^\infty \mathrm{d}\tau' \int_0^\infty \mathrm{d}\tau'' \int_0^\infty \mathrm{d}\tau''' \cdot \cos(K_n v\tau')\cdot\cos(K_n v\tau''')\cdot(u\pi^{\frac{1}{2}})^{-1}\cdot \\ \cdot \exp(-v^2/u^2)\cdot \exp\{-(\gamma+\mathrm{i}\omega_{ab}-\mathrm{i}\omega_n)\tau' - \gamma_a\tau''\}\cdot \\ \cdot [\exp\{-(\gamma+\mathrm{i}\omega_{ab}-\mathrm{i}\omega_n)\tau'''\} + \exp\{-(\gamma+\mathrm{i}\omega_n-\mathrm{i}\omega_{ab})\tau'''\}] + \\ + \text{same with } \gamma_b \text{ replacing } \gamma_a + \\ + \text{complex conjugates}. \tag{9.144}$$

We now carry out the integration over v. To do this we substitute complex exponentials for the cosine terms in the above, so obtaining integrals of the form

$$\frac{1}{2}\int_{-\infty}^{+\infty}\mathrm{d}v \cdot \exp(-v^2/u^2)\cdot[\exp\{K_n v(\tau'-\tau''')\} + \exp\{K_n v(\tau'+\tau''')\}], \tag{9.145}$$

which describe Fourier transforms of the velocity distribution, and which have already been discussed in (§ 6). They may accordingly be replaced by

$$\tfrac{1}{2}(u\pi^{\frac{1}{2}})\cdot[\exp\{-\tfrac{1}{4}K^2u^2(\tau'-\tau''')^2\}+\exp\{-\tfrac{1}{4}K^2u^2(\tau'+\tau''')^2\}], \quad (9.146)$$

so that, after also performing the integration over τ'', (9.144) becomes

$$\begin{aligned}P_n^{(3)}(t) = {} & (\mathrm{i}\mathscr{P}^4/32\hbar^3\gamma_a)\cdot\overline{N}(t)\cdot E_n^3\cdot\exp\{-\mathrm{i}\omega_n t-\mathrm{i}\phi_n(t)\}\cdot\\ & \cdot\int_0^\infty \mathrm{d}\tau'\int_0^\infty \mathrm{d}\tau'''[\exp\{-\tfrac{1}{4}K^2u^2(\tau'-\tau''')^2\}+\exp\{-\tfrac{1}{4}K^2u^2(\tau'+\tau''')^2\}]\cdot\\ & \cdot[\exp\{-(\gamma+\mathrm{i}\omega_{ab}-\mathrm{i}\omega_n)\tau'''\}+\exp\{-(\gamma+\mathrm{i}\omega_n-\mathrm{i}\omega_{ab})\tau'''\}]\cdot\\ & \cdot\exp\{-(\gamma+\mathrm{i}\omega_{ab}-\mathrm{i}\omega_n)\tau'\}+\\ & +\text{same with } \gamma_b \text{ replacing } \gamma_a+\\ & +\text{complex conjugates.} \qquad (9.147)\end{aligned}$$

In order to continue in the solution of the above equation, we need to go to the limiting case where $Ku \gg \gamma$ (i.e. where the Doppler linewidth is very much greater than the homogeneous linewidth). This is known as the "Doppler limit". In this case

$$\exp\{-\tfrac{1}{4}K^2u^2(\tau'-\tau''')^2\}$$

acts approximately as a delta function of $(\tau'-\tau''')$, so that we have the form

$$\int_0^\infty \mathrm{d}\tau'\int_0^\infty \mathrm{d}\tau'''\cdot G(\tau',\tau''')\cdot\exp\{-\tfrac{1}{4}(Ku)^2(\tau'''-\tau')^2\}\approx\frac{2\pi^{\frac{1}{2}}}{Ku}\int_0^\infty \mathrm{d}\tau'\, G(\tau',\tau'). \quad (9.148)$$

Terms involving $\exp\{-\tfrac{1}{4}(Ku)^2(\tau'''+\tau')^2\}$ in (9.147) may be neglected in comparison, since this latter Gaussian factor does not have its full peak in the range of integration. In the Doppler limit, we may then perform the integration over τ' in (9.147) to finally obtain

$$\begin{aligned}P_n^{(3)}(t) = {} & (\mathrm{i}\pi^{\frac{1}{2}}/16\hbar^3)\cdot\mathscr{P}^4\cdot(\gamma/Ku)\cdot(\gamma_a\cdot\gamma_b)^{-1}\cdot\overline{N}(t)\cdot\\ & \cdot\{\mathscr{D}(0)+\mathscr{D}(\omega_{ab}-\omega_n)\}\cdot E_n^3\cdot\exp\{-\mathrm{i}\omega_n t-\mathrm{i}\phi_n(t)\}+\\ & +\text{complex conjugate.} \qquad (9.149)\end{aligned}$$

From this expression we obtain the following in-phase and in-quadrature components for substitution into the self-consistency equations (9.21) and (9.22)

References on page 259

$$C_n^{(3)}(t) = (\pi^{\frac{1}{2}}/8\hbar^3)\cdot \mathscr{P}^4 \cdot \overline{N}(t)\cdot (Ku\gamma_a\gamma_b)^{-1}\cdot(\omega_{ab}-\omega_n)\cdot\gamma\cdot\mathscr{L}(\omega_{ab}-\omega_n)\cdot E_n^3; \tag{9.150}$$

$$S_n^{(3)}(t) = (\pi^{\frac{1}{2}}/8\hbar^3)\cdot \mathscr{P}^4 \cdot \overline{N}(t)\cdot (Ku\gamma_a\gamma_b)^{-1}\cdot\{1+\gamma^2\mathscr{L}(\omega_{ab}-\omega_n)\}\cdot E_n^3, \tag{9.151}$$

where

$$\mathscr{L}(\omega_{ab}-\omega_n) = \{\gamma^2+(\omega_{ab}-\omega_n)^2\}^{-1}. \tag{9.152}$$

The equation describing the amplitude of the mode is, therefore,

$$\dot{E}_n = \{\tfrac{1}{2}(\omega/\varepsilon_0)\cdot(\mathscr{P}^2/\hbar Ku)\cdot\overline{N}\cdot Z_i(\omega_n-\omega_{ab})-\tfrac{1}{2}(\omega/Q_n)\}E_n- -(\pi^{\frac{1}{2}}\omega/16\hbar^3\varepsilon_0)\cdot\mathscr{P}^4\cdot\overline{N}\cdot(Ku\gamma_a\gamma_b)^{-1}\{1+\gamma^2\mathscr{L}(\omega_{ab}-\omega_n)\}E_n^3. \tag{9.153}$$

By looking for the steady state solution to the above ($\dot{E}_n = 0$), we can determine the intensity of the radiation field in the cavity at which the laser saturates; i.e. the intensity for which the decrease in the population inversion, brought about by the influence of the radiation field itself on the populations of the upper and lower levels, is sufficient to reduce the gain of the active medium until it just equals the cavity loss.

It is usual to express (9.153) in terms of a "relative excitation" defined by

$$\mathscr{N} = \overline{N}/\overline{N}_{\mathrm{T}}, \tag{9.154}$$

where $\overline{N}_{\mathrm{T}}$ is the population inversion (or excitation) required to produce threshold oscillation at the line centre ($\Omega_n = \omega_{ab}$) and as such is given by first order theory (9.127) as

$$\overline{N}_{\mathrm{T}} = \varepsilon_0\hbar Ku/\{Q_n\cdot\mathscr{P}^2\cdot Z_i(0)\}. \tag{9.155}$$

We may, therefore, write (9.153) as

$$\dot{E}_n = (\omega/2Q_n)\cdot\left\{\frac{Z_i(\omega_n-\omega_{ab})}{Z_i(0)}\mathscr{N}-1\right\}E_n- -(\pi^{\frac{1}{2}}\mathscr{P}^2\cdot\omega)/\{Z_i(0)\cdot 16\hbar^2\cdot Q_n\cdot\gamma_a\gamma_b\}\cdot\mathscr{N}\cdot\{1+\gamma^2\mathscr{L}(\omega_{ab}-\omega_n)\}\cdot E_n^3,$$

so that the field intensity at saturation is given by

$$E_n^2 = 8\cdot Z_i(0)\cdot\hbar^2\cdot\gamma_a\gamma_b/(\pi^{\frac{1}{2}}\mathscr{P}^2)\cdot\left\{\frac{Z_i(\omega_n-\omega_{ab})}{Z_i(0)}-\mathscr{N}^{-1}\right\}\cdot \cdot\{1+\gamma^2\mathscr{L}(\omega_{ab}-\omega_n)\}^{-1}. \tag{9.156}$$

In this case, where $\gamma \ll Ku$, the plasma dispersion function $Z_i(\omega_n-\omega_{ab})$

may be expanded out (appendix N), so that (9.156) becomes

$$E_n^2 = (8\hbar^2\gamma_a\gamma_b/\mathscr{P}^2)\cdot[\exp\{-(\omega_n-\omega_{ab})^2/(Ku)^2\}-\mathscr{N}^{-1}]\cdot \cdot\{1+\gamma^2\mathscr{L}(\omega_{ab}-\omega_n)\}^{-1}. \quad (9.157)$$

The form of (9.157) is discussed in § 7.4.

For convenience (9.156) may be written in terms of two coefficients α_n and β_n as

$$E_n^2 = \alpha_n/\beta_n, \quad (9.158)$$

where

$$\alpha_n = \tfrac{1}{2}(\omega/Q_n)\{[Z_i(\omega_n-\omega_{ab})/Z_i(0)]\mathscr{N}-1\} \quad (9.159)$$

and

$$\beta_n = \tfrac{1}{16}\pi^2(\omega/Q_n)[\mathscr{N}\mathscr{P}^2/(\hbar^2\gamma_a\gamma_b Z_i(0))]\cdot[1+\gamma^2\mathscr{L}(\omega_n-\omega_{ab})]. \quad (9.160)$$

The frequency behaviour in the presence of saturation may be determined by substitution of (9.150) and (9.123) into the self-consistency equation (9.22), to obtain a solution of the form (ϕ_n again put equal to zero)

$$\omega_n = \Omega_n+\sigma_n+\rho_n E_n^2, \quad (9.161)$$

where

$$\sigma_n = \tfrac{1}{2}(\omega/Q_n)\mathscr{N}Z_r(\Omega_n-\omega_{ab})/Z_i(0) \quad (9.162)$$

and

$$\rho_n = \tfrac{1}{16}\pi^{\frac{1}{2}}(\omega/Q_n)\mathscr{N}[\mathscr{P}^2/(\hbar^2\gamma_a\gamma_b Z_i(0))]\cdot\gamma\cdot(\Omega_n-\omega_{ab})\cdot\mathscr{L}(\Omega_n-\omega_{ab}). \quad (9.163)$$

In the expressions for the two coefficients σ_n and ρ_n, the approximation of replacing ω_n by Ω_n has been made. At threshold ($\mathscr{N} = 1$), (9.161) reduces to (9.130), the expression for frequency pulling given by first order theory. The final term in (9.161), which depends on the radiation field intensity inside the cavity, describes a frequency "pushing" effect in that, as examination of (9.163) shows, its influence is to move the resonance frequency further away from the line centre. Since the first term describes a frequency "pulling" effect, which we have already discussed, whether the resonance frequency moves towards the line centre or away from the line centre as the excitation increases depends on the relative magnitudes of the two terms. Examination of (9.161), (9.158), (9.162) and (9.163) readily shows that if

$$[\gamma(\omega_{ab}-\Omega_n)Z_i(\Omega_n-\omega_{ab})/Z_r(\Omega_n-\omega_{ab})] > [(\Omega_n-\omega_{ab})^2+2\gamma^2] \quad (9.164)$$

as the excitation increases, the resonance frequency moves away from the line centre.

References on page 259

7.3 *Population terms*

It is apparent that (9.97) may be used to determine the influence of the optical radiation field on the populations of the upper and lower laser states. Integration of (9.97) over the time of excitation of the state (t_0) and over the velocity distribution may be performed as discussed previously to give an expression for the total population associated with the upper (or lower) state:

$$\rho_{aa}(z, t) = \int_{-\infty}^{+\infty} \mathrm{d}v \int_{-\infty}^{+\infty} \mathrm{d}t_0 \int \mathrm{d}z_0 \cdot \delta(z - z_0 - vt + vt_0) \times \\ \times \sum_{\alpha = a, b} \lambda_\alpha(z_0, t_0, v) \cdot \rho_{aa}(\alpha, z_0, t_0, v, t). \qquad (9.165)$$

Substituting from (9.97) for $\rho_{aa}^{(2)}(\alpha, z_0, t_0, v, t)$ in (9.165) and evaluating the various integrals, leads to the following for the case of single frequency operation

$$\rho_{aa}(z, t) = [\Lambda_a(z, t)/\gamma_a] - \{[\Lambda_a(z, t)/\gamma_a] - \\ - [\Lambda_b(z, t)/\gamma_b]\}\tfrac{1}{4}(\mathscr{P}E)^2(\gamma_a Ku)^{-1} Z_i(\omega_{ab} - \omega_n). \qquad (9.166)$$

In the case of two frequency operation, as well as the d.c. changes in population due to each mode independently and of the form given by (9.166), there is also a "pulsating" component in the population change due to the optical radiation field, which is at a frequency close to the frequency difference, Δ, between the two modes. In the case when $\gamma \sim \gamma_a \gg \Delta \ll Ku$ (9.166) becomes

$$\rho_{aa}(z, t) = [\Lambda_a(z, t)/\gamma_a] - \{[\Lambda_a(z, t)/\gamma_a] - [\Lambda_b(z, t)/\gamma_b]\} \times \\ \times [\alpha E_1^2 + \beta E_2^2 + \gamma E_1 E_2 \sin(\Delta \cdot t)], \qquad (9.167)$$

where the magnitude of the pulsating component relative to the d.c. components is given by

$$\frac{\gamma E_1 E_2}{\alpha E_1^2 + \beta E_2^2} = \frac{2 E_1 E_2}{(E_1^2 + E_2^2)} \left(\frac{\gamma_a}{\Delta}\right) \cdot \cos(\pi z/L). \qquad (9.168)$$

7.4 *Lamb dip and hole burning*

We return now to a consideration of the behaviour of the field intensity at saturation as described by (9.156) or (9.158). The coefficient, α_n, which forms the numerator of (9.158), has a maximum value at the line centre ($\Omega_n = \omega_{ab}$), but so also has the coefficient, β_n, which forms the denominator of (9.158). By comparison with (9.124) it will be apparent that α_n is just the linear gain profile. The width of this profile is determined by that part of the Doppler profile lying above threshold, and is hence dependent on the relative excitation. The denominator describes the saturation behaviour and its width is determined essentially by the homogeneous linewidth (γ). It is therefore apparent that, in general, the curve displaying intensity of oscillation as a function of cavity detuning from the line centre will in general have a flattened peak at the line centre, and for a sufficiently large relative excitation may exhibit a central dip lying between two maxima. This dip is referred to as the 'Lamb dip'. It may be shown using (9.158) that the conditions for the double peak to occur are: firstly, that the detuning from the line centre required to stop oscillation, $(\Omega_n^* - \omega_{ab})$, satisfies the following

$$\tfrac{1}{2}\{1-\exp[-(\Omega_n^*-\omega_{ab})^2/(Ku)^2]\} > (\gamma/Ku)^2; \tag{9.169}$$

and secondly, that the relative excitation exceeds a certain level given by

$$\mathcal{N} > 1/\{1-2[\gamma/(Ku)]^2\}. \tag{9.170}$$

It is apparent from both these expressions that the larger the Doppler width compared to the homogeneous linewidth, the more readily will the 'Lamb dip' occur. In fig. 9.2 the form of the "'Lamb dip'" is illustrated. In ch. 8 the phenomenon of "hole burning" has been discussed. This phenomenon is associated with the curve of population difference $(\rho_{aa}-\rho_{bb})$ as a function of velocity. In the presence of the optical radiation field this profile differs from the normal (Doppler) curve associated with the unperturbed atoms by having, in the case of single frequency operation, two holes burned into it due to saturation effects. The positions of the two holes are given by $Ku = \pm(\omega_n - \omega_{ab})$. If in the second order solutions for ρ_{aa} and ρ_{bb} discussed in § 7.3, the integration over velocity had not been performed and the population difference as a function of velocity displayed, then this hole burning effect in the velocity profile would have become apparent. The reason two holes are burnt in the profile is a consequence of the standing wave nature of the optical radiation field in the cavity. If the perturbation term for a moving atom (9.92) is expressed in terms of two travelling waves instead of a standing wave, it is apparent that an atom moving with velocity component u sees a radiation field at two frequencies $(\omega_n \pm Ku)$. The interaction between the atom and the radiation field becomes significant when one of these frequencies lies, approximately, within the

References on page 259

homogeneous linewidth of the resonance frequency of the atom (ω_{ab}), and therefore there are two velocities for which the associated atoms interact strongly with the radiation field, and hence have their populations modified by it, namely $Ku = \pm(\omega_n - \omega_{ab})$. In the case when the resonance frequency of the cavity lies well away (greater than the homogeneous linewidth) from the centre of the Doppler profile, the two holes are independent corresponding to different populations. However as the cavity

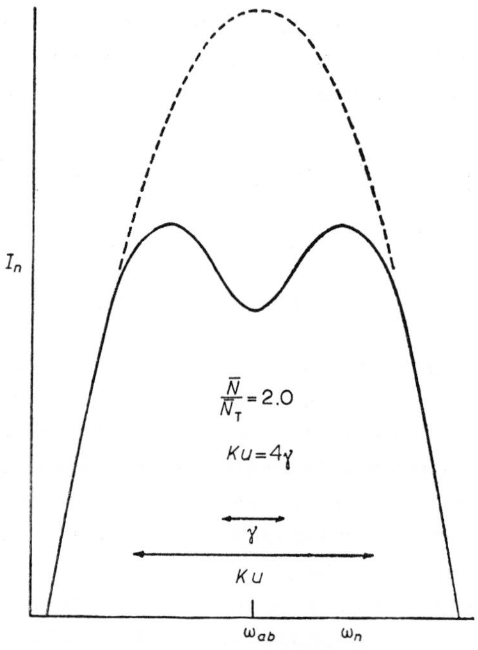

Fig. 9.2. Relative intensity of oscillation as a function of detuning from line centre. The solid curve is drawn for parameters $\bar{N} = 2\bar{N}_T$ and $Ku = 4\gamma$, and represents (9.157). The dotted curve indicates the Doppler gain profile of the numerator of (9.157) (from LAMB [1964a] by courtesy of the American Physical Society).

resonance moves towards the line centre, the two holes begin to overlap, so that, in general, the fraction of the population inversion coupled to the radiation field decreases, and hence so also does the intensity of the field at saturation. The Lamb dip in power output at the line centre may therefore be interpreted in terms of the merging of the holes burnt in the velocity profile.

8. Multimode operation

It is not our intention to derive in detail the solutions to (9.139) in the case of multimode operation, and the reader is referred to Lamb's original paper (Lamb [1964a]) for a full discussion of this aspect. However, for the sake of completeness, we summarize briefly in this section some of the results of this analysis.

In the case of two mode oscillation, the differential equations derived from (9.139), describing the amplitudes, E_1 and E_2, of the modes are

$$\begin{aligned} \dot{E}_1 &= \alpha_1 E_1 - \beta_1 E_1^3 - \theta_{12} E_1 E_2^2, \\ \dot{E}_2 &= \alpha_2 E_2 - \beta_2 E_2^3 - \theta_{21} E_2 E_1^2. \end{aligned} \tag{9.171}$$

The coefficients α_1, α_2 and β_1, β_2 in the above are each associated with only one of the modes and are of the form given by (9.159) and (9.160) for the case of single mode operation. The coefficients θ_{12} and θ_{21} each depend on the frequencies of both the modes, and describe the manner in which the presence of one mode modifies the saturation intensity of the other (i.e. the coupling between the modes). It will be realized that apart from the final term on the right hand sides of the differential equations (9.171), these equations are analogous to those obtained in the case of single frequency operation.

If we write

$$X = E_1^2 \qquad \text{and} \qquad Y = E_2^2, \tag{9.172}$$

then (9.171) may be written

$$\begin{aligned} \dot{X} &= 2X(\alpha_1 - \beta_1 X - \theta_{12} Y), \\ \dot{Y} &= 2Y(\alpha_2 - \beta_2 Y - \theta_{21} X). \end{aligned} \tag{9.173}$$

The conditions for steady state oscillation are obtained from (9.173) by putting $\dot{X} = 0$, $\dot{Y} = 0$. Obviously two possible solutions correspond to single mode operation, namely

$$\begin{aligned} X &= 0, \qquad Y = \alpha_2/\beta_2, \\ Y &= 0, \qquad X = \alpha_1/\beta_1, \end{aligned} \tag{9.174}$$

while the third possible solution is given by

$$\begin{aligned} L_1&: \beta_1 X + \theta_{12} Y = \alpha_1, \\ L_2&: \beta_2 Y + \theta_{21} X = \alpha_2. \end{aligned} \tag{9.175}$$

References on page 259

Each of the equations (9.175) describes a straight line (L_1 and L_2) in the $X-Y$ plane, and the point of intersection of the two lines, if it lies in the upper, right hand quadrant of the plane, describes a possible two mode solution. In figs. 9.3–9.5, these solutions are displayed for different values of the parameters α, β, θ_{12} etc. Under certain conditions (fig. 9.3) only single mode oscillation is possible, whereas under other conditions (figs. 9.4 and 9.5) either single mode or multimode oscillation is possible.

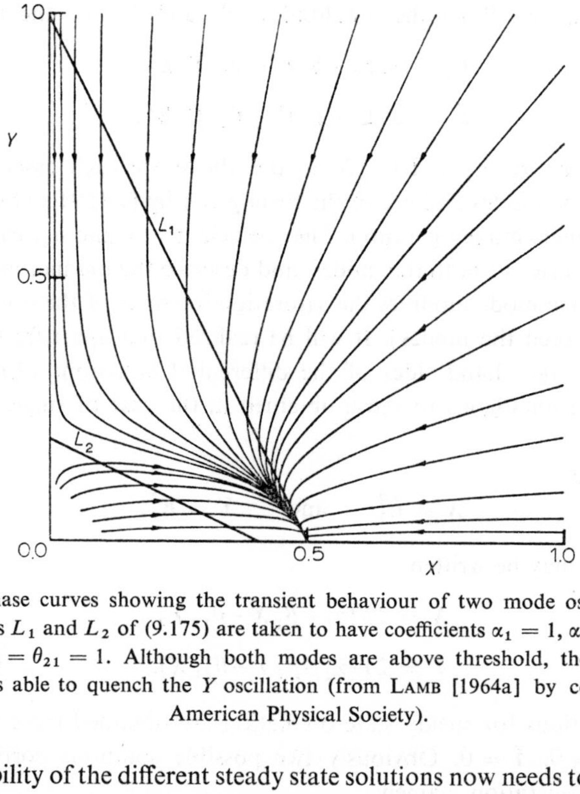

Fig. 9.3. Phase curves showing the transient behaviour of two mode oscillation. The straight lines L_1 and L_2 of (9.175) are taken to have coefficients $\alpha_1 = 1$, $\alpha_2 = 0.4$, $\beta_1 = \beta_2 = 2$, $\theta_{12} = \theta_{21} = 1$. Although both modes are above threshold, the favoured X oscillation is able to quench the Y oscillation (from LAMB [1964a] by courtesy of the American Physical Society).

The stability of the different steady state solutions now needs to be examined to decide which of them represent stable states of oscillation. By performing graphical integration of the differential equations (9.173), Lamb has been able to illustrate the stability of the different steady state solutions in a very clear fashion. The results of this analysis are displayed by means of phase curves, marked with arrows in figs. 9.3–9.5. These curves represent the transient behaviour of the state of oscillation, and are such that they describe the path (in time) of an initial state of oscillation, other than a possible steady state, as it moves towards steady state conditions.

In so far as all the phase curves in fig. 9.3 converge on the steady state solution ($Y = 0$, $X = 0.5$) corresponding to single mode oscillation, this means that no matter what the past history of the state of oscillation, this solution represents the only possible stable oscillation. In the case of fig. 9.4, a similar behaviour is also seen, but now the stable solution represents two mode oscillation. Under the conditions of fig. 9.5, however, it will be seen that the phase curves converge on two of the steady state solutions, both of

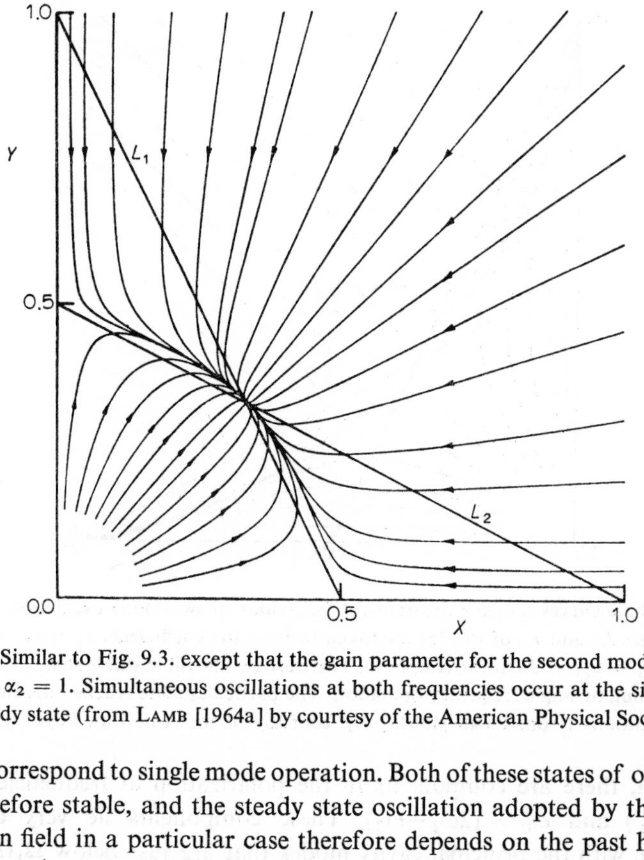

Fig. 9.4. Similar to Fig. 9.3. except that the gain parameter for the second mode has been raised to $\alpha_2 = 1$. Simultaneous oscillations at both frequencies occur at the single stable steady state (from LAMB [1964a] by courtesy of the American Physical Society).

which correspond to single mode operation. Both of these states of oscillation are therefore stable, and the steady state oscillation adopted by the optical radiation field in a particular case therefore depends on the past history of the system (i.e. where abouts in the X–Y plane the transient state originates).

The oscillation frequencies in the two mode case are given by

$$\begin{aligned} \nu_1 &= \Omega_1+\sigma_1+\rho_1 E_1^2+\tau_{12} E_2^2, \\ \nu_2 &= \Omega_2+\sigma_2+\rho_2 E_2^2+\tau_{21} E_1^2. \end{aligned} \tag{9.176}$$

As in the case of single mode operation the coefficients σ_1, σ_2 and ρ_1, ρ_2

References on page 259

each refer only to a single mode and are of the form given by (9.162) and (9.163). The coupling between the modes is described by the last term on the right-hand side of (9.176), the coefficients θ_{12} and θ_{21} each depending on the frequencies (Ω_1 and Ω_2) of both the modes.

Examination of (9.139) shows that the third order polarization of the active medium has components that oscillate at all possible frequencies of the form $(\omega_\mu - \omega_\rho + \omega_\sigma)$, and hence even for the case of two frequency

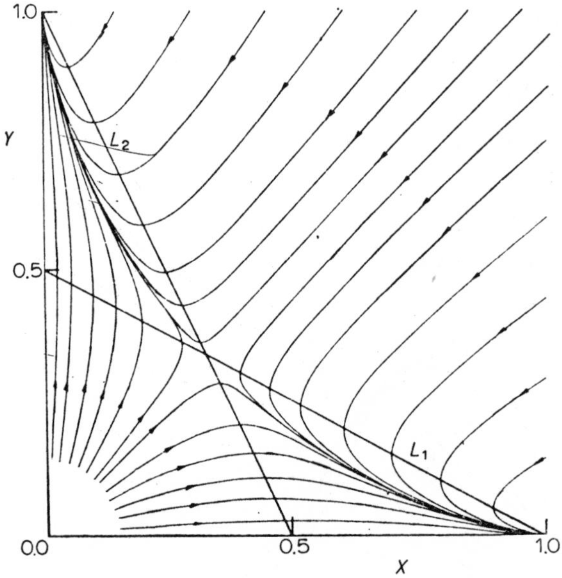

Fig. 9.5. Phase curves showing the transient behaviour of two-mode oscillation when the straight lines L_1 and L_2 of (9.175) are taken to have the coefficients $\alpha_1 = \alpha_2 = 1$, $\beta_1 = \beta_2 = 1$, $\theta_{12} = \theta_{21} = 2$ (strong coupling). There are two possible stable steady states, each corresponding to single frequency operation. The particular state reached depends on the initial conditions (from Lamb [1964a] by courtesy of the American Physical Society).

operation, there are components in the polarization at frequencies $\omega_3' = (2\omega_2 - \omega_1)$ and $\omega_0' = (2\omega_1 - \omega_2)$. These components lie very close to resonance with the principal cavity modes that are just below $\Omega_1(\Omega_0)$ and just above $\Omega_2(\Omega_3)$. In so far as these components exist in the polarization and lie close to principal cavity resonances, Maxwell's equations imply that they also exist as radiation fields inside the cavity. These are the so called "combination tones". Hence even in the case when the linear gain associated with a third mode, say α_3, is negative, quasi-three-frequency operation may occur. For example, in order to observe the combination tone ω'_3, one would adjust the cavity tuning so that the cavity resonance Ω_2 was slightly above the atomic resonance frequency, ω_{ab}, since this would ensure that as the excitation of the medium was increased oscillation corresponding to the cavity mode Ω_1, reached threshold before that corresponding to cavity mode Ω_3. The combination tone lying close to the cavity resonance Ω_3 could then be observed, before the linear gain associated with this resonance became positive and true three mode oscillation occurred.

References

ALPERT, S. S. and A. D. WHITE, 1963, Proc. I.E.E.E. **51**, 1665–6.

BOERSCH, H., G. HERZIGER, H. LINDNER and G. MAKOSCH, 1967, Phys. Lett. **24A**, 227–8.

BOLWIJN, P. T., 1964, Phys. Lett. **13**, 311.

BOLWIJN, P. T., 1965, Phys. Lett. **19**, 384–5.

BOLWIJN, P. T., 1966, J. Appl. Phys. **37**, 4487–92.

CULSHAW, W. and J. KANNELAUD, 1966, Phys. Rev. **145**, 257–67.

CULSHAW, W., 1967, Phys. Rev. **164**, 329–39.

DURAND, G., 1966, I.E.E.E. J. of Quant. Electron. **QE–2**, 448–55.

D'YAKONOV, M. I. and A. S. FRIDRIKHOV, 1966, Usp. Fiz. Nauk. **90**, 565, Eng. trans. Sov. Phys. – Usp. **9**, 837 (1967).

FORK, R. L. and M. A. POLLACK, 1965, Phys. Rev. **139**, 1408–14.

FORK, R. L. and M. SARGENT, III, 1966, Proc. of the International Conference on the Physics of Quantum Electronics, ed. Kelley, P. L., B. Lax and P. E. Tannenwald (McGraw-Hill, New York) pp. 611–19.

GYORFFY, B. L., M. BORENSTEIN and W. E. LAMB, Jr., 1968, Phys. Rev. **169**, 340–359.

HAISMA, J. and G. BOUWHUIS, 1964, Phys. Rev. Lett. **12**, 287–9.

HEER, C. V. and R. D. GRAFT, 1965, Phys. Rev. **140**, A1088.

LAMB, W. E., Jr., 1960, Lectures in Theoretical Physics, ed. Brittin, W. E. and B. W. Downes (Interscience) pp. 435–83.

LAMB, W. E., Jr., 1964a, Phys. Rev. **134**, A1429–50.

LAMB, W. E., Jr., 1964b, International School of Physics 'Enrico Fermi', Course XXXI, Quantum Electronics and Coherent Light (Academic Press) pp. 78–110.

LAMB, W. E., Jr., 1965, "Quantum Optics and Electronics, Les Houches, 1964", ed. de Witt, C., A. Blandin and C. Cohen-Tannoudji (Gordon and Breach) pp. 331–81.

MCFARLANE, R. A., W. R. BENNETT, Jr. and W. E. LAMB, Jr., 1963, Appl. Phys. Lett. **2**, 189–190.

SARGENT, M., III, W. E. LAMB, Jr. and R. L. FORK, 1967a, Phys. Rev. **164**, 436–49.

SARGENT, M., III, W. E. LAMB, Jr. and R. L. FORK, 1967b, Phys. Rev. **164**, 450–65.

SETTLES, R. A. and C. V. HEER, 1968, Appl. Phys. Lett. **12**, 350–2.

SMITH, P. W., 1966, J. of Appl. Phys. **37**, 2089–93.

SZOKE, A. and A. JAVAN, 1963, Phys. Rev. Lett. **10**, 521–4.

SZOKE, A. and A. JAVAN, 1966, Phys. Rev. **145**, 137–47.

TOMLINSON, W. J. and R. L. FORK, 1967, Phys. Rev. **164**, 466–83.

UCHIDA, T., 1967, I.E.E.E. J. of Quant. Electron. **QE–3**, 7–16.

UEHARA, K. and K. SHIMODA, 1965, Japan. J. Appl. Phys. **4**, 921.

VAN HAERINGEN, W., 1967, Phys. Rev. **158**, 256–72.

CHAPTER 10

COHERENCE

1. Introduction

In recent years conventional optical coherence theory has developed considerably in an attempt to define more precisely concepts such as "coherence", and also to describe a wider range of coherence phenomena. This has been brought about both by unconventional experiments involving thermal sources, and by the advent of the laser which generates radiation having a different statistical nature to that from thermal sources. Optical coherence can be studied by classical or quantum mechanical methods. However, in a quantum treatment, a fuller specification of the system (the density matrix) is required than in the classical approach Although considerable progress has been made in developing the quantum theory of coherence, the classical approach is of value in predicting results rather more easily.

This chapter, therefore, begins by providing an introduction to modern (classical) theory. The more precise terminology introduced is illustrated by a consideration of certain well understood, and familiar, coherence experiments. The main differences between laser and thermal radiation, and the effects of these differences on the outcome of the more sophisticated coherence experiments, are then discussed.

2. Elementary principles and definitions

The concept of temporal coherence can be introduced by considering the operation of the Michelson Interferometer (fig. 10.1). Here a steady beam of light from a small source (S) is split (amplitude division) into two beams by a partially silvered mirror (M). These then travel along paths of different length before being reunited at P. If the path difference introduced in the beams is less than a certain value, s_c, with which is associated a relative time delay, $\tau_c = s_c c^{-1}$, interference fringes are observed in the plane (P).

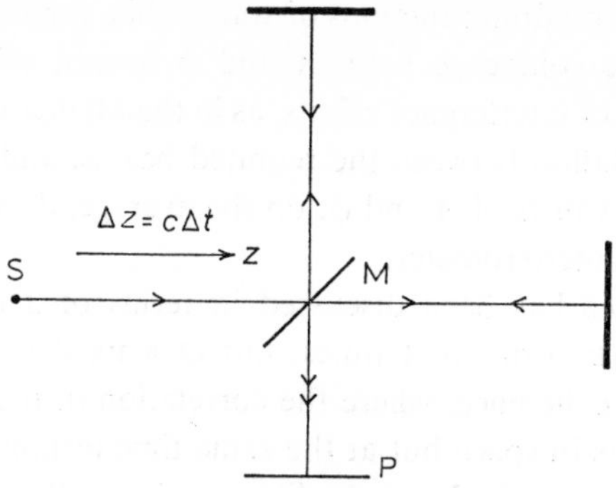

Fig. 10.1. Michelson interferometer.

The observation of fringes implies that at a fixed point in the light beam, there is a correlation between the optical disturbance at a time t_1 and the optical disturbance at a time t_2, provided that $|t_1 - t_2| \leqslant \tau_c$. The time delay, τ_c, is known as the coherence time of the light and the path difference, s_c, the coherence length.

The total intensity of the light may be considered to be the sum of the different frequency components in the range ν to $(\nu+\Delta\nu)$. The coherence time can be simply related to the effective bandwidth of the light, $\Delta\nu$, as follows.

The light is temporally coherent at a certain point only over a time interval, Δt, such that the number of maxima passing with frequency ν does not exceed by more than unity, say, the number of maxima passing with frequency $(\nu+\Delta\nu)$, i.e. Δt is the time over which the various intensity maxima remain in step. Therefore,

$$(\nu+\Delta\nu)\Delta t - \nu\Delta t \leqslant 1,$$

$$\Delta\nu \cdot \Delta t \leqslant 1. \tag{10.1}$$

The equality in (10.1) defines the coherence time, τ_c. The light is said to be quasimonochromatic provided that

$$\Delta\nu/\nu \ll 1. \tag{10.2}$$

A quasimonochromatic light beam (assumed linearly polarized) can be represented by a real function of the form

$$E(\boldsymbol{r}, t) = A(\boldsymbol{r}, t) \cos [\phi(\boldsymbol{r}, t) - 2\pi\nu t]. \tag{10.3}$$

Over time intervals long compared with the coherence time, A and ϕ

References on page 299

must be considered random functions of time, while for time intervals short compared with the coherence time, A and ϕ remain effectively constant.

The observation of interference effects, as in the Michelson Interferometer, implies a phase relation between the reunited beams, and hence only small fluctuations in the values of A and ϕ, on the average, during the time delay introduced in the interferometer.

So far, coherence has been discussed in terms of a correlation in the radiation field, at two different times, but at a fixed point in space. The concept of spatial coherence, where the correlation in the radiation field at two different points in space but at the same time instant is investigated, is now discussed in terms of a Young's slits experiment (fig. 10.2). Suppose an extended luminous source is placed at a distance, R, large compared to the

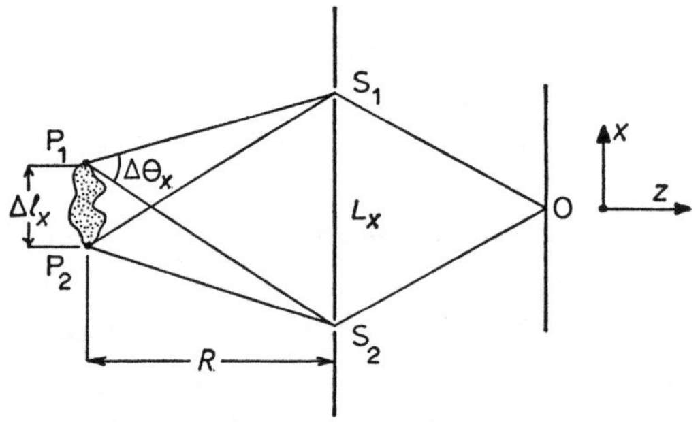

Fig. 10.2. Young's slits experiment.

extent of the source, Δl_x, from a screen containing two sampling apertures S_1 and S_2 a distance L_x apart. (The apertures sample across the wavefront.) To investigate the spatial coherence of the radiation field, the correlation between the fields at S_1 and S_2 is examined at a fixed time. This is most conveniently carried out by examining the interference fringes formed by the reuniting fields at the point O, which is equidistant from S_1 and S_2. The appearance of fringes is taken as evidence of coherence.

By moving O from the symmetric position, the correlation between the radiation field at the point S_1 at time t_1 and the radiation field at the point S_2 at time t_2 can be examined, where

$$|t_2 - t_1| = c^{-1} \cdot |\mathrm{OS}_2 - \mathrm{OS}_1|. \qquad (10.4)$$

The condition for spatial coherence between the radiation at S_1 and S_2 is deduced as follows. The source is considered to be made up of a collection of independent oscillators, each of which radiates to both slits; for a particular oscillator, P_1, there is a path difference $(P_1S_1 - P_1S_2)$ between the radiation reaching S_1 and S_2. The condition for spatial coherence between S_1 and S_2 is that the change in this path difference on moving from any one element in the source to another is less than the effective wavelength of the radiation, for only under this condition will there remain a phase relation between the fields at S_1 and S_2 on summing over all the sources. Now

$$(P_1S_1)^2 = R^2 + (L_x - \Delta l_x)^2/4, \qquad (P_1S_2)^2 = R^2 + (L_x + \Delta l_x)^2/4. \tag{10.5}$$

Therefore,

$$|P_1S_1 - P_1S_2| \approx \Delta l_x L_x/R. \tag{10.6}$$

The path difference for the symmetric position is zero, and therefore, for spatial coherence at the slits

$$\Delta l_x \cdot L_x/R \leqslant \lambda, \qquad \Delta l_x \cdot \Delta\theta_x \leqslant \lambda. \tag{10.7}$$

So far, temporal and spatial coherence have been discussed in terms of a wave formulation of the radiation field. It is useful to consider the same experiments in terms of photons. The uncertainty relationships for photons are required, and may be written as follows

$$\Delta p_x \cdot \Delta x \sim h, \qquad \Delta p_y \cdot \Delta y \sim h, \qquad \Delta p_z \cdot \Delta z \sim h. \tag{10.8}$$

The axes are defined in figs. 10.1 and 10.2 for the two experiments (the z-axis is taken as the direction of propagation).

We consider, firstly, Young's slits experiment, which was used to define spatial coherence. When radiation from the source is observed, this amounts to localizing the position of a photon to an accuracy given by the source dimensions $(\Delta l_x, \Delta l_y)$. The first two relationships can now be used to find the uncertainty in the x and y components of the momentum of the photon, and with these is associated an angular uncertainty in its direction of emission, which, considering only the one dimensional case, can be seen to be

$$\Delta\theta_x \sim \lambda/\Delta l_x, \tag{10.9}$$

References on page 299

since

$$p_z = h/\lambda. \tag{10.10}$$

If the slits lie within this angular uncertainty, it is not possible to locate the photon as passing through one slit as opposed to the other. It has already been shown that under the above condition, interference fringes are observed and the radiation field is spatially coherent at the two slits. If the slits are separated by more than the above angular uncertainty, it is possible to decide through which slit a particular photon has passed, and this has likewise been shown to correspond to the case when the radiation field is spatially incoherent at the two slits.

The Michelson interferometer experiment can also be discussed in terms of the uncertainty principle. The third relationship is written in a slightly different form, using

$$\Delta E = \Delta p_z/c = h \cdot \Delta \nu \tag{10.11}$$

and expressing Δz in terms of a time uncertainty

$$c \cdot \Delta t = \Delta z, \tag{10.12}$$

substitution into the third uncertainty relationship gives

$$\Delta t \cdot \Delta \nu \sim 1. \tag{10.13}$$

The value of $\Delta \nu$ is determined by the nature of the source (and can be related to the lifetime of the radiating species or the stability of the oscillator). If the experiment does not involve relative time delays between the two paths which exceed the reciprocal of the bandwidth of the source, the uncertainty principle prevents the photon from being localized in one path of the interferometer as opposed to the other (without additional experiments being performed, which would, in fact, destroy the effect). The photon must, therefore, be considered to traverse both paths. It has already been shown that a similar relationship to (10.13) is the condition for the formation of fringes and for temporal coherence. Once the path differences are so great that it becomes possible to localize the photon in one or another of them, then fringes can no longer be observed and the temporal coherence disappears.

From the results of the two experiments described, it is possible to make the following general statement: two points (space or time) in a radiation field are coherent only if, through the operation of the uncertainty principle, it is not possible to localize a particular photon at one point as opposed to the other. "Each photon, then, interferes only with itself. Intererfence between two different photons never occurs" (DIRAC [1958]).

Using the uncertainty relationships, it is possible to introduce the concept of a cell in the six dimensional phase space of the photons, the volume of which is given by

$$\Delta p_x \cdot \Delta p_y \cdot \Delta p_z \cdot \Delta x \cdot \Delta y \cdot \Delta z = h^3. \tag{10.14}$$

The cell is a measure of the finest detail in phase space that can be seen by any experiment, the limit being determined by the uncertainty principle. It is possible to locate a photon within a particular cell, but not to define its position within the cell. Related to the cell in phase space is the "volume of coherence"

$$\Delta V_c = c \cdot \tau_c \cdot A_c, \tag{10.15}$$

where τ_c is the coherence time, and A_c is the coherence area (which can be defined by an extension of the one dimensional argument so far employed). The "volume of coherence" corresponds to the spatial volume of an elementary cell in the phase space of the photons.

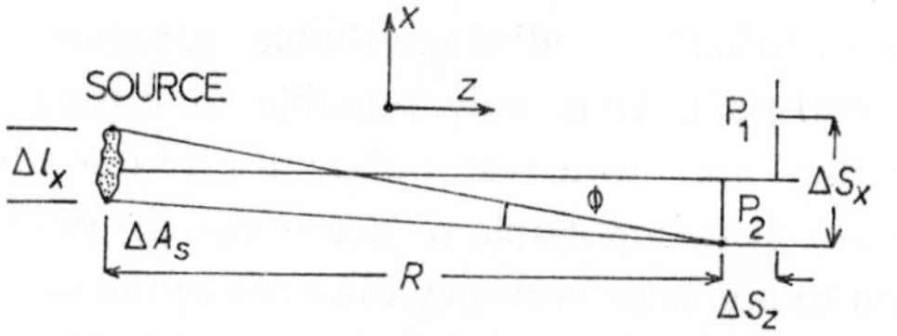

Fig. 10.3. The relationship between "volume of coherence" and the elementary cell of phase space.

To demonstrate this, consider the experimental arrangement illustrated in fig. 10.3. At the observation points (P_1 and P_2) the coherence volume of the radiation from the source of area ΔA_s is found by substituting from (10.1) and (10.7) into (10.15), and is

$$\Delta V_c = (\lambda^2 R^2 c)/(\Delta A_s \cdot \Delta \nu). \tag{10.16}$$

We now apply the uncertainty principle to the measurements carried out at the two observation points (P_1 and P_2), which are separated by (Δs_x, Δs_y, Δs_z). The location of a photon at one of the observation points corresponds to measuring its x (or y) component of momentum to an uncertainty

$$\Delta p_x = p_z \phi = (h \cdot \Delta l_x)/(\lambda R). \tag{10.17}$$

The uncertainty in the momentum in the z-direction is as before

$$\Delta p_z = h \cdot \Delta\nu/c. \tag{10.18}$$

Therefore, in the observation plane

$$\Delta p_x \Delta p_y \Delta p_z = (h^3 \cdot \Delta A_s \cdot \Delta\nu)/(\lambda^2 \cdot R^2 \cdot c). \tag{10.19}$$

The spatial volume at the observation points associated with an elementary cell of phase space is found by substituting from (10.19) into (10.14),

$$\Delta x \cdot \Delta y \cdot \Delta z = (\lambda^2 \cdot R^2 \cdot c)/(\Delta A_s \cdot \Delta\nu). \tag{10.20}$$

Comparison of (10.16) and (10.20) confirms the relationship between the volume of coherence and the spatial volume of an elementary cell. For the radiation to be coherent at the two points, then we have

$$\Delta s_x \Delta s_y \Delta s_z < \Delta V_c = \Delta x \cdot \Delta y \cdot \Delta z. \tag{10.21}$$

The number of photons in one cell of phase space (which corresponds to the number of photons in the volume of coherence in real space) is known as the "degeneracy parameter" (δ) of the radiation.

Photons are intrinsically indistinguishable particles (and hence obey Bose-Einstein statistics). It is only possible to distinguish between two photons when they are located at different positions in space, or have different momenta. It is impossible to determine the position and momentum of a photon to a greater accuracy than the cell in which it is located in phase space. Hence, two photons that are located within the same cell of phase space are completely indistinguishable, and no experiment is capable of discriminating between them. It is therefore possible to extend the previous generalization to the following: each cell in phase space interferes only with itself.

For thermal sources the largest values of the degeneracy parameter that can be obtained are of the order of 10^{-3}, while for laser sources degeneracy parameters of the order of 10^{14} are possible. When $\delta \ll 1$ the radiation is non-degenerate, and when $\delta \gg 1$, it is highly degenerate.

3. Classical wave theory of coherence

The formalism for describing coherence phenomena by the classical wave theory was developed, in the main, by WOLF [1955] who used the analytic signal representation (explained below), due to GABOR [1946], to describe the real field. We treat only the scalar formulation of this theory, but do so in terms of the analytic signal representation since this is the approach usually employed in the more advanced literature (although for most simple problems the real field representation is adequate).

We assume that the radiation field at any point, $\boldsymbol{r}$, and time, t, can be described by a real, scalar function, $V^{\mathrm{r}}(\boldsymbol{r}, t)$ which, for example, might be the magnitude of one component of the electric or magnetic field. In fact, a vector function is often required to completely represent the field, but providing we consider radiation with only a small angular spread, and are not concerned with polarization aspects, the scalar theory is satisfactory for developing most of the fundamental concepts of classical coherence theory. For a vector formulation the reader is referred to MANDEL and WOLF [1965].

We introduce a complex, analytic function (the analytic signal), $V(\boldsymbol{r}, t)$, associated with $V^{\mathrm{r}}(\boldsymbol{r}, t)$, in the following manner. Suppose that $V^{\mathrm{r}}(t)$ (for the time being we omit the space variable), has a Fourier transform, $\tilde{V}^{\mathrm{r}}(\nu)$ then

$$V^{\mathrm{r}}(t) = \int_{-\infty}^{+\infty} \tilde{V}^{\mathrm{r}}(\nu) \cdot \exp(-2\pi i \nu t) \mathrm{d}\nu \qquad (10.22)$$

and from the inversion theorem

$$\tilde{V}^{\mathrm{r}}(\nu) = \int_{-\infty}^{+\infty} V^{\mathrm{r}}(t) \exp(2\pi i \nu t) \mathrm{d}t. \qquad (10.23)$$

If we replace ν by $(-\nu)$ in this last equation, then

$$\tilde{V}^{\mathrm{r}}(-\nu) = \int_{-\infty}^{+\infty} V^{\mathrm{r}}(t) \exp(-2\pi i \nu t) \mathrm{d}\nu = [\tilde{V}^{\mathrm{r}}(\nu)]^*, \qquad (10.24)$$

where the asterisk denotes the complex conjugate. This shows that the negative frequency components of $\tilde{V}^{\mathrm{r}}(\nu)$ do not carry any information in addition to that contained in the positive frequency components (i.e. all the "information" about a real function is carried in the positive, or negative, frequencies alone).

Consequently, we may omit the negative components of $\tilde{V}^{\mathrm{r}}(\nu)$ without loss of information about the field, and define the complex, analytic signal by

$$V(t) = \int_{0}^{\infty} \tilde{V}(\nu) \cdot \exp(-2\pi i \nu t) \mathrm{d}\nu, \qquad (10.25)$$

where

$$\tilde{V}(\nu) = 2\tilde{V}^{\mathrm{r}}(\nu). \qquad (10.26)$$

References on page 299

Suppose that $\tilde{V}(\nu)$ is of the form

$$a(\nu)\exp[\mathrm{i}\Phi(\nu)], \tag{10.27}$$

where $a(\nu)$ and $\Phi(\nu)$ are real functions. Then if we replace ν by $(-\nu)$ in (10.22)

$$V^{\mathrm{r}}(t) = \int_{-\infty}^{+\infty} [\tilde{V}^{\mathrm{r}}(\nu)]^{*}\exp(2\pi\mathrm{i}\nu t)\cdot \mathrm{d}\nu \tag{10.28}$$

and add together this and the original equation, we find

$$V^{\mathrm{r}}(t) = \int_{0}^{\infty} a(\nu)\cos[-2\pi\nu t+\Phi(\nu)]\cdot \mathrm{d}\nu. \tag{10.29}$$

This demonstrates the usefulness of the analytic signal representation. The equation above describes the real field function as a superposition of different frequency components (real) which have different amplitudes and phases, and is the most general representation of a polychromatic radiation field. The analytic signal representation enables a complex exponential function to be associated with the real cosine function, thereby extending a useful technique normally associated with monochromatic radiation to polychromatic radiation (in the case of monochromatic radiation, $a_0\exp(2\pi\mathrm{i}\nu t)$ is the analytic signal representation of the real signal $a_0\cos 2\pi\nu t$). Also the problem of negative frequencies (which carry redundant information) is avoided. As will be shown later, the analytic signal is particularly convenient for quasi-monochromatic radiation.

There is a simple relation between the analytic signal and the real signal. If we further define another real function

$$V^{\mathrm{i}}(t) = \int_{0}^{\infty} a(\nu)\sin[-2\pi\nu t+\Phi(\nu)]\mathrm{d}\nu, \tag{10.30}$$

then it can be seen by combining (10.29) and (10.30) that

$$V(t) = V^{\mathrm{r}}(t)+\mathrm{i}V^{\mathrm{i}}(t). \tag{10.31}$$

The real function is therefore the real part of the analytic signal. In order to derive the analytic signal from a knowledge of the real signal, we use the following transform

$$V(t) = V^{\mathrm{r}}(t)+\mathrm{i}H[V^{\mathrm{r}}(t)], \tag{10.32}$$

where $H[V^{\mathrm{r}}(t)]$ is the Hilbert transform of $V^{\mathrm{r}}(t)$

$$V^{\mathrm{i}}(t) = H[V^{\mathrm{r}}(t)] = \frac{1}{\pi}P\int_{-\infty}^{+\infty}\frac{V^{\mathrm{r}}(t')\mathrm{d}t'}{(t'-t)}, \tag{10.33}$$

where P denotes the Cauchy principle value of the integral at $t' = t$ (see appendix A for a discussion of Hilbert transforms).

The inversion theorem for the analytic signal is

$$\tilde{V}(\nu) = \int_{-\infty}^{+\infty} V(t) \exp(2\pi i \nu t) dt \qquad \text{for} \quad \nu > 0,$$
$$= 0 \qquad \text{for} \quad \nu < 0. \tag{10.34}$$

By using (10.25), (10.29), (10.30) and Parseval's theorem, it can be shown that

$$\int_{-\infty}^{+\infty} [V^r(t)]^2 dt = \int_{-\infty}^{+\infty} [V^i(t)]^2 dt = \frac{1}{2}\int_{-\infty}^{+\infty} V(t) \cdot V^*(t) dt = \frac{1}{2}\int_0^{\infty} |\tilde{V}(\nu)|^2 d\nu. \tag{10.35}$$

The analytic signal has a particularly convenient form for quasi-monochromatic radiation ($\Delta\nu \ll \bar{\nu}$). Suppose we write the analytic signal as

$$V(t) = a(t) \cdot \exp[-2\pi i \bar{\nu} t + i\Phi(t)], \tag{10.36}$$

where $a(t)$ and $\Phi(t)$ are real. The associated real signal is

$$V^r(t) = a(t) \cos[-2\pi \bar{\nu} t + \Phi(t)]. \tag{10.37}$$

Using (10.25), we have

$$a(t) \cdot \exp[i\Phi(t)] = \int_0^{\infty} \tilde{V}(\nu) \cdot \exp[-2\pi i(\nu - \bar{\nu})t] d\nu. \tag{10.38}$$

Replace $(\nu - \bar{\nu})$ by μ and let $g(\mu) = \tilde{V}(\mu + \bar{\nu})$, then

$$a(t) \cdot \exp[i\Phi(t)] = \int_{-\bar{\nu}}^{\infty} g(\mu) \cdot \exp(-2\pi i \mu t) \cdot d\mu. \tag{10.39}$$

The condition for quasi-monochromaticity is that $g(\mu)$ is only significant for values of $|\mu| \leqslant \Delta\nu$, where $\Delta\nu \ll \bar{\nu}$. In this case, it is apparent that the expression on the left hand side of the above equation is made up from components with frequencies very much less than the mean frequency of the field ($\bar{\nu}$). The function, $a(t)$, can be regarded, therefore, as an amplitude envelope, which varies only slowly over times comparable with $\bar{\nu}^{-1}$ and which modulates a wave of frequency $\bar{\nu}$. When $a(t)$ contains frequency components comparable with $\bar{\nu}$, the envelope concept is no longer a useful one, although the relationships (10.36) and (10.39) are still valid. The phase function, $\Phi(t)$, is also a slowly varying function of time in this approx-

References on page 299

imation. The following relationships enable the real signal to be derived from the analytic signal

$$
\begin{aligned}
a(t) &= \sqrt{(VV^*)} = |V|, \\
\Phi(t) &= 2\pi\bar{\nu}t + \tan^{-1}\left(\mathrm{i}\cdot\frac{V^*-V}{V^*+V}\right).
\end{aligned}
\tag{10.40}
$$

Over times comparable with $\bar{\nu}^{-1}$, we have shown that $a(t)$ and $\Phi(t)$ change only slowly. However, it is apparent that they will start to change appreciably over times comparable to $\Delta\nu^{-1}$ since they contain components with frequencies of the order of $\Delta\nu$. We have associated with $\Delta\nu$ a coherence time, τ_c, where

$$\tau_c \sim \Delta\nu^{-1},$$

and therefore the coherence time represents an upper limit to the interval over which the amplitude and phase of a quasi-monochromatic wave are effectively constant.

Before going on to consider classical coherence theory in terms of the analytic signal, it is necessary to consider more carefully the validity of Fourier analysis in this context. It has been assumed in (10.34) that the function $V^r(t)$ is defined for all values of t, whereas, in fact, it will only be finite (or will only be observed) for some time interval $-T \leqslant t \leqslant T$. In most cases T is usually large compared to the physically significant times such as $\bar{\nu}^{-1}$ and τ_c, and so it is convenient to take the limit $T \to \infty$. When this limit is taken, however, quantities such as the time average of the intensity over the interval must also approach a finite limit. In other words,

$$\lim_{T\to\infty} \cdot \frac{1}{2T}\int_{-T}^{+T} [V^r(t)]^2\,\mathrm{d}t \tag{10.41}$$

must be finite, when it is apparent that $\int_{-\infty}^{+\infty}[V^r(t)]^2\mathrm{d}t$ diverges. However, a condition for a function to be Fourier analysable is that it be square integrable. In order to overcome this difficulty, and to be able to use the techniques of Fourier analysis, we define a truncated function

$$
\begin{aligned}
V_T^r(t) &= V^r(t) \qquad \text{for} \quad |t| \leqslant T, \\
&= 0 \qquad\quad\ \ \text{for} \quad |t| > T,
\end{aligned}
\tag{10.42}
$$

which has a Fourier integral

$$V_T^r(t) = \int_{-\infty}^{+\infty} \tilde{V}_T(\nu)\cdot\exp(-2\pi\mathrm{i}\nu t)\mathrm{d}\nu. \tag{10.43}$$

Similarly, we introduce a truncated analytic signal

$$V_T(t) = \int_0^{+\infty} \tilde{V}_T(\nu) \exp(-2\pi i\nu t) d\nu, \tag{10.44}$$

where $\tilde{V}_T(\nu) = 2\tilde{V}_T^r(\nu)$ as before.

Relationships similar to (10.35) hold for the truncated function

$$\frac{1}{2T}\int_{-\infty}^{+\infty} [V_T^r(t)]^2 dt = \frac{1}{4T}\int_{-\infty}^{+\infty} V_T(t)V_T^*(t) dt = \frac{1}{2}\int_0^{\infty} G_T(\nu) \cdot d\nu, \tag{10.45}$$

where

$$G_T(\nu) = \frac{\tilde{V}_T(\nu)\tilde{V}_T^*(\nu)}{2T}. \tag{10.46}$$

The limit $T \to \infty$ is taken as the final step. In many cases of interest, it is found that when this limit is taken the function $G_T(\nu)$ does not tend to a definite value, but oscillates. To overcome this further difficulty, a "smoothing" process is applied to $G_T(\nu)$, which has its basis in the essentially statistical nature of the radiation field. We leave further consideration of this process until we have introduced the statistical ideas that lie at the root of classical coherence theory. This we do in the next section. In subsequent sections, we will not always work explicitly in terms of the truncated function, but the corresponding truncated function expressions should be readily derivable from those given.

4. The stochastic description of the classical radiation field

Since classical coherence theory was originally formulated to describe light from thermal sources, we stress in this section the differences between such light and that from the laser. In this way the limitations of some of the procedures adopted in the description of thermal light, will become apparent. We leave to a later section a discussion on how the conventional theory is modified so as to attempt a description of the different statistical nature of laser light.

Now, light from thermal sources can be regarded as being made up from a sequence of random wavetrains (different wave trains being emitted by the different independent radiators that make up the source). Therefore, since $V^r(t)$ is the superposition of a large number of Fourier components that are independent of each other, it is a fluctuating function of time (Einstein [1915], Root and Pitcher [1955], Janossy [1957, 1959]), and is represented by a Gaussian random time function of zero mean (MANDEL [1963]). (For the time being we consider only the real function, but will generalize to the analytic signal later.)

In the case of the light from a laser, the Fourier components will not be completely independent, since the atomic sources of the field are coupled to one another (stimulated emission), but there will be some random fluctuations because of the presence of spontaneous emission. In § 12 we develop this further when we consider laser light as the superposition of harmonic components (modes) of fixed amplitudes and random phases, with Gaussian components representing spontaneous emission (and thermal and mechanical instabilities).

In general, therefore, the wave amplitude $V^{\mathrm{r}}(t)$ will fluctuate in a random fashion which is never completely predictable. There are two ways in which we can arrive at an average value for $V^{\mathrm{r}}(t)$. We can consider it as a typical member of an ensemble consisting of all possible values of the field ampli-de. Each member possesses an associated probability density

$$p[V^{\mathrm{r}}(t)] \cdot \mathrm{d}[V^{\mathrm{r}}(t)], \tag{10.47}$$

which is the probability that at time, t, the field amplitude (at a particular point) has a value lying between $V^{\mathrm{r}}(t)$ and $\{V^{\mathrm{r}}(t)+\mathrm{d}[V^{\mathrm{r}}(t)]\}$. For thermal radiation this probability density is of Gaussian form. The statistical (ensemble) average of $V^{\mathrm{r}}(t)$ is

$$\langle V^{\mathrm{r}}(t)\rangle_{\mathrm{e}} = \int_{-\infty}^{+\infty} V^{\mathrm{r}}(t) \cdot p[V^{\mathrm{r}}(t)] \cdot \mathrm{d}[V^{\mathrm{r}}(t)]. \tag{10.48}$$

Alternatively, we can take the time average of $V^{\mathrm{r}}(t)$

$$\langle V^{\mathrm{r}}(t)\rangle_{\mathrm{t}} = \lim_{T\to\infty} \cdot \frac{1}{2T}\int_{-T}^{+T} V^{\mathrm{r}}(t) \cdot \mathrm{d}t. \tag{10.49}$$

In the case of thermal light, the probability density is independent of the choice of the origin of time (such fields are called stationary fields), and if we assume ergodicity, then

$$\langle V^{\mathrm{r}}(t)\rangle_{\mathrm{e}} = \langle V^{\mathrm{r}}(t)\rangle_{\mathrm{t}}. \tag{10.50}$$

It is possible to generalize these statistical considerations, and define joint probability densities, such that

$$p[V_1(t_1), V_2(t_2), \cdots V_n(t_n)] \cdot \mathrm{d}[V_1(t_1)] \cdot \mathrm{d}[V_2(t_2)] \cdots \mathrm{d}[V_n(t_n)] \tag{10.51}$$

is the probability that the field at point $\boldsymbol{r}_1$ and time t_1 has a value between $V_1(t_1)$ and $\{V_1(t_1)+\mathrm{d}[V_1(t_1)]\}$, at point $\boldsymbol{r}_2$ and time t_2 has a value between $V_2(t_2)$ and $\{V_2(t_2)+\mathrm{d}[V_2(t_2)]\}$, and so on. If $F[V_1(t_1), V_2(t_2) \ldots, V_n(t_n)]$ is some function of the fields at these different space-time points, we can introduce average values of this function in the same way as before

$$\langle F\rangle_{\mathrm{e}} = \int_{-\infty}^{+\infty} \cdots \int_{-\infty}^{+\infty} F[V_1(t_1) \cdots V_n(t_n)] \cdot p[V_1(t_1) \cdots V_n(t_n)] \cdot \cdot \mathrm{d}[V_1(t_1) \cdots V_n(t_n)], \tag{10.52}$$

$$\langle F\rangle_{\mathrm{t}} = \lim_{T\to\infty} \int_{-T}^{+T} \frac{F[V_1(t_1) \cdots V_n(t_n)]}{2T} \cdot \mathrm{d}t. \tag{10.53}$$

5. Second order coherence – the mutual coherence function

From the preliminary discussion on the relation between the coherence properties of light and the formation of interference fringes in different forms of interferometer, it can be seen that, traditionally, coherence is related to the ability of two light beams to give rise to interference fringes on superposition. The two light beams might originate from a single space point at two different times (Michelson Interferometer), from two space points at the same time (symmetrical position in Young's slits experiment), or from different space and time points together (general position in Young's slits experiment). Coherence effects of this kind are now classified as being of second order, since they are described by second order probability densities†.

Both spatial and temporal, second order coherence effects are characterized by a single function, the mutual coherence function due to WOLF [1955], it being found that, except in special cases, the two effects are not independent. If the analytic signal representing the light disturbance at a point $\boldsymbol{r}_1$ at time $(t+\tau)$ is $V_1(t+\tau)$, and at point $\boldsymbol{r}_2$ and time t is $V_2(t)$, then the mutual coherence function is defined by

$$\Gamma_{12}(\tau) = \langle V_1(t+\tau)V_2^*(t)\rangle_{\mathrm{t}} = = \lim_{T\to\infty} \cdot \frac{1}{2T} \cdot \int_{-T}^{+T} V_1(t+\tau)V_2^*(t)\mathrm{d}t. \tag{10.54}$$

† We adopt here the classification due to WOLF [1963]. Some authors designate these coherence effects as being of first order.

Assuming that the system is ergodic and stationary, we also have the equivalent definition

$$\Gamma_{12}(\tau) = \langle V_1(t+\tau)V_2^*(t)\rangle_e =$$
$$= \int_{-\infty}^{+\infty}\int_{-\infty}^{+\infty} V_1(t+\tau)V_2^*(t)\cdot p[V_1(t+\tau), V_2(t)]\cdot \mathrm{d}[V_1(t+\tau)]\cdot \mathrm{d}[V_2(t)], \tag{10.55}$$

which involves a second order probability density.

Associated with the mutual coherence function there is a normalized function, known as the "complex degree of coherence" or "complex coherence factor", which is defined by

$$\gamma_{12}(\tau) = \frac{\Gamma_{12}(\tau)}{\sqrt{[\Gamma_{11}(0)]}\cdot\sqrt{[\Gamma_{22}(0)]}}\,. \tag{10.56}$$

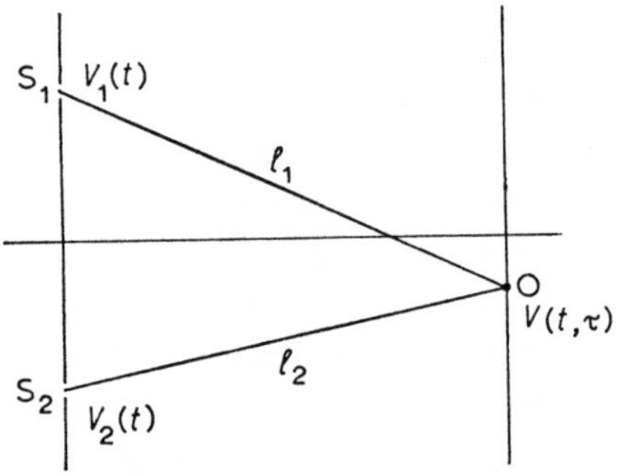

Fig. 10.4. Young's slits experiment—general arrangement.

In order to demonstrate the importance of the mutual coherence function, we consider its operational definition in terms of Young's slits experiment, and for the case of quasimonochromatic radiation, relate the values that it can assume to the appearance of the observed fringes. We now consider the fringes at a general point, O, on the observation screen (fig. 10.4). Let the field at one slit (S_1) be $V_1(t)$, and at the other slit (S_2) be $V_2(t)$. The instantaneous field at the point, O, on the observation screen, due to the radiation from the two slits is then

$$V(t, \tau) = K_1 V_1(t-t_1)+K_2 V_2(t-t_2), \tag{10.57}$$

where $t_1 = l_1/c$, $t_2 = l_2/c$, $\tau = (t_2-t_1)$, and K_1 and K_2 depend on the geometry.

The instantaneous intensity is taken as $V(t, \tau)V^*(t, \tau)$. This is not always strictly proportional to the square of the real signal. (However, it can readily be seen that, for quasi-monochromatic radiation, the instantaneous intensity is twice the average, over a few mean periods, of the square of the real signal. This is, of course, what is seen at optical frequencies by any feasible square law detector. When the instantaneous intensity is time averaged, (10.35) shows its equivalence to the time average of the square of the real signal.) We have, therefore, that

$$I(t,\tau) = V(t,\tau)V^*(t,\tau) =$$
$$= |K_1|^2 I_1(t-t_1)+|K_2|^2 I_2(t-t_2)+2\,\mathcal{R}e\,[K_1 K_2^* V_1(t-t_1)V_2(t-t_2)]. \tag{10.58}$$

The time average of the instantaneous intensity at the point, O, is

$$I(\tau) = |K_1|^2 \cdot I_1+|K_2|^2 \cdot I_2+2\,\mathcal{R}e\,[K_1 K_2^* \Gamma_{12}(\tau)], \tag{10.59}$$

and the intensities can now be associated with the square of the real signal. In the above time average, we have assumed that the field is stationary.

The first two terms in (10.59) are just the intensities at the observation point due to the two slits separately, and it is therefore only when $|\Gamma_{12}(\tau)|$ is non-zero that the superposition of the beams gives rise to interference effects. The modulus of the complex coherence factor γ lies between 0 and 1; the former value representing what is conventionally understood as incoherence, and the latter value complete coherence. Intermediate values designate conditions of partial coherence. Temporal and spatial coherence are described by $|\gamma_{11}(\tau)|$ and $|\gamma_{12}(0)|$ respectively.

In arriving at (10.59), we have time averaged the instantaneous values over all time. This is the signal seen by a detector with a response time long compared to the mean period and coherence time of the radiation. If the response time is short compared to these times, then transient interference effects may be observed (§ 6).

We investigate further the form of $\Gamma_{12}(\tau)$ by expanding the analytic signals into their frequency components. Here it is appropriate to work explicitly in terms of the truncated functions, so that

$$\Gamma_{12}(\tau) = \lim_{T\to\infty} \frac{1}{2T}\left\{\int_{-\infty}^{+\infty} V_{2T}^*(t)\left[\int_0^{\infty} \tilde{V}_{1T}(\nu)\cdot \exp\left(-2\pi i\nu\overline{t+\tau}\right)d\nu\right] dt\right\}. \tag{10.60}$$

Interchanging the order of integration, we then obtain

$$\Gamma_{12}(\tau) = \lim_{T\to\infty} \frac{1}{2T}\int_0^{\infty} \tilde{V}_{1T}(\nu)\cdot \tilde{V}_{2T}^*(\nu)\cdot \exp\left(-2\pi i\nu\tau\right)d\nu. \tag{10.61}$$

References on page 299

We have already pointed out (§ 3) that

$$\lim_{T\to\infty}\left[\frac{\tilde{V}_{1T}(\nu)\cdot\tilde{V}^*_{2T}(\nu)}{2T}\right]$$

does not always tend to a definite value and may oscillate. To avoid this we carry out a smoothing procedure, such as taking the ensemble average over the random functions $V_1(t)$ and $V_2(t)$ before taking the limit. In this case we can write

$$\Gamma_{12}(\tau)=\int_0^\infty G_{12}(\nu)\cdot\exp\left(-2\pi i\nu\tau\right)\cdot d\nu, \tag{10.62}$$

where

$$G_{12}(\nu)=\lim_{T\to\infty}\left[\frac{\langle\tilde{V}_{1T}(\nu)\cdot\tilde{V}^*_{2T}(\nu)\rangle_e}{2T}\right]. \tag{10.63}$$

In particular

$$\Gamma_{11}(\tau)=\int_0^\infty G(\nu)\cdot\exp\left(-2\pi i\nu\tau\right)\cdot d\nu, \tag{10.64}$$

where

$$G(\nu)=\lim_{T\to\infty}\left[\frac{\langle\tilde{V}_{T}(\nu)\tilde{V}^*_{T}(\nu)\rangle_e}{2T}\right], \tag{10.65}$$

which is the function we have already introduced in § 3.

In the terminology of stochastic processes, $\Gamma_{11}(\tau)$ is known as the auto-correlation function of $V_1(t)$, $\Gamma_{12}(\tau)$ as the cross-correlation function of $V_1(t)$ and $V_2(t)$, $G(\nu)$ as the power spectrum (spectral density) of $V_1(t)$, and $G_{12}(\nu)$ as the cross (or mutual) spectral density of $V_1(t)$ and $V_2(t)$. If the analytic signal, $V_1(t)$, is identified with the electric field, then $G(\nu)$ is the electric energy spectrum of the radiation.

From the preceeding equations, it can be seen that $\Gamma_{11}(\tau)$ and $G(\nu)$, and $\Gamma_{12}(\tau)$ and $G_{12}(\nu)$ form Fourier transform pairs. Instead of using smoothing procedures to define $G(\nu)$ and $G_{12}(\nu)$ they can therefore be defined as the Fourier inverses of $\Gamma_{11}(\tau)$ and $\Gamma_{12}(\tau)$ respectively:

$$\begin{aligned}G_{12}(\nu)&=\int_{-\infty}^{+\infty}\Gamma_{12}(\tau)\exp\left(2\pi i\nu\tau\right)d\tau &&\text{for}\quad \nu>0,\\&=0 &&\text{for}\quad \nu<0;\end{aligned} \tag{10.66}$$

$$\begin{aligned}G(\nu)&=\int_{-\infty}^{+\infty}\Gamma_{11}(\tau)\exp\left(2\pi i\nu\tau\right)d\tau &&\text{for}\quad \nu>0,\\&=0 &&\text{for}\quad \nu<0.\end{aligned} \tag{10.67}$$

Expressions (10.62) and (10.66) with the definitions for $G_{12}(\nu)$ and $\Gamma_{12}(\tau)$ are the Wiener-Khintchine theorem.

If we examine (10.62) more closely, we can see that the mutual coherence function takes a particularly simple form for quasi-monochromatic radiation, since in this case $G_{12}(\nu)$ is only appreciable over a narrow frequency interval, $\Delta\nu$, centered on $\bar{\nu}$ (where $\Delta\nu \ll \bar{\nu}$). The exponential factor, therefore, can be taken outside the integral, so that in general

$$\Gamma_{12}(\tau_2) = \Gamma_{12}(\tau_1) \cdot \exp\left[2\pi i\bar{\nu}(\tau_2-\tau_1)\right] \qquad \text{for} \quad |(\tau_2-\tau_1)| \ll \tau_c. \tag{10.68}$$

Similarly for the complex degree of coherence

$$\gamma_{12}(\tau_2) = \gamma_{12}(\tau_1) \cdot \exp\left[2\pi i\bar{\nu}(\tau_2-\tau_1)\right] \qquad \text{for} \quad |(\tau_2-\tau_1)| \ll \tau_c. \tag{10.69}$$

We now give the complex degree of coherence an operational significance by relating it to the visibility of the fringes that are observed. We write (10.59) in the form

$$I(\tau) = I_1^{(0)}+I_2^{(0)}+2[I_1^{(0)}]^{\frac{1}{2}}[I_2^{(0)}]^{\frac{1}{2}}\, \mathcal{R}e\, [\gamma_{12}(\tau)], \tag{10.70}$$

where we have combined $|K_1|^2$ with I_1 so that $I_1^{(0)}$ is the intensity at the point, O, due to the source at S_1, etc. Let us now write $\gamma_{12}(\tau)$ in the form

$$\gamma_{12}(\tau) = |\gamma_{12}(\tau)| \cdot \exp\left[i\alpha_{12}(\tau) - 2\pi i\bar{\nu}\tau\right], \tag{10.71}$$

where

$$\alpha_{12}(\tau) = \arg\left[\gamma_{12}(\tau)\right] + 2\pi\bar{\nu}\tau, \tag{10.72}$$

and $\bar{\nu}$ is the effective frequency of the light.

Substitution into (10.70) gives

$$I(\tau) = I_1^{(0)}+I_2^{(0)}+2[I_1^{(0)}]^{\frac{1}{2}}[I_2^{(0)}]^{\frac{1}{2}} \cdot |\gamma_{12}(\tau)| \cdot \cos\left[\alpha_{12}(\tau)-2\pi\bar{\nu}\tau\right]. \tag{10.73}$$

Over a small region about the observation point, O, $I_1^{(0)}$ and $I_2^{(0)}$ will change only slowly. As was shown in the previous section, this also applies to $|\gamma_{12}(\tau)|$ and $\alpha_{12}(\tau)$ provided that the light is quasi-monochromatic and that changes in τ resulting from changes in the position of O are small compared to the coherence time, τ_c. However, the cosine term in (10.73) will change rapidly because of the frequency term involving $\bar{\nu}$. Therefore the maximum and minimum intensities around the point, O are

$$\begin{aligned} I(\tau)_{\max} &= I_1^{(0)}+I_2^{(0)}+2[I_1^{(0)}]^{\frac{1}{2}}[I_2^{(0)}]^{\frac{1}{2}} \cdot |\gamma_{12}(\tau)|, \\ I(\tau)_{\min} &= I_1^{(0)}+I_2^{(0)}-2[I_1^{(0)}]^{\frac{1}{2}}[I_2^{(0)}]^{\frac{1}{2}} \cdot |\gamma_{12}(\tau)|. \end{aligned} \tag{10.74}$$

References on page 299

The visibility of the fringes is defined by

$$\mathscr{V} = \frac{I(\tau)_{\max} - I(\tau)_{\min}}{I(\tau)_{\max} + I(\tau)_{\min}} \tag{10.75}$$

and is, therefore

$$\mathscr{V} = \frac{2[I_1^{(0)}]^{\frac{1}{2}}[I_2^{(0)}]^{\frac{1}{2}}}{[I_1^{(0)} + I_2^{(0)}]} \cdot |\gamma_{12}(\tau)|. \tag{10.76}$$

The modulus of the complex coherence function, determines the fringe visibility. When the intensities at the observation point due to each source separately, are equal, then the visibility is just $|\gamma_{12}(\tau)|$. For completely coherent light, the maximum value of $|\gamma_{12}(\tau)|$, with respect to τ, is unity. The argument of $\gamma_{12}(\tau)$ can readily be given an operational significance in terms of the location of the intensity maxima in the fringe pattern, but we will not consider this further here.

6. Transient coherence

Suppose we consider the form of the mutual coherence function for two monochromatic sources of different frequencies ν_1 and ν_2 then

$$\Gamma_{12}(\tau) = \lim_{T\to\infty} \cdot \frac{1}{2T} \int_{-T}^{+T} a_1 a_2^* \exp[2\pi i(\nu_1 - \nu_2)t] \cdot \exp[2\pi i \nu_1 \tau] dt. \tag{10.77}$$

When the limit is taken, the value of $\Gamma_{12}(\tau)$ tends to zero. However, it can be argued that a definite phase relationship exists between the two sources, and that, therefore, they are coherent, although of different frequency. The reason the mutual coherence function vanishes is that, as defined above, it is a measure of only the linear dependence between two variables, and although the phases are linearly related, this is not the case for the corresponding exponential terms. If (10.77) is integrated over a finite time interval

$$\Gamma_{12}(\tau, T, t) = \frac{1}{2T} \int_{t-T}^{t+T} V_1(t'+\tau) V_2^*(t') dt', \tag{10.78}$$

then for the monochromatic sources

$$\Gamma_{12}(\tau, T, t) = \frac{\sin[2\pi(\nu_2 - \nu_1)T]}{2\pi(\nu_2 - \nu_1)T} \cdot \exp[2\pi i \nu_1 \tau] \cdot \exp[2\pi i(\nu_2 - \nu_1)t]. \tag{10.79}$$

If the observation time, T, is less than the reciprocal of the frequency separation, $\Gamma_{12}(\tau, T, t)$ is finite, and transient interference effects may be observed. If, instead of monochromatic, we had considered quasi-monochromatic sources, then there would be the additional requirement that the observation time be less than the reciprocal of the bandwidth.

With the advent of gas lasers such effects have become observable, since it is possible to arrange two separate sources with bandwidths of the order of a couple of hundred cycles, and mean frequencies about a kilocycle apart. Transient interference effects can then be observed for times of the order of a millisecond (it is possible to detect such effects not only photoelectrically, but photographically as well). With thermal sources, the bandwidth limitation by itself usually makes it difficult to observe transient interference effects, since very fast detectors are required.

7. Propagation of the mutual coherence function

It has been demonstrated by Wolf [1955] that the mutual coherence function in vacuo obeys two wave equations of the form

$$\nabla_i^2 \Gamma = \frac{1}{c^2} \frac{\partial^2 \Gamma}{\partial \tau^2} \qquad (i = 1, 2). \tag{10.80}$$

where the subscript i denotes differentiation with respect to the coordinates of one of the space points ($\boldsymbol{r}_1$ or $\boldsymbol{r}_2$). Using this equation it is possible to follow the behaviour of the mutual coherence function throughout an optical field. The propagation of the mutual coherence function explains why it is not generally possible to treat temporal and spatial coherence independently, since one type of coherence can be transformed into the other through the propagation of the field. Also, spatially incoherent light may become partially coherent through the process of propagation. One example of this is the way in which starlight, originating from a collection of independent radiators making up the star and therefore initially spatially incoherent, becomes spatially coherent in propagating to the Earth, forming sharp diffraction fringes in a telescope. (This is illustrated by the expression for coherence volume derived in § 2, where it can be seen that the coherence volume expands with distance of propagation.) In § 9 we consider how a passive laser cavity imparts complete spatial coherence to quasi-monochromatic light, initially completely spatially incoherent, through the processes of propagation and diffraction alone.

In order to do this we need to make use of a particularly convenient expression derived from (10.80) by Parrent [1959] which describes the

References on page 299

propagation of the mutual coherence function from a finite plane source (σ) and which is therefore useful for dealing with problems involving plane apertures

$$\Gamma_\rho(P_1, P_2, \tau) = \frac{1}{(2\pi)^2}\int_\sigma\int \frac{\cos\theta_1 \cdot \cos\theta_2}{R_1^2 \cdot R_2^2} \cdot$$

$$\cdot \mathscr{D}\Gamma_\sigma\left(S_1, S_2, \tau - \frac{R_1 - R_2}{c}\right) \cdot \mathrm{d}S_1 \cdot \mathrm{d}S_2, \qquad (10.81)$$

where

$$\mathscr{D} = 1 + \frac{(R_1 - R_2)}{c} \cdot \frac{\partial}{\partial\tau} - \frac{R_1 R_2}{c^2}\frac{\partial^2}{\partial\tau^2}. \qquad (10.82)$$

The points S_1 and S_2 are two points on the plane source (σ), while P_1 and P_2 are two points in the wavefield (ρ). The other quantities are defined in fig. 10.5.

If σ is a thermal source, then we can assume that the radiations emitted from different elements of the source are completely incoherent, in other words

$$\Gamma_\sigma(S_1, S_2, \tau) \sim \delta(S_1 - S_2) \cdot f(\tau, S_1). \qquad (10.83)$$

If we further assume that the radiation is quasi-monochromatic, we can use (10.68), to obtain

$$\Gamma_\sigma(S_1, S_2, \tau) \sim I(S) \cdot \delta(S_1 - S_2) \cdot \exp(-2\pi i\bar{\nu}\tau) \qquad \text{for} \quad |\tau| \ll \tau_c, \qquad (10.84)$$

$I(S)$ being the averaged intensity per unit area at S.

If θ_1 and θ_2 are small and

$$|\tau - (R_1 - R_2)c^{-1}| \ll \tau_c,$$

then (10.81) can be simplified to

$$\Gamma_\rho(P_1, P_2, \tau) \sim \left(\frac{\bar{k}}{2\pi}\right)^2 \cdot \exp(-2\pi i\bar{\nu}\tau) \cdot \int_\sigma \frac{I(S)}{R_1 R_2} \cdot \exp[-i\bar{k}(R_1 - R_2)] \cdot \mathrm{d}S. \qquad (10.85)$$

This equation is similar in form to the Huygens-Fresnel formula used in the scalar theory of diffraction and is, in fact, a mathematical formulation of the Van Cittert-Zernike theorem (VAN CITTERT [1936], ZERNIKE [1938]). We shall use (10.85) in § 9 to demonstrate the development of spatial coherence within a laser cavity. Firstly, however, we need to consider other equivalent ways of defining coherence in a radiation field.

8. Coherent light

In considering radiation from a laser, we are particularly concerned with light that has a high degree of spatial and temporal coherence. In terms of the complex coherence function, completely coherent light must satisfy the condition that

$$\max_{\tau} \cdot |\gamma_{12}(\tau)| = 1 \tag{10.86}$$

for all pairs of points ($\boldsymbol{r}_1$ and $\boldsymbol{r}_2$) within the light beam. We take the maximum value of the modulus with respect to τ, since in general it is a function of τ, and it is this maximum value that determines the visibility of the fringes.

It can be shown that the above definition can also be written in two other equivalent forms:

(i) For any pair of points, $\boldsymbol{r}_1$ and $\boldsymbol{r}_2$, within a completely coherent radiation field, there exists a series of relative time delays, τ_n, such that

$$\frac{V_1(t+\tau_n)}{\sqrt{I_1}} - \frac{V_2(t)}{\sqrt{I_2}} = W(t, \tau_n), \tag{10.87}$$

where the square of the modulus of the difference function, $W(t, \tau_n)$ has a time average that approaches zero as the averaging interval is increased, i.e.

$$\lim_{T\to\infty} \frac{1}{2T} \int_{-T}^{+T} |W(t, \tau_n)|^2 \cdot \mathrm{d}t = 0. \tag{10.88}$$

The equivalence of this condition to condition (10.86) can be seen by multiplying both sides of (10.87) by their respective complex conjugates and then taking the limit of their time averages, as the averaging interval becomes infinitely large.

This condition is particularly useful in that it illustrates that for complete coherence, the disturbance need not necessarily be strictly periodic, nor need it have an exceedingly narrow spectral range. We shall consider this condition again later when we have obtained forms of the analytic signal describing laser radiation.

(ii) For completely coherent light, the mutual coherence function is of the form

$$\Gamma_{12}(\tau) = U_1 U_2^* \exp(-2\pi i\bar{\nu}\tau) \tag{10.89}$$

for a range of values of τ short compared to the coherence time τ_c. Here U_1 and U_2 are functions only of the coordinates of the points $\boldsymbol{r}_1$ and $\boldsymbol{r}_2$ respectively, and are independent of τ. We shall use this condition again

References on page 299

later in our consideration of the spatial coherence developed in a light beam by its propagation through a passive cavity. For a fuller discussion of this form of the mutual coherence function for completely coherent light see BERAN and PARRENT [1964].

9. The development of spatial coherence within a laser cavity

We now consider the manner in which quasi-monochromatic radiation, which is initially spatially incoherent, achieves complete spatial coherence after a sufficiently large number of transits within a passive cavity. The development of spatial coherence within a cavity can be studied by using a simplified form of (10.81) which describes the propagation of the mutual coherence function from a finite plane region (σ).

If the radiation is assumed to be quasi-monochromatic, then

$$\Gamma(S_1, S_2, \tau) \sim \Gamma(S_1, S_2, 0) \exp(-2\pi i\bar{\nu}\tau). \tag{10.90}$$

Substituting this expression into (10.85) and assuming that all dimensions are large compared to the mean wavelength of the radiation and that the angles θ_1 and θ_2 are small, we obtain

$$\Gamma_\rho(P_1, P_2) \sim \left(\frac{\bar{k}}{2\pi}\right)^2 \int_\sigma\int \Gamma_\sigma(S_1, S_1) \cdot \frac{1}{R_1 R_2} \cdot \exp[i\bar{k}(R_1 - R_2)] \cdot dS_1 \cdot dS_2, \tag{10.91}$$

where the various parameters have already been defined (fig. 10.5).

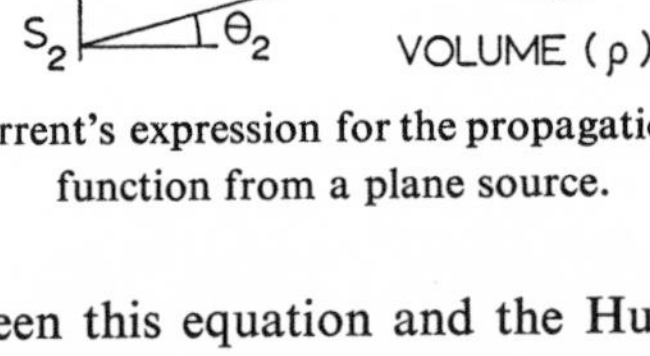

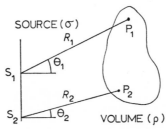

Fig. 10.5. Notation for Parrent's expression for the propagation of the mutual coherence function from a plane source.

The similarity between this equation and the Huygens-Fresnel equation used by Fox and Li to investigate the modes associated with laser cavities (ch. 6 § 1) suggests the method for obtaining steady-state solutions.

We consider the case of a plane parallel Fabry-Pérot cavity, and use the transmission line analogue (ch. 5 § 5) which replaces the cavity by a periodic structure of identical, equidistant, opaque screens, spaced by the separation of the cavity mirrors, and having apertures with the dimensions of the cavity mirrors (fig. 10.6). We now use (10.91) to follow the propagation of the mutual coherence function from one plane aperture to the next.

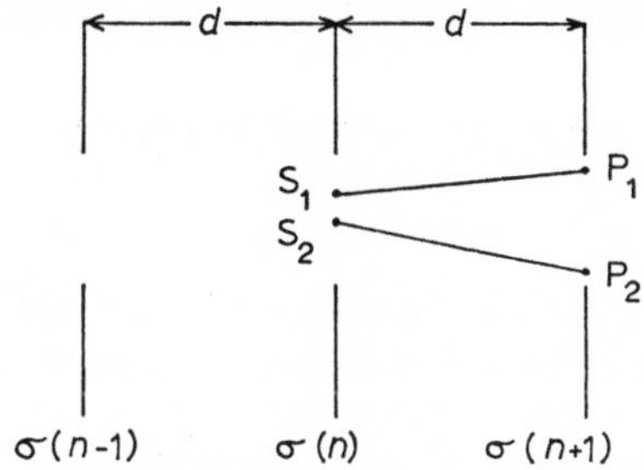

Fig. 10.6. Cavity analogue for propagation of mutual coherence function.

In the Fox and Li treatment, the light was assumed to be completely monochromatic, and necessarily completely coherent. The influence of propagation and diffraction on the coherence of the radiation, as it travelled through the cavity, was not, therefore, considered. In the present case we assume that the light is quasi-monochromatic, and that it is initially spatially incoherent.

The mutual coherence function after propagation to aperture $(n+1)$ is given in terms of the mutual coherence function at aperture (n) by

$$\Gamma_{(n+1)}(P_1, P_2) \sim \left(\frac{\bar{k}}{2\pi}\right)^2 \int_{\sigma(n)}\int \Gamma_{(n)}(S_1, S_2) \cdot \frac{1}{R_1 R_2} \exp[i\bar{k}(R_1 - R_2)] \mathrm{d}S_1 \mathrm{d}S_2. \tag{10.92}$$

We look for steady state solutions for the mutual coherence function after propagation through a large number of apertures, in the sense that for equivalent pairs of points on two adjacent apertures

$$\Gamma_{(n+1)}(P_1, P_2) = \left(\frac{1}{\alpha}\right) \cdot \Gamma_{(n)}(P_1, P_2), \tag{10.93}$$

where α is a constant (possibly complex) representing the attenuation, phase shift, etc. of the light as it travels from one aperture to the next. These solutions correspond to the Fox and Li solutions where the pattern of the wave disturbance is required to reproduce itself from one aperture to the next.

References on page 299

Substitution of (10.93) into (10.92) gives

$$\Gamma(P_1, P_2) = \alpha \cdot \left(\frac{\bar{k}}{2\pi}\right)^2 \int_\sigma \int \Gamma(S_1, S_2) \cdot \frac{1}{R_1 R_2} \exp\left[\mathrm{i}\bar{k}(R_1 - R_2)\right] \mathrm{d}S_1 \, \mathrm{d}S_2 \,. \tag{10.94}$$

It can be shown that the above equation has eigenfunctions and eigenvalues of the form

$$\Gamma_{ij}(P_1, P_2) = U_i(P_1) \cdot U_j^*(P_2), \tag{10.95}$$

$$\alpha_{ij} = \beta_i \beta_j^*,$$

where the U's and β's are the eigenfunctions and eigenvalues respectively of the following equation (which also appears in the Fox and Li treatment)

$$U(P) = \beta \int_\sigma U(S) \cdot \frac{\exp(\mathrm{i}\bar{k}R)}{R} \cdot \mathrm{d}S. \tag{10.96}$$

The solutions form a complete, orthogonal set, and so the initial mutual coherence function may be expanded in terms of these eigenfunctions

$$\Gamma_{(1)}(P_1, P_2) = \sum_{i,j} \lambda_{ij} U_i(P_1) \cdot U_j(P_2). \tag{10.97}$$

Substitution of (10.97) into the iterative equation (10.93) yields the following solution

$$\Gamma_{(n)}(P_1, P_2) = \sum_{i,j} \lambda_{ij} (\alpha_{ij})^{(1-n)} \cdot \Gamma_{ij}(P_1, P_2). \tag{10.98}$$

If α_{11} is taken as the lowest eigenvalue (i.e. the one whose modulus has the lowest value) then for a sufficiently large value of n,

$$\Gamma_{(n)}(P_1, P_2) \sim \lambda_{11} (\beta_1 \beta_1^*)^{1-n} \cdot U_1(P_1) \cdot U_1^*(P_2). \tag{10.99}$$

We therefore see that irrespective of the initial form of the mutual coherence function, after a sufficiently large number of transits of the radiation through the cavity, it factorizes into the product of two functions, each of which depends on the coordinates of only one of the points. We have already pointed out (§ 8) that this is a description of a completely coherent field, and so have shown that propagation and diffraction of radiation through a passive cavity imparts spatial coherence to the field. The high degree of spatial coherence found in laser radiation is a consequence of the cavity properties alone. The active medium, through the process of stimulated emission, maintains the radiation field against the losses, as the radiation propagates from one mirror to the other in the cavity, but does so without altering the spatial coherence of the field (neither destroying nor enhancing the coherence).

10. Higher order coherence effects for thermal radiation

So far in our treatment of coherence effects, we have used time averages in calculating functions such as the mutual coherence function. It was pointed out in § 4 that this method of averaging is equivalent to taking an ensemble average, provided that the system is stationary and ergodic. Time averaging is particularly appropriate when the radiation is composed of a finite number of periodic components (as in the case of quasi-monochromatic radiation) or can be represented in time as a stationary, random, time series. However, in this section, where we consider higher order coherence effects, we will see that it is often convenient to work partly in terms of ensemble averages.

To carry out an ensemble average, it is necessary to know the probability densities associated with the different states of the system. We have already pointed out that thermal radiation is described by a Gaussian random time function of zero mean, which we now give explicitly as

$$p\{V^{r}(t)\} = (\pi\langle I\rangle)^{-\frac{1}{2}} \exp\{-[V^{r}(t)]^{2}/\langle I\rangle\}, \qquad (10.100)$$

where $\langle I\rangle$ is the average value of the wave intensity associated with the analytic signal of $V^{r}(t)$. It has been stressed that this is not an appropriate description for laser radiation. In this section, therefore, we consider only thermal radiation, and will show that for such radiation, higher order coherence effects can be described in terms of the second order mutual coherence function. This comes about because the higher order probability densities are multivariate Gaussian distributions, and as such are completely characterized by their second order members. (This will become clearer when we consider a specific example later in this section.) In other words, all the statistical information about thermal radiation is contained within second order theory. Higher order coherence experiments do not yield any information in addition to that obtainable from second order experiments, although they may provide more convenient ways of measuring the mutual coherence function. In particular we consider the correlation functions for intensities, and for intensity fluctuations, since these are associated with well known forms of interferometer.

In § 13, we shall develop the statistical approach to laser radiation. We shall show that although the mutual coherence function is still adequate for a description of second order experiments (of the type we have already described), this is no longer the case for higher order experiments. In particular, we contrast the different results to be expected in the correlation of intensity fluctuations as observed by the Hanbury-Brown and Twiss interferometer.

We firstly investigate the correlation in the intensities of the light at two different space-time points in a thermal radiation field. We define the intensity correlation function in terms of the instantaneous intensities at each point, as

$$\langle I_1(t+\tau)I_2(t)\rangle_e = \langle V_1(t+\tau)V_1^*(t+\tau)V_2(t)V_2^*(t)\rangle_e. \tag{10.101}$$

Since $V^r(t)$ is a Gaussian random process, and $V^i(t)$ is obtained from it by a Hilbert transformation (which is a linear transformation), it follows that $V^i(t)$ is also a Gaussian random process with the same mean (zero). It can be shown (MANDEL [1963]) that the autocorrelation functions of $V^r(t)$ and $V^i(t)$ are equal, and that at the same instant of time $V^r(t)$ and $V^i(t)$ are uncorrelated (the so-called random phase approximation of radio theory). Expansion of (10.101) in terms of the real signal therefore gives

$$\begin{aligned}\langle I_1(t+\tau)I_2(t)\rangle_e &= \langle[V_1^r(t+\tau)]^2[V_2^r(t)]^2\rangle_e+\\ &+\langle[V_1^r(t+\tau)]^2[V_2^i(t)]^2\rangle_e+\langle[V_1^i(t+\tau)]^2[V_2^r(t)]^2\rangle_e+\\ &+\langle[V_1^i(t+\tau)]^2[V_2^i(t)]^2\rangle_e.\end{aligned} \tag{10.102}$$

In order to evaluate the various terms in (10.102) we require the bivariate Gaussian distribution for two, not necessarily independent, variates, x_1 and x_2, say. When the two variates both have zero mean, its normalized form is (see appendix O),

$$\begin{aligned}p(x_1, x_2) &= [2\pi\sigma_1\sigma_2(1-\rho^2)]^{-1}\cdot\\ &\cdot\exp\{-(1-\rho^2)^{-1}[x_1^2/(2\sigma_1^2)+x_2^2/(2\sigma_2^2)-\rho x_1x_2/(\sigma_1\sigma_2)]\},\end{aligned} \tag{10.103}$$

where

$$\sigma_1^2 = \langle x_1^2\rangle, \qquad \sigma_2^2 = \langle x_2^2\rangle$$

and

$$\rho = \langle x_1x_2\rangle/(\sigma_1\sigma_2).$$

The parameter ρ is a measure of the correlation between the two variates (in probability theory it is known as the "correlation coefficient") and it is apparent that this parameter is related to the mutual coherence function when the variates are associated with V^r (or V^i). Similarly, the variances are related to the mean intensities. Each of the terms in expression (10.102) leads to an integral of the form

$$\int_{-\infty}^{+\infty}\int_{-\infty}^{+\infty} x_1^2 x_2^2 p(x_1, x_2)\mathrm{d}x_1\,\mathrm{d}x_2 \tag{10.104}$$

which has a value

$$\sigma_1^2\sigma_2^2(1+2\rho^2).$$

If we now associate with the variates, x_1 and x_2 either V^r or V^i as appropriate, the various parts of (10.102) can be evaluated, giving an expression of the form

$$\langle[V_1^r(t+\tau)]^2[V_2^r(t)]^2\rangle_e = \tfrac{1}{4}\langle I_1\rangle\langle I_2\rangle+2[\langle V_1^r(t+\tau)V_2^r(t)\rangle]^2. \tag{10.105*}$$

The real correlation functions can be related to the mutual coherence function (appendix K) by the following expressions

$$\begin{aligned}\langle V_1^r(t+\tau)V_2^r(t)\rangle &= \langle V_1^i(t+\tau)V_2^i(t)\rangle = \tfrac{1}{2}\,\mathscr{Re}\,[\Gamma_{12}(\tau)],\\ \langle V_1^r(t+\tau)V_2^i(t)\rangle &= -\langle V_1^i(t+\tau)V_2^r(t)\rangle = -\tfrac{1}{2}\mathscr{I}[\Gamma_{12}(\tau)].\end{aligned} \tag{10.106}$$

Substitution from (10.105) and (10.106) into (10.102) gives as a final result

$$\langle I_1(t+\tau)I_2(t)\rangle = \langle I_1\rangle\langle I_2\rangle+|\Gamma_{12}(\tau)|^2. \tag{10.107}$$

We have, therefore, demonstrated that the mutual coherence function characterizes intensity correlations provided that the radiation is described by a Gaussian random process. If intensity fluctuations are defined by

$$\Delta I = I-\langle I\rangle, \tag{10.108}$$

it can readily be seen from expression (10.107) that the correlation in the intensity fluctuations is

$$\langle\Delta I_1(t+\tau)\Delta I_2(t)\rangle = |\Gamma_{12}(\tau)|^2. \tag{10.109}$$

In principle, therefore, by measuring correlations in intensities or intensity fluctuations, it is possible to evaluate the mutual coherence function without recourse to more conventional second-order interference experiments based on the observation of fringes. Measurement of correlations in intensity fluctuations is the basis of the Hanbury-Brown and Twiss interferometer. The interferometer was initially used to investigate intensity correlations in radio frequency radiation from astronomical sources (HANBURY-BROWN

* We no longer distinguish between ensemble and time averages, see § 4.

and TWISS [1957a]), but was later extended to the optical region of the spectrum (HANBURY-BROWN and TWISS [1957b]).

The optical arrangement for this experiment is illustrated in fig. 10.7. Two photomultiplier tubes, for which the electrical response is proportional to the intensity of the incident light, were used to detect the radiation from a filtered mercury arc after it had been split into two beams by a half-silvered mirror. One of the tubes could be moved across the beam, and, using the appropriate signal processing circuits, the correlations in the intensity fluctuations were measured for various positions. These were compared to the theoretical values for $|\Gamma_{12}(\tau)|^2$ derived from the geometry of the arrangement, the nature of the spectral source, etc. Reasonable agreement was found between theory and experiment, but the experimental results were

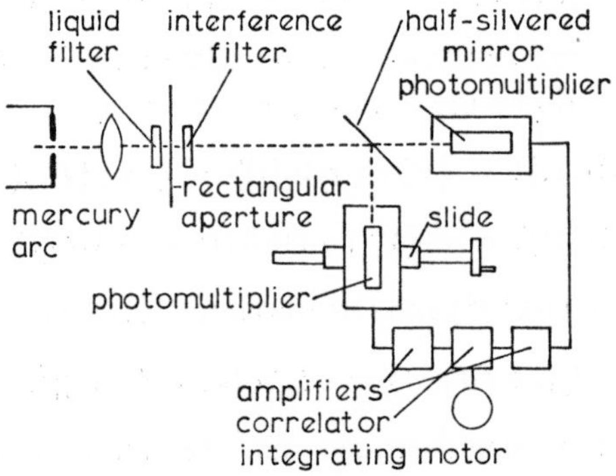

Fig. 10.7. Hanbury-Brown and Twiss experiment.

subject to large statistical errors. Two aspects that have not been considered in this simple theoretical treatment are: the intensity fluctuations normally involve frequency components that are too rapid to be followed by the electronic correlator, so that what is really being measured is not $\langle \Delta I_1(t+\tau)\Delta I_2(t)\rangle$ but the corresponding correlation function where the instantaneous intensities are replaced by their short-time averages; the quantum nature of the radiation field (photon noise) and fluctuations in the emission of photoelectrons in the photomultiplier (shot noise) both lead to additional fluctuations in the photo-currents going to the correlator.

11. Laser noise

In ch. 1 § 6 where the Schawlow-Townes expression describing the spectral linewidth of a (single mode) laser oscillator, above threshold, was first introduced, it was pointed out that the finite linewidth was a consequence of the intensity and phase fluctuations brought about by spontaneous emission. In ch. 4 this expression was derived using simple arguments relating to the quality factor of an active cavity. In the final sections of this chapter, we shall consider, in more detail, fluctuations in both amplitude and phase of a single-mode laser oscillator, and will relate such fluctuations to the coherence functions describing the classical field.

In the case of fluctuations associated with the laser oscillator, it is important to consider non-linear effects. For example, amplitude fluctuations are subject to the non-linear phenomenon of gain saturation in the active medium, which tends to minimize such fluctuations when the oscillator is operating well above threshold. In other words, in order to describe the noise in the output from a laser, it is necessary to examine in detail the processes occurring in the oscillator which modify the injected noise signal.

We begin by considering how the equation describing a damped, simple harmonic oscillator

$$\frac{\mathrm{d}^2E}{\mathrm{d}t^2}+\gamma\frac{\mathrm{d}E}{\mathrm{d}t}+\omega_0^2E=0 \tag{10.110}$$

may be modified in order to describe a laser oscillator. We have the following additional phenomena to take into consideration: the small signal gain of the active medium, describable by a coefficient α; and gain saturation effects which reduce the gain of the active medium under oscillation conditions, by an amount dependent on the intensity of the radiation field, and describable by $(-\beta E^2)$. Inclusion of these additional terms in (10.110) leads to

$$\frac{\mathrm{d}^2E}{\mathrm{d}t^2}+(\gamma-\alpha)\frac{\mathrm{d}E}{\mathrm{d}t}+\beta\left(E^2\frac{\mathrm{d}E}{\mathrm{d}t}\right)_{lf}+\omega_0^2E=0, \tag{10.111}$$

where γ describes the losses associated with the oscillator. The form of the saturation term in (10.111) has been justified by Lamb [1964], the bracket indicating that only low frequency terms (ω_0) are to be retained, the high frequency terms $(3\omega_0)$ being discarded.

The noise which starts the laser oscillator, and subsequently perturbs it, is now introduced on the right hand side of (10.111) as an inhomogeneous driving term to give

$$\frac{\mathrm{d}^2E}{\mathrm{d}t^2}+(\gamma-\alpha)\frac{\mathrm{d}E}{\mathrm{d}t}+\beta\left(E^2\frac{\mathrm{d}E}{\mathrm{d}t}\right)_{lf}+\omega_0^2E=\omega_0^2N(t). \tag{10.112}$$

References on page 299

Only statistical information is available regarding $N(t)$ (namely the average noise power and its frequency spectrum).

The above equation was originally derived by VAN DER POL [1927] in connection with the triode oscillator, but has been extensively applied to many different oscillators (e.g. the quartz crystal clock). LAMB [1964] has applied it to a study of gas lasers, and its derivation and solution with regard to intensity fluctuations in the laser are discussed by ARMSTRONG and SMITH [1967].

In this section we work in terms of the real signal, and look for solutions to (10.112) of the form

$$E(t) = E_0 \cos[\omega_0 t + \phi(t)] + e_n(t). \tag{10.113}$$

The amplitude, E_0, describing the coherent output, is time-independent, and is not subject to fluctuations. The function $\phi(t)$ is a random function describing the phase fluctuations, and is such that its root mean square value changes only slowly compared to $(\omega_0 t)$. The function $e_n(t)$, which has a frequency centered on ω_0, describes the intensity fluctuations and its root mean square value is assumed to be much smaller than E_0.

In the case of narrow band noise, such as that due to spontaneous emission (where the bandwidth is determined by the linewidth of the optical transition), the noise term in (10.112) may be written in the form

$$N(t) = n_1(t)\cos\omega_0 t + n_2(t)\sin\omega_0 t, \tag{10.114}$$

where n_1 and n_2 are random functions.

Substitution of (10.114) and (10.113) into (10.112) leads to the following expressions for E_0, ϕ and e_n if terms of the order $\ddot{\phi}$, $(\dot{\phi})^2$, etc. are neglected:

$$E_0^2 = \frac{4(\alpha-\gamma)}{\beta} - 4\langle e_n^2\rangle_t; \tag{10.115}$$

$$\ddot{e}_n + (\gamma - \alpha + \tfrac{1}{2}\beta E_0^2)\dot{e}_n + \omega_0^2 e_n = \omega_0^2 n_2(t)\sin\omega_0 t; \tag{10.116}$$

$$\dot{\phi} = -\omega_0 n_1(t)/2E_0. \tag{10.117}$$

Using (10.115), (10.116) may be written in terms of the mean square amplitude fluctuations as

$$\ddot{e}_n + (\alpha - \gamma - 2\beta\langle e_n^2\rangle_t)\dot{e}_n + \omega_0^2 e_n = \omega_0^2 n_2(t)\sin\omega_0 t. \tag{10.118}$$

Equations (10.116) and (10.117) are valid both above ($E_0 > 0$) and below ($E_0 = 0$) threshold. Below threshold (10.115) is no longer valid, and hence (10.118) applies to above threshold operation only.

The first equation (10.115) describes the amplitude of the coherent output, and demonstrates how its energy is decreased by the presence of the noise term. Since eq. (10.118) is linear, we can take its Fourier transform to obtain the spectral density (§ 5) of the amplitude fluctuations above threshold

$$G_{e_n}(\omega) = \frac{G_N}{4\omega_0^2} \frac{1}{(\omega-\omega_0)^2+(b/2)^2}, \tag{10.119}$$

where

$$b = \alpha-\gamma-2\beta\langle e_n^2\rangle_t \tag{10.120}$$

and G_N is the spectral density of the driving noise, which is assumed flat over the spectrum of e_n. This is a reasonable approximation in so far as the width of the spectral density of the driving noise is the linewidth of the laser transition. The spectrum of the amplitude noise is thus described by a Lorentzian function, the width of which depends on the mean square value of the amplitude fluctuations.

Below threshold $E_0 = 0$, and from (10.118) we see that the bandwidth of the noise spectrum is still described by a Lorentzian function, but now its width is given by

$$b = (\gamma-\alpha). \tag{10.121}$$

As the laser is adjusted so that threshold is approached from below, α increases, and hence b decreases; that is to say the bandwidth of the amplitude noise below threshold decreases as threshold is approached. This effect is known as "gain narrowing" (ch. 8 § 3).

We now return to operation above threshold. We can relate the bandwidth of the noise (associated with the amplitude fluctuations) to its total mean square amplitude, since

$$\langle e_n^2\rangle_t = \frac{1}{2\pi}\int_0^\infty W_{e_n}(\omega)\cdot d\omega, \tag{10.122}$$

so that using (10.119) we obtain

$$P_n = \langle e_n^2\rangle_t = \frac{W_N}{8\omega_0^2 b}. \tag{10.123}$$

We have another expression relating b and $\langle e_n^2\rangle_t$, namely (10.120), so that substituting from (10.123) into (10.120) we obtain the following expressions for the bandwidth, b, and the power, P_n, of the amplitude noise

References on page 299

$$b = \tfrac{1}{2}\beta P_{\text{coh}} - \frac{W_{\text{N}}}{\omega_0^2 P_{\text{coh}}}\,; \tag{10.124}$$

$$P_n = W_{\text{N}}/2\beta\omega_0^2 P_{\text{coh}}, \tag{10.125}$$

where P_{coh} is the coherent output of the oscillator in the absence of noise, i.e.

$$P_{\text{coh}} = 2(\alpha-\gamma)/\beta. \tag{10.126}$$

As may be seen from (10.125), the mean noise power decreases as the coherent power from the oscillator increases (i.e. as the gain of the active medium is increased). This equation also illustrates the influence of the saturation parameter, β; the greater the saturation effects in the active medium, the smaller is the mean noise power at a given coherent power level. We saw from (10.121) that below threshold the bandwidth of the noise decreased as threshold was approached. Above threshold the bandwidth depends on the coherent power level of the oscillator, and above a certain coherent power level, the bandwidth is proportional to the coherent power.

We now consider the phase fluctuations described by (10.117). We require to evaluate the mean square phase fluctuation as a function of time, i.e. $\langle\Delta\phi^2\rangle$†.

We can write (10.117) in the form

$$\delta\phi(t') = -\{\omega_0 n_1(t')/2E_0\}\delta t', \tag{10.127}$$

and using

$$\delta[\phi^2(t')] = 2\phi(t')\delta\phi(t')$$

we then have

$$\delta[\phi^2(t')] = 2\left(\frac{\omega_0}{2E_0}\right)^2 \int_0^{t'} n_1(t'')\mathrm{d}t''\, n_1(t')\delta t'.$$

Integration over t' from 0 to t enables the mean square phase fluctuation as a function of t to be evaluated as

$$\langle\Delta\phi^2\rangle = 2\left(\frac{\omega_0}{2E_0}\right)^2 \int_0^{t}\int_0^{t'} \langle n_1(t'')n_1(t')\rangle \mathrm{d}t''\mathrm{d}t'. \tag{10.128}$$

In so far as the function $n_1(t)$ describes a stationary random process, its autocorrelation function, $\langle n_1(t'')n_1(t')\rangle$ is symmetrical about the $t'' = t'$ axis,

† In so far as we are treating a stationary, random process here, the average $\langle\ \rangle$ may be either an ensemble or time average.

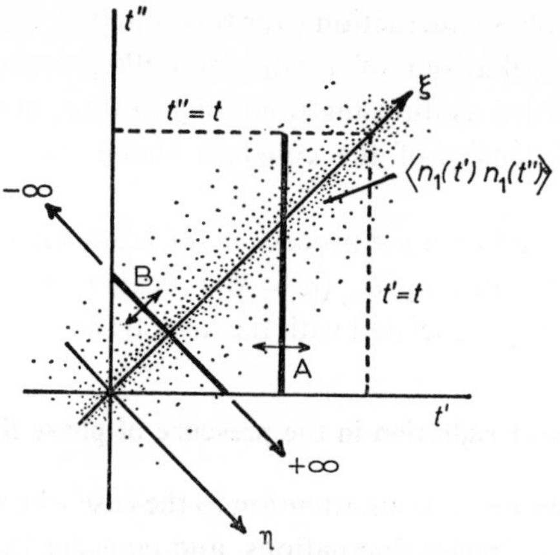

Fig. 10.8. Evaluation of the integral (10.128). The integration over t' and t'' is indicated by A, and over ξ and η by B. The density of dots roughly indicates the magnitude of the autocorrelation function in different regions of the plane of integration.

and reference to (fig. 10.8) shows that (10.128) may therefore be written

$$\langle \Delta\phi^2 \rangle = \left(\frac{\omega_0}{2E_0}\right)^2 \int_0^t \int_0^t \langle n_1(t'')n_1(t')\rangle \mathrm{d}t''\mathrm{d}t'. \tag{10.129}$$

We next transform to variables defined by

$$\xi = t'+t'',$$
$$\eta = t''-t',$$

so that (10.129) becomes

$$\langle \Delta\phi^2 \rangle = \left(\frac{\omega_0}{2E_0}\right)^2 \int_0^t \int_{-\xi}^{+\xi} \left\langle n_1\left(\frac{\xi-\eta}{2}\right) n_2\left(\frac{\xi+\eta}{2}\right)\right\rangle \mathrm{d}\eta \cdot \mathrm{d}\xi. \tag{10.130}$$

The auto-correlation function is only significant in the region where $\eta \sim 0$ (i.e. $t' \sim t''$), and provided t is greater than the correlation time of the noise, the limits of the intergration over η can be extended to run from $-\infty$ to $+\infty$ (Fig. 10.8 illustrates the arguments used above.). Reference to (10.67) shows that the integral over η now defines the spectral density at zero frequency of the noise process $n_1(t)$, so that (10.130) may be evaluated to give

$$\langle \Delta\phi^2 \rangle = \frac{\omega_0^2}{4E_0^2} \cdot G_{n_1}(\omega = 0) \cdot t. \tag{10.131}$$

References on page 299

The mean square phase fluctuation over time interval, t, is proportional to t, which is just the behaviour of a random walk process. The phase of the laser may be regarded as diffusing randomly in time, at a rate which is dependent on the intensity of the coherent signal as well as on the perturbing noise.

In so far as the random noise process, $n_1(t)$, is associated with spontaneous emission, the spectral density, $G_{n_1}(\omega = 0)$ is proportional to the spontaneous transition probability associated with the laser transition.

12. Statistics of laser radiation in the presence of phase fluctuations alone

In this section we restrict our attention to the case where the laser radiation is perturbed only by phase fluctuations, and consider their influence on the mutual coherence function and spectral density describing such radiation. The description that we shall develop will be applicable, therefore, only under those circumstances where gain saturation has limited the influence of the amplitude fluctuations.

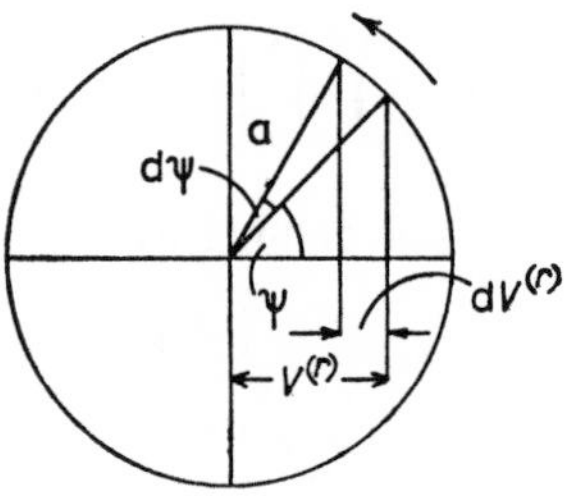

Fig. 10.9. Representation of laser output on vector diagram.

For reasons that will shortly be apparent, it is convenient to work in this section in terms of the analytic signal. The expression (10.113) describing the radiation field in terms of the real signal is, therefore, replaced by

$$\begin{aligned} V(t) &= a \exp\left(-2\pi i\nu_0 t - i\phi(t)\right) \\ &= a \exp\left(-i\psi\right), \end{aligned} \tag{10.132}$$

where a is constant, and where ϕ (and hence ψ) fluctuates randomly. In the complex plane, the signal can be represented by a vector of length a, which takes up an angle ψ relative to the real axis. The projections of a on to the real and imaginary axes are the real and imaginary parts of the analytic signal respectively (fig. 10.9). We now derive the probability distribution of the real signal. The probability that the phasor makes an angle between ψ and $(\psi+d\psi)$ with the real axis is

$$d\psi/2\pi, \tag{10.133}$$

since all phase angles are equally probable. The real signal associated with the phase angle ψ is

$$V^r = a \cos \psi \tag{10.134}$$

and therefore

$$\begin{aligned} dV^r &= a \sin \psi \cdot d\psi = \\ &= d\psi[a^2-(V^r)^2]^{\frac{1}{2}}. \end{aligned} \tag{10.135}$$

The probability density for the real signal is then readily seen to be

$$p(V) = \pi^{-1}[\langle I\rangle-(V^r)^2]^{-\frac{1}{2}}, \tag{10.136}$$

where I is the instantaneous intensity. In fig. 10.10 the probability densities of thermal and laser light are contrasted.

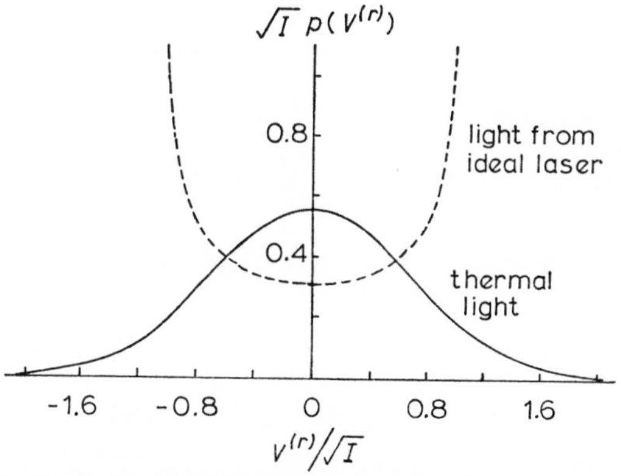

Fig. 10.10. Probability distributions relating to thermal light and to light from an ideal laser (after MANDEL [1963]).

It is also apparent that V^r and V^i are no longer statistically independent, since it may be seen from fig. 10.9 that the joint probability density can be expressed mathematically as a delta function of the form

$$p(V^r, V^i) = \pi^{-1}\delta[(V^r)^2+(V^i)^2-\langle I\rangle]. \tag{10.137}$$

Since the probability distribution has been derived on the basis of an amplitude stabilized signal, it is apparent that the intensity fluctuation correlation function (associated with the Hanbury-Brown-Twiss interferometer)

References on page 299

is zero, and will only depart from zero for real laser light in so far as amplitude fluctuations are present (§ 13). In § 11 we demonstrated that the mean square phase fluctuations follow a random walk process, in that for a time interval t the mean square fluctuation is (10.131)

$$\langle \Delta\psi^2 \rangle = \tau\lambda, \tag{10.138}$$

where

$$\lambda = \frac{\omega_0^2}{4E_0^2} \cdot G_{n_1}(\omega = 0). \tag{10.139}$$

In so far as the phase fluctuations follow a Gaussian random process, their distribution density is given by

$$p(\Delta\psi) = [\sqrt{(2\pi)} \cdot \sigma]^{-1} \cdot \exp[-(\Delta\psi)^2/(2\sigma^2)], \tag{10.140}$$

where the variance, σ, is

$$\sigma = [\tau\lambda]^{\frac{1}{2}}. \tag{10.141}$$

We use this distribution function to derive the autocorrelation function

$$\Gamma_\phi(\tau) = \langle V(t+\tau)V(t)\rangle_e \;^\dagger$$

by substituting from (10.140) and (10.141) into (10.52) to obtain

$$\Gamma_\phi(\tau) = a^2 \cdot \exp(-2\pi i\nu_0\tau) \cdot \langle \exp(-i\Delta\psi)\rangle_e =$$

$$= \frac{a^2}{\sqrt{(2\pi)} \cdot \sigma} \cdot \exp(-2\pi i\nu_0\tau) \cdot \int_{-\infty}^{+\infty} \exp\{-i\Delta\psi - (\Delta\psi)^2/(2\sigma^2)\} d(\Delta\psi), \tag{10.142}$$

therefore finding that

$$\Gamma_\phi(\tau) = a^2 \exp\{-2\pi i\nu_0\tau - \lambda\tau/2\}. \tag{10.143}$$

For $\tau \ll \lambda^{-1}$, the autocorrelation function takes a form that has previously been associated with quasi-monochromatic radiation. By using (10.67) the spectral density of laser radiation subject to phase fluctuations may be derived as

$$G_\phi(\nu) \propto \frac{1}{[4\pi^2(\nu-\nu_0)^2 + (\lambda/2)^2]}, \tag{10.144}$$

where λ is given by (10.139). The linewidth is inversely proportional to the intensity of the coherent output.

† An ensemble average is required here, since although $\Delta\psi$ follows a Gaussian random process, this is no longer true of $V(t)$.

13. Statistics of laser radiation with amplitude fluctuations

In this section we will again be working in terms of the analytic signal representation and so we replace our expression for the output signal (10.113) by the corresponding analytic signal

$$V(t) = E_0 \exp\{-2\pi i \nu_0 t - i\phi(t)\} + e_n'(t), \qquad (10.145)$$

where $e_n'(t)$ represents the analytic signal associated with $e_n(t)$. We first of all determine the autocorrelation function associated with the amplitude fluctuations, $e_n'(t)$, which is defined by

$$\Gamma_{e_n}(\tau) = \langle e_n'(t+\tau) e_n'^*(t) \rangle. \qquad (10.146)$$

In so far as $e_n'(t)$ represents a stationary, random process the average in (10.146) may be either a time or ensemble average. The above autocorrelation function may be derived from the spectral density of $e_n(t)$, which we have already derived (10.119), by using (10.64)

$$\Gamma_{e_n}(\tau) = \int_0^\infty G_{e_n}(\nu) \cdot \exp(-2\pi i \nu \tau)\, d\nu. \qquad (10.64)$$

Substituting from (10.119) and evaluating the integral in (10.64) we obtain

$$\Gamma_{e_n}(\tau) = 2\langle e_n^2 \rangle \cdot \exp\{-2\pi i \nu_0 \tau - b\tau/2\}. \qquad (10.147)$$

Since we are considering a stationary random process, then we can use (10.107) to determine the intensity correlation function for this process in terms of the autocorrelation function

$$\langle e_n'(t+\tau) e_n'^*(t+\tau) \cdot e_n'(t) e_n'^*(t) \rangle = 4\langle e_n^2 \rangle + |\Gamma_{e_n}(\tau)|^2. \qquad (10.148)$$

The autocorrelation function describing the total signal, given by (10.145), may now be determined since

$$\langle V(t+\tau) V^*(t) \rangle_e = E_0^2 \cdot e^{-2\pi i \nu_0 \tau} \langle e^{-\Delta\phi(\tau)} \rangle_e + \langle e_n'(t+\tau) \cdot e_n'^*(t) \rangle_t, \qquad (10.149)$$

where we have assumed that the phase and amplitude fluctuations are uncorrelated. In the determination of this autocorrelation function an ensemble average must be taken (see § 12), but on expansion this may be replaced by a time average for the second bracket on the right hand side, in accordance with what we have said above about $e_n'(t)$ being a stationary, random function. The first bracket on the right hand side in (10.149) is just the autocorrelation function describing the phase fluctuations which we derived in § 12, while the second bracket is the autocorrelation function for the am-

References on page 299

plitude fluctuations. We can therefore write (10.149) in terms of the complex coherence factors (10.56) as

$$\gamma_V(\tau) = \frac{P_{\text{coh}} \cdot \gamma_\phi(\tau) + P_n \cdot \gamma_{e_n}(\tau)}{(P_{\text{coh}} + P_n)}. \tag{10.150}$$

We now proceed to evaluate the correlation function for intensity fluctuations, defined by

$$\langle \Delta I(t+\tau)\Delta I(t)\rangle_{\mathrm{e}}, \tag{10.151}$$

where

$$\Delta I(t) = I(t) - \langle I \rangle \tag{10.108}$$

and

$$I = VV^*. \tag{10.152}$$

If we substitute from (10.145) for $V(t+\tau)$ into (10.152) we obtain

$$\begin{aligned}\Delta I(t+\tau) = {} & e_n'(t+\tau)e_n'^*(t+\tau) - 2\langle e_n^2\rangle + \\ & + \{E_0 \cdot e_n'(t+\tau) \cdot \exp\left[\mathrm{i}\omega_0(t+\tau) + \mathrm{i}\phi(t+\tau)\right] + c.\ c.\}.\end{aligned} \tag{10.153}$$

Substitution of (10.153) and the equivalent expression for $\Delta I(t)$ into (10.151), and then expanding out, leads to the following, if the amplitude and phase fluctuations are again assumed to be independent

$$\begin{aligned}\langle \Delta I(t+\tau) \cdot \Delta I(t)\rangle_{\mathrm{e}} & = |\Gamma_{e_n}(\tau)|^2 + [\Gamma_{e_n}^*(\tau) \cdot \Gamma_\phi(\tau) + c.\ c.] = \\ & = |\Gamma_{e_n}(\tau)|[|\Gamma_{e_n}(\tau)| + 2 \cdot |\Gamma_\phi(\tau)|].\end{aligned} \tag{10.154}$$

This may be written in terms of the complex coherence factors as

$$\frac{\langle \Delta I(t+\tau) \cdot \Delta I(t)\rangle_{\mathrm{e}}}{\langle I\rangle^2} = \frac{P_n \cdot |\gamma_{e_n}(\tau)|\{P_n \cdot |\gamma_{e_n}(\tau)| + 2P_{\text{coh}} \cdot |\gamma_\phi(\tau)|\}}{(P_n + P_{\text{coh}})^2}. \tag{10.155}$$

If we consider the case when, $\tau \ll b/2$ and $\tau \ll \lambda^{-1}$, the above has a particularly simple form, namely

$$\frac{\langle \Delta I(t+\tau)\Delta I(t)\rangle_{\mathrm{e}}}{\langle I\rangle^2} = \frac{(P_n^2 + 2P_{\text{coh}}P_n)}{(P_n + P_{\text{coh}})^2}. \tag{10.156}$$

The first term describes the noise beating with itself, while the second term describes the noise beating with the coherent signal. When the coherent signal is very large compared to the noise signal ($P_{\text{coh}} \gg P_n$), (10.156) reduces to

$$P_n/P_{\text{coh}},$$

indicating that the correlation in the intensity fluctuations becomes vanishingly small in accordance with our discussion in § 12. The factor 2 arises in (10.156) because we have treated the case where the spectral density of the noise function is Lorentzian. In cases where the spectral density is Gaussian or rectangular this factor may be shown to assume the values $\sqrt{2}$ and 1 respectively. It will be realized by comparison of (10.155) and (10.150) that the relationship between the autocorrelation function and the intensity fluctuation function, that we derived in the case of thermal radiation (10.109), no longer applies in the case of laser radiation.

References

ARMSTRONG, J. A. and A. W. SMITH, 1967, Progress in Optics, Vol. 4 (North-Holland Publishing Co.) p. 213.

BERAN, M. J. and G. B. PARRENT, Jr., 1964, Theory of Partial Coherence (Prentice-Hall, New Jersey) pp. 53–5.

CITTERT, P. H. VAN, 1936, Physica **6**, 1129.

DIRAC, P. A. M., 1958, The Principles of Quantum Mechanics, 4th Edition (Oxford University Press, London) p. 9.

EINSTEIN, A., 1915, Ann. der Phys. **47**, 879.

GABOR, D., 1946, J. Inst. Elec. Eng. **93**, part III, 429.

HANBURY-BROWN, R. and R. Q. TWISS, 1957a, Proc. Roy. Soc. **A242**, 300.

HANBURY-BROWN, R. and R. Q. TWISS, 1957b, Proc. Roy Soc. **A243**, 291.

HODARA, H., 1965, Proc. I.E.E.E. **53** 696.

JANOSSY, L., 1957, Nuovo Cimento **6**, 111.

JANOSSY, L., 1959, Nuovo Cimento **12**, 369.

LAMB, Jr., W. E., 1964, Phys. Rev. **134**, 1429.

MANDEL, L., 1963, Progress in Optics, Vol. 2 (North-Holland Publishing Company, Amsterdam) pp. 181–248.

MANDEL, L., and E. WOLF, 1965, Rev. Mod. Phys. **37**, 231–87.

PARRENT, G. B., Jr. 1959, J.O.S.A. **49** 787.

ROOT, W. L., and T. S. PITCHER, 1955, Ann. Math. Statist. **26**, 313.

VAN DER POL, B., 1927, Phil. Mag. **3**, 65.

WOLF, E., 1955, Proc. Roy. Soc. **A230**, 246.

WOLF, E., 1963, Proc. of Sym. on Optical Masers, Polytechnic Brooklyn, p. 29.

ZERNIKE, F., 1938, Physica **5**, 785.

CHAPTER 11

CAVITY ENGINEERING

1. Introduction

In this chapter we shall discuss some of the aspects of cavity engineering. An enormous effort has been devoted to the establishment of technology for the optimization and control of laser radiation. This effort has been concerned with the interaction of laser radiation with the optical components of the cavity and those outside the cavity and its rapid success is a great tribute to the ingenuity of many scientists. The literature describing this work is very extensive and far too detailed to be adequately treated in a single chapter. However, some description of the principles underlying the cavity "hardware" of lasers is necessary. We aim to cover some of the basic aspects and, by way of some topics which are not necessarily basic, to give some idea of techniques which may be used in cavity engineering and control of beam characteristics.

We shall consider both Brewster angled and perpendicular windows and dispersive cavities. The question of possible alternatives to spherical and plane mirrors for cavities (chs. 4, 5, 6) arises, so corner cube and roof-top reflectors are discussed together with the travelling wave and standing wave modes which they produce. Mode volume and alignment tolerances of mirrors and the active medium are important from many points of view including stability and power optimization. Mode selection and stabilization are also discussed. It is fortunate that the optical laser was sought at a time when there was a well established technology for the production of high reflectivity mirrors by forming multilayer dielectric films. The theory of these mirrors concludes the chapter.

2. Laser windows

2.1 *Brewster angled windows*

The first gas lasers had the cavity mirrors and the gas of the active medium within the same envelope. This is inconvenient and leads to mirror deterioration. It is usual now for mirrors to be quite separate from the active medium. The boundaries of the envelope of the active medium thus introduce refractive index discontinuities at which energy may be lost from the cavity by reflection as described by the Fresnel equations, (D.2), at each transit of the cavity. RIGROD et al. [1962] showed that reflection losses can be reduced to zero for one plane of polarization by having the windows at the ends of the gas discharge tube aligned at the Brewster angle, fig. D.2. Misalignment introduces reflection losses in accordance with (D.2) and as shown in fig. D.2; the amount which can be tolerated depends upon the conditions of oscillation of particular systems. The radiation produced by a laser with Brewster angled windows is plane polarized with the electric vector in the plane of incidence. Windows at angles other than the Brewster angle may result in the laser output being elliptically polarized. A further point which may be important in some applications is that Brewster angled windows introduce some astigmatism into the resonator. This becomes more important at the larger angular spread of the higher order modes.

For maximum power in the laser cavity the material used for the windows should be transparent to the laser radiation and should be Schlieren quality, i.e. free from strains which introduce rotation of the plane of polarization. The material finally selected will depend upon particular requirements. Typically, He–Ne lasers and argon lasers have Brewster windows of Schlieren quality quartz, since in the one case the system is of low gain and in the other the highest possible power is generally required. Infra-red lasers, CO_2–N_2–He (10.6 μ) may have windows of NaCl. Visible lasers with high gain (e.g. pulsed lasers) can operate with Brewster windows made from glass microscope slides.

In the case of crystal lasers such as ruby, antireflection coatings at the ends of the rod may be damaged by the high powers produced, so rods with Brewster angled ends are used for high power systems. A disadvantage of this type of rod is that it is more difficult to align with the other components of the system than is a cylindrical rod with plane ends normal to the rod axis.

Optical components may be inserted into laser cavities with zero loss (apart from absorption and scattering) if the radiation field meets them at the Brewster angle. A commonly used component is a prism to ensure that the laser operates at one frequency (see ch. 11 § 3). Other components include crystals for modulation and cells containing fluids.

A plane parallel piece of glass placed in the cavity may be used to reduce the laser output power or to remove some of the radiation from the cavity.

References on page 338

Insertion of a plane parallel slab of glass into a cavity introduces reflection losses in accordance with (D.2) and fig. D.2 but KOLOMNIKOV et al. [1965] have shown that these losses may be reduced as a result of interference between the beams reflected by the two plate surfaces. In these experiments a He–Ne laser operating at 1.152 μ was used. The arrangement is shown in fig. 11.1 in which the cavity consisted of a spherical mirror, 1, of radius of curvature 4 m and a plane mirror, 2, 2 m apart. The laser tube, 3, had a bore of 10 mm and length 80 cm. Transverse modes were suppressed by the stop, 4. The plane-parallel glass plate, 5, of thickness, $d = 9$ mm and refractive index $\mu = 1.5$, was mounted so that angles of rotation ϕ, could be measured accurately. The laser output was recorded by the photocell, 6.

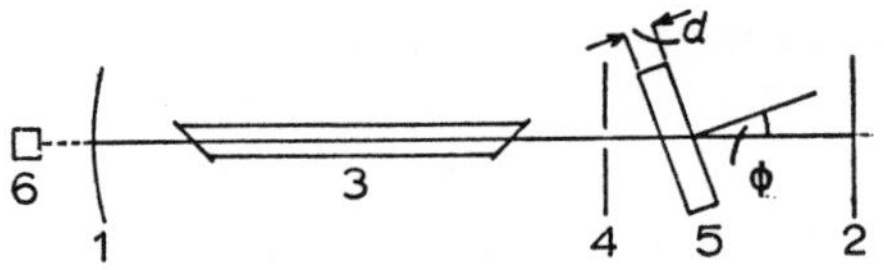

Fig. 11.1. Plane parallel slab of glass inserted into the cavity of a helium-neon laser (after KOLOMNIKOV et al. [1965]).

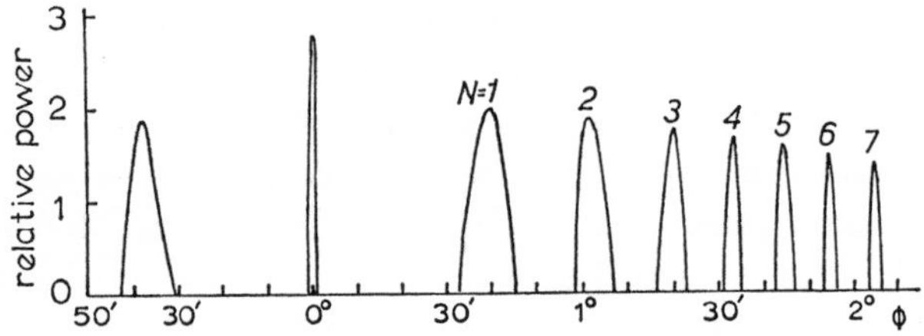

Fig. 11.2. Effect of orientation of a glass plate located in the cavity of a helium-neon laser on the output power of the laser (after KOLOMNIKOV et al. [1965]).

Rotating the plate affected the laser output as shown in fig. 11.2. If the plate was within 1′–1.5′ of the normal to the laser axis, oscillation was obtained. In rotating the plate 6°–7°, up to 50 maxima were obtained. Thicker plates caused the maxima to occur further apart. Reflections from the plate at the laser maxima were not completely extinct and their relative magnitude increased with ϕ. The laser output when $\phi = 0$ was about the same as in the case when the plate was set at the Brewster angle.

The condition for minimum reflection was confirmed by the experiments and is

$$2\mu d[1-(\sin^2 \phi/\mu^2)]^{\frac{1}{2}} = (2k+1)\lambda/2, \qquad k = 0, 1, 2, \cdots \tag{11.1}$$

2.2 *Perpendicular end windows*

Brewster angled windows cause the laser output to be polarized and, because only one plane of polarization is able to oscillate, the available power is reduced. MIELENZ et al. [1964] successfully operated a He–Ne laser with the plane of the windows normal to the discharge tube axis. The laser was 24 cm long. The windows were coated on each side with a dielectric two-layer anti-reflection coating, giving a transmission for each window of 99.5 % at 1.15 μ and were aligned parallel to within ±1′. On analysis with a Nicol prism the mode patterns were found to be polarized. The complete output pattern was a superposition of the polarized ones. The polarization of the off-axis modes was attributed to different phase changes upon reflection from the cavity mirrors for the two planes of polarization at oblique incidence.

3. Dispersive laser cavities

A laser with a non-dispersive cavity may oscillate at several frequencies simultaneously. Both argon and He–Ne lasers do this. In the case of He–Ne, the laser transitions at 6328 Å and 3.39 μ have a common upper level $3s_2$ so there is competition between the oscillations, and suppression of one leads

Fig. 11.3. Prism located in the laser cavity to ensure oscillation at one wavelength only.

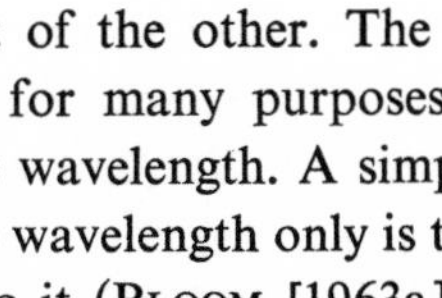

Fig. 11.4.

to increased power output of the other. The argon laser does not have competing transitions but for many purposes it is desirable to have the output confined to a single wavelength. A simple method of ensuring that the laser will operate at one wavelength only is to make the cavity dispersive by introducing a prism into it (BLOOM [1963a]). The prism is designed so that the radiation meets it at the Brewster angle to reduce insertion losses, fig. 11.3.

References on page **338**

The angle of the prism is easily calculated from fig. 11.4. At minimum deviation the beam passes through the prism symmetrically. Angles θ, ϕ are the angles of incidence and refraction, and A is the angle of the prism. The angle θ is to be the Brewster angle given by $\tan\theta = \mu$. By similar triangles, $A/2 = \phi$. Hence, we have

$$\sin\theta/\sin(A/2) = \mu = \tan\theta = \sin\theta/\cos\theta,$$

$$\sin(A/2) = \cos\theta. \tag{11.2}$$

This equation together with $\mu = \tan\theta$ enables the angle of the prism to be calculated. Figure 11.5 shows $(A/2)$ plotted against μ.

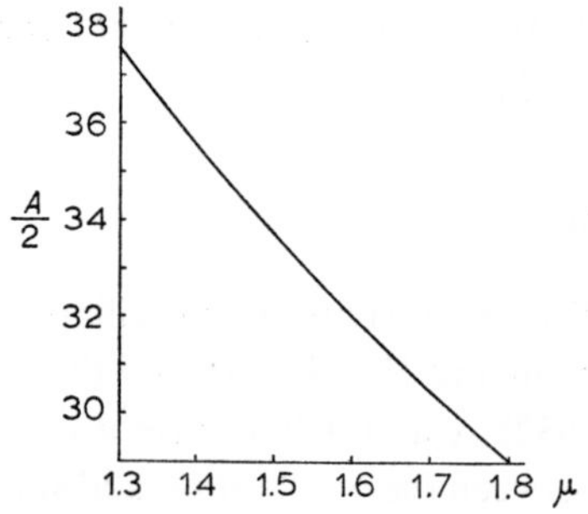

Fig. 11.5. The half-angle of a prism for a dispersive laser cavity as a function of the refractive index of the prism material.

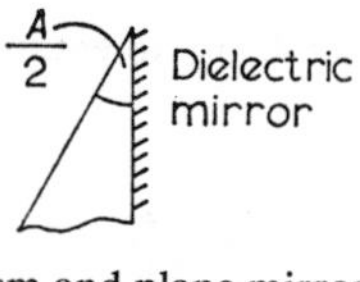

Fig. 11.6. A combination of prism and plane mirror for a dispersive laser cavity.

The dependence is $\sin(A/2) = (\mu^2+1)^{-\frac{1}{2}}$.

A prism and mirror can be incorporated in a single structure as shown in fig. 11.6. In this case the angle of the prism is $A/2$ and has the appropriate dielectric coatings (15 layers, or so) to form the mirror. For fixed wavelength operation this component may be sealed on to the end of the discharge tube, thus combining Brewster window, prism, and mirror to eliminate interfaces at which scattering can occur.

WHITE [1964] has shown that the reflecting surface of the prism may be made spherical to form a confocal cavity but it is necessary to correct for astigmatism by making the refracting face cylindrical as shown in fig. 11.7.

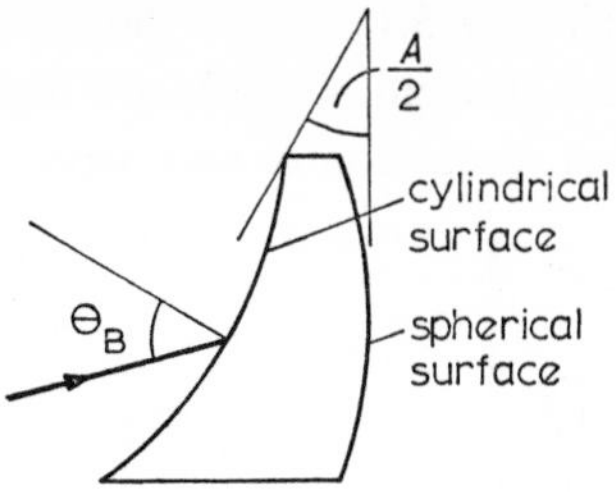

Fig. 11.7. A combination of prism and spherical mirror for a dispersive laser cavity.

The radius is given by

$$R = s/\mu \cos \theta_B, \tag{11.3}$$

where s is the geometrical radius of curvature of the reflecting surface and θ_B is the Brewster angle. The corrected reflecting prism has an "optical" radius of curvature of s/μ.

To align a prism in a cavity is quite straightforward. One method suitable for visible wavelengths is to place a white card with a pinhole of about 2 mm diameter in the cavity at A (Fig. 11.3) so that the laser oscillates. The mirror close to the card is removed and the prism located so that the light emerging from the pinhole passes through the prism with minimum deviation. The mirror is now located at C and adjusted so that the reflected beam coincides with the pinhole. The system usually oscillates when the card at A is removed. For high gain systems such as argon, the prism need only be placed on a level surface and moved about by hand, though obviously for some purposes, it is necessary to have precise mechanical control and stability.

4. Corner-cube and roof-top reflectors

A corner cube prism results from slicing off the corner of a cube. Both corner cube and roof-top prisms are shown in fig. 11.8. The angle ϕ of the roof-top prism usually has a value between 180° and 90°. Both prisms usually rely on total internal reflection to reflect a beam of light back along its path and so may be used to construct a laser cavity. Cavities may be formed from plane-parallel mirrors, corner cube reflectors, roof-top reflectors, or combinations of these. A cavity made up of two corner cubes can be aligned by viewing the system through a beam splitter and rotating the corner cubes about their principal axis until their edges coincide. The longitudinal modes of these cavities are either travelling wave modes or standing wave modes (ch. 5 § 3.1). Since the rays are obliquely incident with the reflecting surfaces, the phase shifts are dependent on the polarization of the light. Detailed discussion of this aspect is given by PECK [1962].

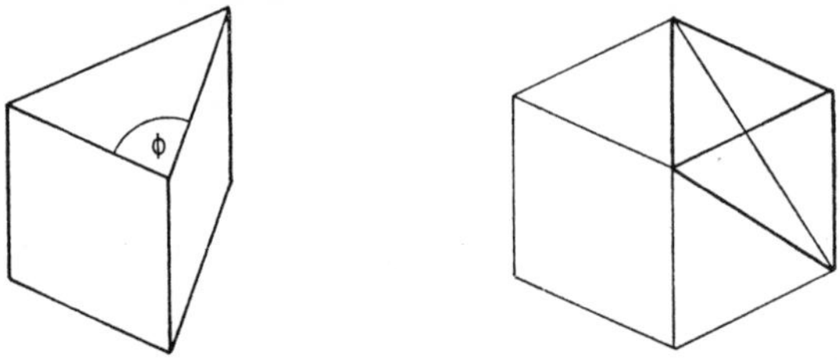

Fig. 11.8. Roof-top prism and corner-cube prism.

We shall use simple geometrical considerations to discuss the establishment of mode systems by multiple reflections and the alignment of the reflectors as described by BOBROFF [1964].

4.1 *Travelling wave modes*

Figure 11.9 represents a corner cube prism as seen by the radiation along the axis of the cavity. This may be represented more conveniently by a circle with sectors. Figure 11.10 shows two such circles representing a

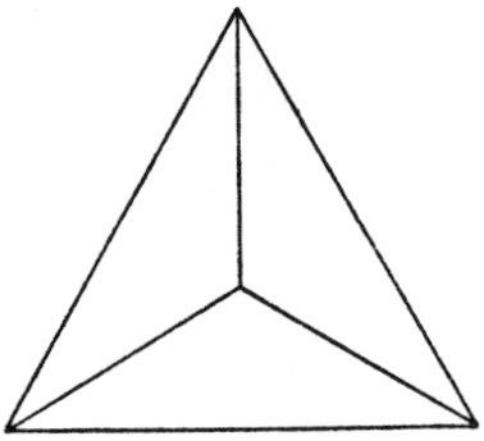

Fig. 11.9. Corner cube prism as seen by radiation along the axis of the cavity.

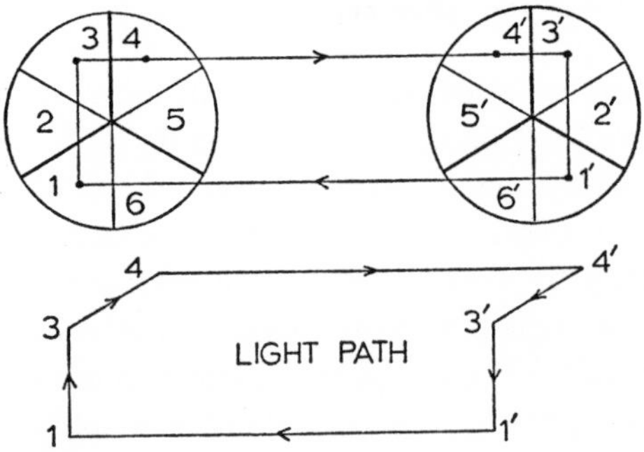

Fig. 11.10. Representation of a corner cube cavity and a light path established in the system.

corner cube cavity. Light incident in sector 1 of cube A is reflected along the closed path 1, 3, 4, 4′, 3′, 1′, 1 between the faces of cubes A and B to form the set of modes for which the phase change around the closed path is $2\pi q$ where q is an integer. The light can travel in either direction around the closed path. When a beam of light is reflected from a surface, its polarized state is altered (appendix D), hence, the modes will be polarized. When the polarization is invariant after a round trip it is called an eigenstate of the cavity. PECK [1962] has considered the problem in detail and shows that plane polarized eigenstates can occur when the prisms are aligned. It is easily shown that there are two other mode systems. Each mode forms a closed path after one complete circuit of the cavity. A 60°-misalignment supports three sets of modes over different paths from the aligned case, which are, therefore, quite different modes, fig. 11.11. When the corner cubes

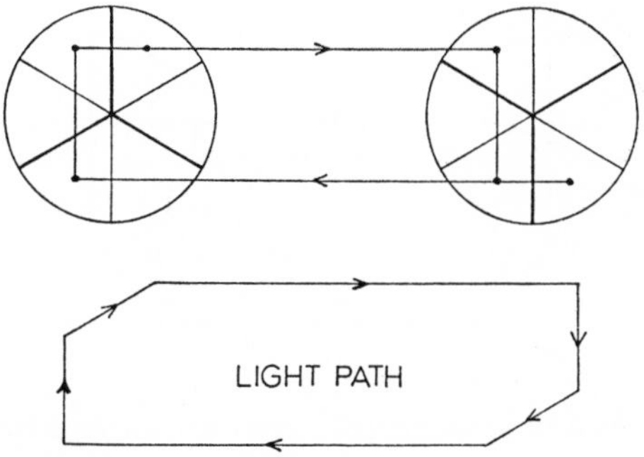

Fig. 11.11. Corner-cube cavity with 60°-misalignment and a resulting light path.

are misaligned as shown in fig. 11.12 all six modes are present, one set maintained by the unshaded area, and the other by the shaded area. Thus, misalignment results in a decrease in the area available for maintenance of a given mode and an increase in the number of modes.

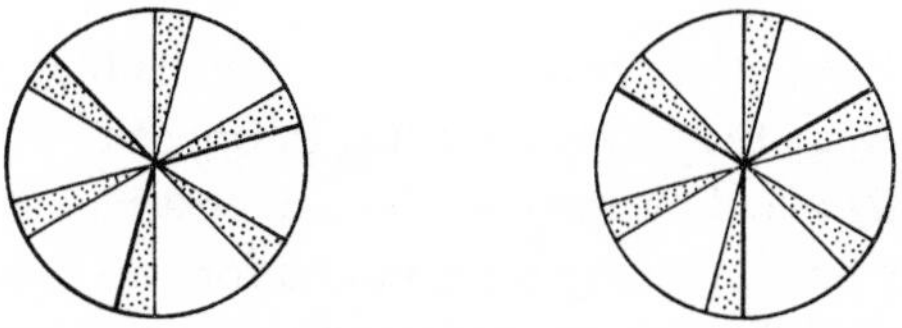

Fig. 11.12. Misaligned corner-cubes.

References on page 338

A cavity consisting of a corner-cube and a roof-top is represented in fig. 11.13 in alignment. The paths of the two sets of modes are shown. The first mode requires two circuits of the cavity to close; the second, only one.

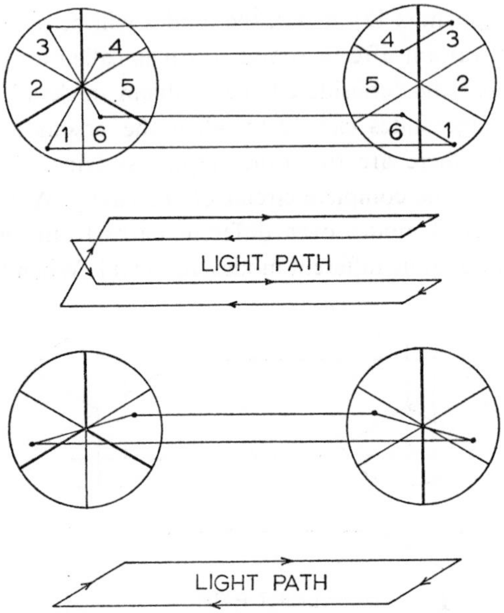

Fig. 11.13. Cavity formed by a roof-top and corner-cube combination; two possible orientations and resulting light paths.

Two roof-tops form a cavity. When they are misaligned, the number of circuits to close the mode is 180° divided by the angle of misalignment. If this is not an integer, the mode does not close on itself and so does not exist. Orientations giving small departures from an integral value result in walk-off losses.

4.2 *Standing wave modes*

If one of the mirrors of the cavity is plane, there is then an amplitude and phase relation between the waves travelling in opposite directions along the same path and only standing wave modes are possible. A cavity formed of two roof tops may have standing wave modes for two polarizations. This is proved in ch. 5 § 3.1. Summarizing, a cavity with a plane mirror has only standing wave modes; a cavity formed by two corner cubes or a corner cube and a roof top has only travelling wave modes; a cavity formed by two roof tops always has standing waves and, for proper orientations of the roof tops, also has travelling waves.

5. Mode volume

The mode volume V may be defined as the product of the spot size at the mirrors and their separation, thus,

$$V = \pi w_s^2 d. \tag{11.4}$$

This definition takes no account of the variation of mode diameter along its axis but is quite adequate for some purposes. The power obtainable in a given mode (TEM_{00}, say) depends upon the volume of active medium (fig. 11.14) in the mode volume and so is an important characteristic of a

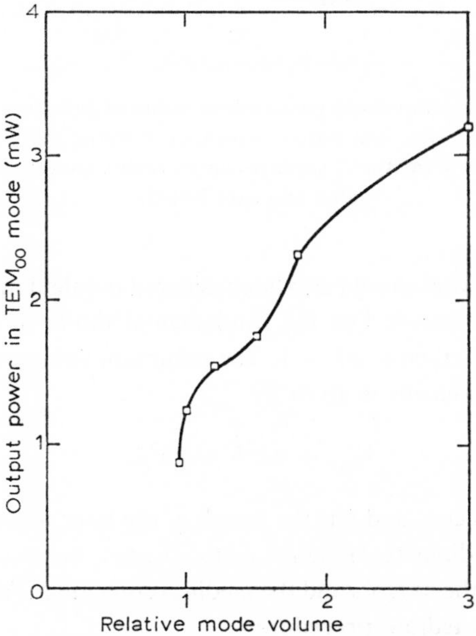

Fig. 11.14. Output power in TEM_{00} mode versus relative mode volume for near-hemispherical cavity (experimental) (reproduced from D. C. Sinclair, Appl. Optics **3**, 1067 (1964) with permission).

cavity (Sinclair [1964]). It can be shown that the confocal system has the smallest mode volume. The effect of mirror radius of curvature on mode volume is shown in fig. 11.15. The higher-order modes increase the effective mode volume. If an active medium within a confocal (or other) cavity is used as an amplifier, the same considerations of matching mode volume and active medium apply.

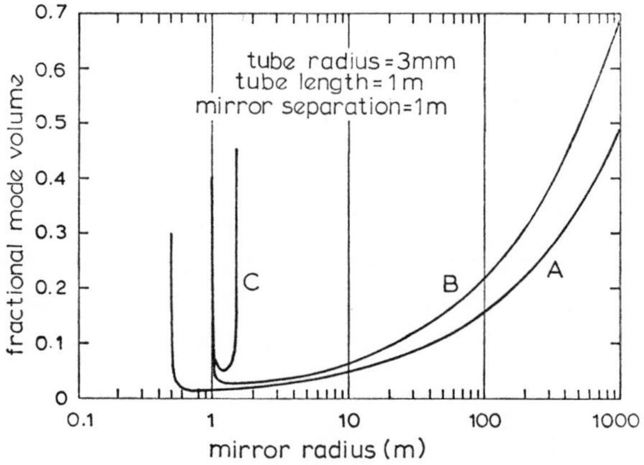

Fig. 11.15. Fractional mode volume versus mirror radius of curvature for a tube of diameter 6 mm, length 1 metre, and mirror separation 1 metre: curve A, double concave cavity; B, plano-concave cavity; C, concave-convex cavity, convex mirror radius, 50 cm (after SINCLAIR [1964]).

The diffraction losses should also be considered in calculating the minimum volume of laser material. For the fundamental mode to have diffraction losses less than 1 %, then $a^2/b\lambda \gtrsim 1$. The minimum volume of laser material fulfilling this requirement is given by

$$V'_{\min} = \pi a^2 b \approx \pi b^2 \lambda, \tag{11.5}$$

where a is the diameter, and b is the length of the laser material.

The smaller the diameter of active material used, the more necessary it is to have the axis of the mirrors and the medium coincident. Alignment of long thin lasers can be a tedious procedure.

Sinclair has calculated mode volumes and alignment tolerances for TEM_{00} modes in gas lasers with cavities consisting of two concave mirrors, a concave and a plane mirror, and a concave and a convex mirror. The mode volume was assumed to be that of the two truncated cones bounded by the spot sizes on the two mirrors and the minimum spot size in the cavity. Figure 11.16 shows the results of the calculations for a wavelength of 0.6 μ.

The fractional mode volume is the actual volume of the mode divided by the tube volume. The mode volume desired must be considered in relation to the consequent alignment tolerances.

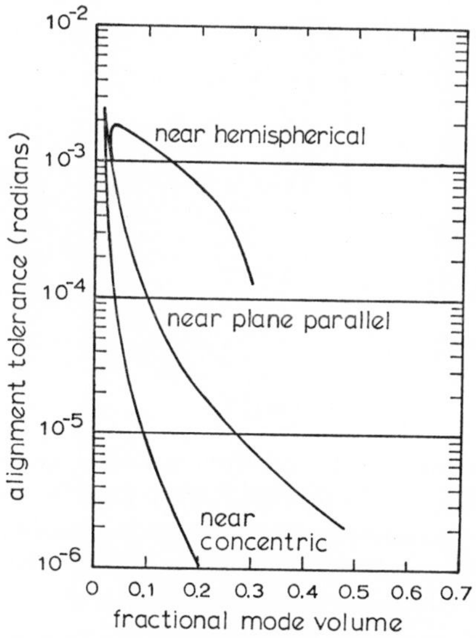

Fig. 11.16. Alignment tolerance versus fractional mode volume for a tube of diameter 6 mm, length 1 metre, and mirror separation 1 metre (after SINCLAIR [1964]).

5.1 *Axial alignment tolerances of gas laser tubes*

The axis of a laser mirror system is the line joining the centres of curvature of the two mirrors. The alignment of a gas laser requires that, given perfectly aligned mirrors, the axis passes through the laser tube and that it is far enough away from the tube wall to allow the TEM_{00} mode to oscillate. The aperture stops of the system used to calculate the alignment tolerances are located at the ends of the tube and are of radius $r_t - w_{st}$ where r_t is the tube radius and w_{st} is the spot size at the tube end (in fact this is taken as the spot size at the mirrors). Figure 11.15 shows that for a given mode volume, the near-hemispherical cavity is the easiest of all spherical mirror cavities to align. Its maximum fractional mode volume is $\frac{1}{3}$. The effect of mirror radius of curvature is shown in fig. 11.17.

A simple method of aligning long thin lasers where tube flexibility may cause trouble is to use a piece of cotton stretched along the axis of the tube.

References on page 338

The tube is clamped, the cotton removed, and the windows sealed on.

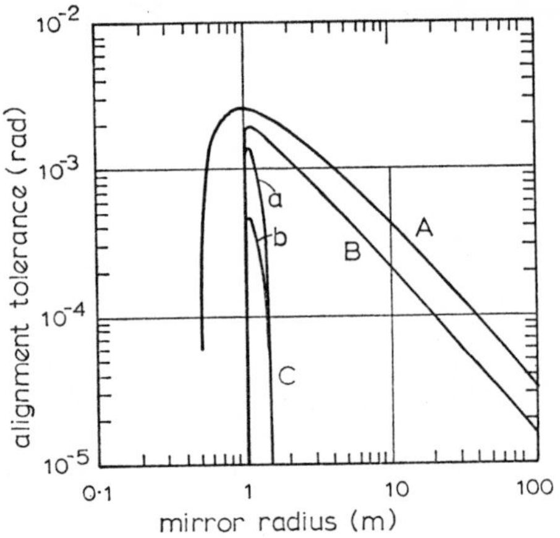

Fig. 11.17. Alignment tolerance versus radius of concave mirror for a tube of diameter 6 mm, length 1 metre, and mirror separation 1 metre: curve A, double concave cavity; B, plano-concave cavity; C, concave-convex cavity with curve *a* for convex mirror and curve *b* for concave mirror, convex mirror radius 50 cm (after SINCLAIR [1964]).

6. Mirror alignment tolerances

Mirror alignment may be considered from three points of view which involve quite different degrees of rigor but which are nevertheless very useful.

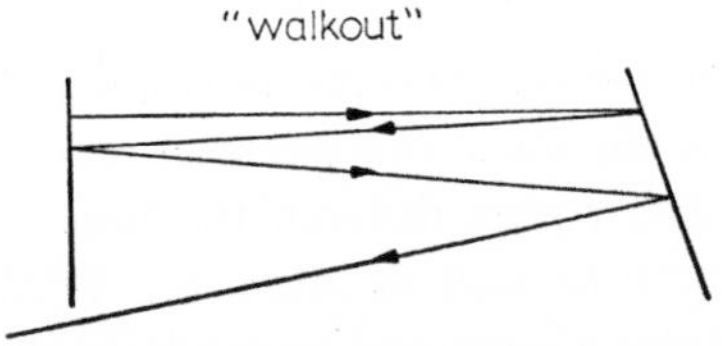

Fig. 11.18. Misaligned plane mirrors giving rise to "walkout".

The simplest idea is that the radiation should be kept within the boundaries of the reflecting surfaces while traversing the active medium many times, i.e. it should not "walk-out", fig. 11.18. Consideration of walk-out using simple ray diagrams shows that spherical mirrors, forming a concave cavity because of their focussing action, need not be aligned as accurately as plane parallel mirrors.

BLOOM [1963b] has considered the problem from the point of view of maintaining the fundamental mode. In a perfectly aligned system, this mode always lies symmetrically along the line joining the centres of the two mirrors (i.e. the laser axis). As a mirror is tilted, the diffraction losses of the mode increase rapidly and the spot diameter decreases but remains symmetrically distributed about the laser axis. We may regard the fundamental mode volume of the perfectly aligned system as defining the limits of a region in

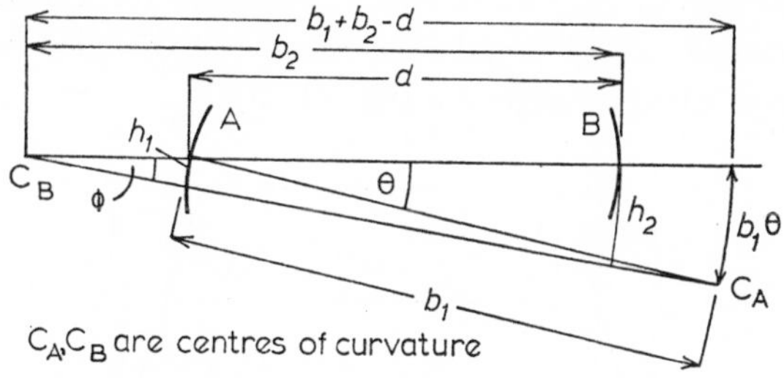

Fig. 11.19. Geometry of misaligned concave mirrors (after BLOOM [1963b]).

which the laser axis can be orientated. Figure 11.19 shows the mirror system to be discussed. The angles θ and ϕ are small, therefore

$$b_1\theta = (b_1+b_2-d)\phi, \qquad (b_2-d)\phi = h_1, \qquad b_2\phi = h_2.$$

Hence, we have,

$$h_1 = b_1(b_2-d)\theta/(b_1+b_2-d)$$

and

$$h_2 = b_1 b_2\theta/(b_1+b_2-d).$$

Which of the values, h_1 or h_2, is the greater depends upon which mirror is tilted. The angle, θ, for which $h_{1,2} = w_{1,2}$ is taken as being the largest angle θ_m of tilt at which the fundamental mode may be expected to oscillate; θ_m is given by

$$d = w_1/\theta = h_2/\theta_m. \tag{11.6}$$

Examples

1. Confocal mirrors. In this case $h_1 = 0$; $h_2 = b\theta$. For $b = 1$ metre, $w \approx 0.5$ mm we get $\theta_m \approx 1.5$ minutes of arc.

2. Large-radius mirrors. In this case $b_1 = b_2 \gg d$; $\theta_m = 2w/b$. This gives $\theta_m \approx 10$ seconds of arc.

References on page 338

3. Hemispherical resonator.
Flat mirror tilted ($b_1 = \infty$). Then $h_1 = (b_2 - d)\theta$; $h_2 = b_2\theta$.
Spherical mirror tilted ($b_2 = \infty$), then, $h_1 = b_1\theta$.
The alignment sensitivity is equal to that of the confocal resonator.

Of the various possible resonator geometries, the confocal and hemispherical are the least sensitive.

The third approach is that of Fox and Li [1963], who used the iterative technique described in ch. 4 § 2 in which conditions for a steady state were sought by repeatedly calculating the wavefront as it travelled back and forth between mirrors tilted at angle δ/a (fig. 11.20).

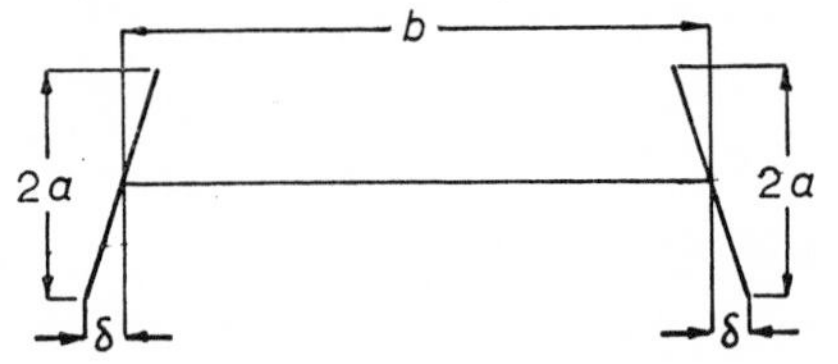

Fig. 11.20. Misaligned plane mirrors.

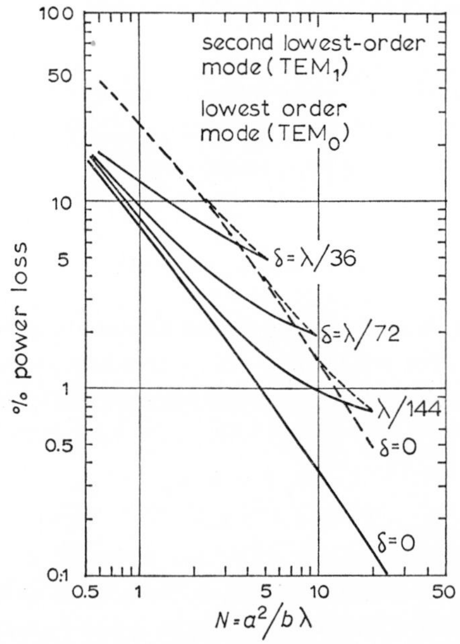

Fig. 11.21. The power loss per transit for tilted infinite-strip mirrors versus Fresnel number for several values of tilt. (After Fox and Li [1963].)

The angle of tilt was assumed to be small and the same for both mirrors. In this technique, the relative magnitudes of the eigenvalues determine the speed with which the iterated solution converges. Fox and Li found that, if the losses of the higher order modes are high, they will decay rapidly. If the losses of the two lowest-order modes, say, are approximately equal and much less than the other modes, these two modes will beat together after the other modes have died. The relative phase and magnitude of the eigenvalues of the two remaining modes determine the beat characteristics. They found that tilting tended to equalize the losses of the two lowest order modes and produce this beating effect. For large values of Fresnel number and/or tilt, δ/a, several modes had losses close together and gave complicated beat patterns. The power loss per transit for tilted infinite-strip mirrors is shown in fig. 11.21. (The three dimensional problem of rectangular mirrors can be reduced to a two-dimensional problem of infinite strip mirrors.)

7. Frequency effects

7.1 *Causes of frequency variations*

The theoretical limit of stability of an optical laser is ~ 1 in 10^{14}. There are several factors which cause the stability attained in practice to fall short of this.

1. Many longitudinal and transverse modes may oscillate simultaneously in a frequency band determined by the spontaneous line profile of the transition which is oscillating. This is because the cavity resonances are much narrower than the atomic resonance. The multimode output is thus of a range of frequencies almost as broad as that of a spontaneous emitter.

2. Since the cavity resonance is much narrower than the atomic resonance, its frequency stability is dependent on the stability of the cavity dimensions. The cavity length varies because of vibrations and thermal fluctuations. Obviously, the longer the cavity, the greater this problem will be. A change in mirror spacing of 4 Å in a 20 cm cavity of a laser operating at 6328 Å causes a frequency shift of 1 MHz.

3. The refractive index of the active medium, μ, determines the optical length, μl, of the cavity. A change in optical length of $\lambda/2$ is caused by a variation of refractive index of $\lambda/2l$. This causes a shift in the resonant frequency of the cavity of $c/2l$. Such changes in μ are possible even in low pressure gases. In practice, many laser systems operate with the fraction x of the optical path between the mirrors exposed to the atmosphere. Dust in

References on page 338

this atmosphere introduces noise into the laser cavity. Long term drifts result from changes in the ambient temperature and pressure. Some figures given by WHITE [1965] illustrate this. The frequency shift $\Delta\nu$ due to pressure and temperature changes is approximately

$$|\Delta\nu| = \nu_0(\alpha\Delta T + 3.63\times 10^{-7}\Delta p x), \tag{11.7}$$

where α is the linear expansion coefficient of the cavity spacers, ΔT is the temperature change (°C), Δp is the pressure change (in torr), and ν_0 is the centre frequency of the resonance. Typically, uncompensated invar cavities with $x = 0.1$ have a drift of 500 MHz/°C and 20 MHz/torr for 6328 Å.

White has estimated that detuning caused by air convection and sound waves in the optical path is probably less than 0.5 MHz in a laboratory.

Another cause of frequency variation is mode jumping. Usually, several modes oscillate simultaneously within the profile of the particular transition. Oscillation of a mode depletes the population maintaining the oscillation and, since the pumping process is usually quite random in its selection of atoms, the excited atoms needed to maintain a given mode may not be supplied and thus the mode extinguishes. By this means, modes are established and extinguished quite randomly.

4. Many gas discharge lasers operate in conditions in which magnetic fields greater than a few gauss are present and thus cause Zeeman splitting. The flash tubes of optically pumped lasers also create magnetic fields which can lead to undesirable Zeeman effects.

5. Pulsed lasers have a bandwidth of at least $1/\Delta\tau$, where $\Delta\tau$ is the pulse width.

7.1.1. *Refractive index of gas discharges* (*due to density of free electrons*)

In the theory of laser cavities the mode structures have been considered on the basis of a homogeneous medium which determines the optical path (refractive index × mirror spacing) between the mirrors. One of the factors determining the refractive index, μ, is the electron density, thus

$$\mu = [1-(4\pi e^2 n/m\omega^2)]^{\frac{1}{2}}, \tag{11.8}$$

where ω is the frequency, n the mean electron density, and e and m are the electron charge and mass, respectively. Variation of electron density causes variation of the optical spacing of the mirrors and hence the wavelengths at which a given cavity will resonate also alter. The wavelength shift can be calculated from (11.8) and (4.3a).

7.2 *Influence of active medium on mode formation*

The modes of oscillation of optical cavities have generally been studied on the assumption that the cavity is filled with a homogeneous, passive medium. This assumption is not satisfied in practice.

If the cavity contains an active medium, the losses may be compensated by the gain of the medium and this increases the effective Q of the cavity ch. 4 § 5. When the gain is greater then the total losses, the system oscillates (i.e. it operates as a laser). The lower order modes have the smaller diffraction losses and so these oscillate first. The higher order modes start to oscillate when the gain is increased further. The lowest order mode (usually TEM_{00q}) increases in intensity until it saturates. In saturation, the field distribution changes from a Gaussian (ch. 7) to give increased field at the wings relative to the centre, and thus relatively greater diffraction losses. Other higher order modes start to oscillate because they now have lower losses than the saturated lower order modes. This coupling of power into other modes is due to the non-linearity of the medium.

The gain of a laser medium depends upon the pumping process. In optically pumped lasers the surface of the laser medium receives more photons than do the inner regions because of absorption of the pump light as it travels through the medium. In gas discharges (pumping ultimately due to electrons) the electron density and temperature varies across the cross-section of the tube, being generally higher at the centre and lower at the periphery because of wall losses. Since the pumping is generally non-uniform across the laser cross-section, the gain is also non-uniform.

In addition to non-uniform pumping, there is also the problem of gain saturation. For a laser to oscillate, it is necessary for the inverted population of the active medium to provide gain equal to or greater than the losses. The amplitude of the oscillation can increase only at the expense of this population which is maintained by pumping. Hence, when the inverted population reaches the steady state representing balance between pumping and losses, the gain has saturated. The amplitudes of modes varies considerably depending on location in the cavity and the degree of gain saturation varies accordingly.

The effect of gain saturation in deforming modes has been investigated theoretically by STATZ and TANG [1965].

7.3 *Beat frequencies between modes*

We shall consider a cavity consisting of two concave mirrors of radii of curvature b_1 and b_2 respectively, distance d apart such that the con-

References on page 338

figuration is stable, see ch. 5 §§ 4, 6. The resonant frequencies for the different modes in Cartesian coordinates have been shown in ch. 6 § 3 to be given by

$$\nu = [q+(1+m+n)f]c/2d, \tag{11.9}$$

where

$$f = \frac{1}{\pi}\cos^{-1}\left[\left(1-\frac{d}{b_1}\right)\left(1-\frac{d}{b_2}\right)\right]^{\frac{1}{2}}. \tag{11.10}$$

The longitudinal modes are designated by q and the transverse ones by m and n. When the value of $(m+n)$ is fixed we have seen ch. 4 § 2 that $\delta\nu$, the separation between longitudinal modes corresponding to $\Delta q = 1$, is given by $\delta\nu = c/2d$. When q is fixed, the frequency separation $\Delta\nu$ between transverse modes corresponding to $\Delta(m+n)$ is given by

$$\Delta\nu = f\Delta(m+n)c/2d. \tag{11.11}$$

For stable systems, f varies between 0 and $\frac{1}{2}$, For each value of $(m+n)$ there is a longitudinal set of frequencies (modes); for each value of q, there is a transverse set.

In a typical gas laser the Fresnel numbers are $\sim$ 50, and, for these values, the diffraction losses are small and the approximations involved in (11.9) are valid. For resonators operating near the boundaries of stability, larger Fresnel numbers are required for (11.9) to remain valid. A Fabry-Perot, being plane parallel, has the value $f = 0$ and so, in this case, $(m+n)$ does not influence frequency. However, GOLDSBOROUGH [1964] points out that extrapolation of the curves of FOX and LI [1961] gives $f = 4\times10^{-3}$ for $N = 60$ and that with $d = 1$ m, $\delta\nu = 150$ MHz, the splitting between the TEM_{00q} and the TEM_{01q} modes is (4×10^{-3}) $(150\times10^{6}) = 0.6$ MHz, which agrees with the experimental value obtained by JAVAN et al. [1961].

Irregularities in the resonator cause removal of degeneracy resulting in small splittings in the mode frequencies but the departure from the TEM_{mnq} modes of a lossless resonator are small.

In terms of cylindrical coordinates, BOYD and KOGELNIK [1962] give the resonant frequencies as

$$\nu = [q+(1+2p+l)f]c/2d. \tag{11.12}$$

The mode frequencies are properties of the laser system and so obviously are independent of the coordinates used. YARIV and GORDON [1963] point out that a linear combination of modes with the same value of $(m+n)$ can be found which is identical with a mode in cylindrical coordinates with $(2p+l) = m+n$. If the resonator is symmetrical and lossless it does not matter whether Cartesian or cylindrical coordinates are used. In a real resonator there are imperfections in the mirrors, the windows, etc. which make the resonator asymmetric in the azimuthal direction and then Cartesian coordinates are used.

To observe beat frequencies between the various modes which may oscillate simultaneously within the gain profile of a given laser, one may arrange the laser beam to fall on the cathode of a photo-multiplier whose output is fed to a spectrum analyser (see, for example JAVAN et al. [1961] or GOLDSBOROUGH [1964]). One aspect of interest which illustrates the general approach is described by Goldsborough. Figure 11.22 shows TEM_{00q} and

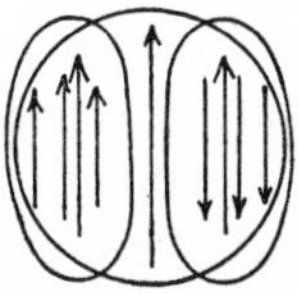

Fig. 11.22. Superimposed modes TEM_{10q}, TEM_{00q}. The fields in the two halves of the TEM_{10q} mode are in opposite directions and so produce beats with the TEM_{00q} mode which are out of phase with each other and which, therefore cancel in the photomultiplier. To observe the beats it is necessary to block off half the laser beam down the centre line shown (after GOLDSBOROUGH [1964]).

TEM_{10q} modes superimposed. Referring to the phase reversals occurring in transverse modes discussed in ch. 6 § 1, we see that the field in one half of the asymmetric mode is reversed, thus causing the beats with the symmetric mode also to be reversed in phase to produce a net result of zero at the photo-multiplier.

A beat frequency is observed if half the pattern is screened from the photo-multiplier. This is generally true of beats between transverse modes.

8. Mode selection

8.1 *Introduction*

The populations of the energy levels of an active medium may be sufficiently inhomogeneously broadened to allow independent oscillation of a number of longitudinal and transverse modes which saturate independently

References on page 338

so that a laser output may (and generally does) consist of many discrete frequencies. The modes are, in most cases, uncoupled (to a good approximation). It is sometimes desirable to restrict the number of modes and the methods used vary in sophistication and depend upon the tolerances of the particular experiment. Important fields where mode suppression and control are vital are those of optical communication and holography.

The basic problem in devising a method of mode selection is to ensure that selection is obtained with the minimum loss of usable power.

8.2.1. *Selection of transverse modes by tilting mirrors*

When a laser is set up it is very soon discovered that tilting the mirrors produces fascinating changes in the mode patterns observed. The axis of the resonator is defined as the line through the two centres of curvature of the two mirrors, and the axis of the active medium generally is arranged to coincide with this, so that the mode volume in the active medium is maximized ch. 11 § 5. Tilting the mirrors moves the axis of the cavity, and so may affect gain by reducing the volume of active medium available to one mode, and increasing that available to another; gain is also affected by the fact that different regions of the active medium will be used, and since there are generally gradients of population inversion across the section of an active medium, this can produce interesting effects; different areas of the mirrors and the Brewster angled windows form the cavity and so any irregularities of reflectivity, transmission, etc. cause variation of Q. Dust in the atmosphere also affects the modes and can be a troublesome source of noise in the output beam.

8.2.2. *Selection of modes by a circular aperture*

A screen with a circular aperture of diameter equal to the beam diameter of the TEM_{00} mode suppresses the higher order modes while introducing very little loss to the fundamental. The screen may be placed inside or outside the cavity. If inside, it helps to reduce effects due to competing modes; if outside, the beam may be passed through a lens with the aperture located at the focus (such a system may be used to "clean" the beam for holography, or schlieren photography).

RIGROD [1964] has studied the isolation of axi-symmetrical TEM_{pl} transverse modes of a He–Ne laser by adjusting the diameter of the aperture when it is accurately located on the optic axis at either mirror. He found that the fundamental (TEM_{00}) mode is the first to oscillate but as the aperture is gradually increased, the mode distorts and is replaced in a smooth transition by the next higher order mode. This process is repeated for each of the circular-symmetric modes in turn. He found that the laser oscillated in the highest-order mode permitted by the diffraction losses of the iris and the available laser gain. This behaviour cannot be explained in terms of linear mode theory. Rigrod isolated modes with azimuthal periodicity ($l = 1, 2, 3, 4$) by means of two straight wires crossing the optic axis at appropriate angles as suggested by Rigden and White.

Li [1963] has made a theoretical study of mode selection in an aperture limited concentric laser cavity in which there is no active medium (no gain). He solved the integral equations iteratively.

8.3 *Mode selection in long lasers*

It is easy to suppress the transverse modes of a laser by using a diaphragm § 8.2.2 so that only one transverse mode and the longitudinal modes remain. If no further precautions are taken, the gain profile of the active medium will support oscillation at a number of frequencies separated by $c/2d$, ch. 4 § 2. The problem is to obtain single frequency oscillation without reducing d (since this reduces the power available).

Smith [1965] has proposed two methods of mode selection, one is external to the cavity and the other is internal.

8.3.1. *External mode selection (Smith)*

The multifrequency output from a gas discharge laser is passed through a spherical Fabry-Perot interferometer whose longitudinal resonances are

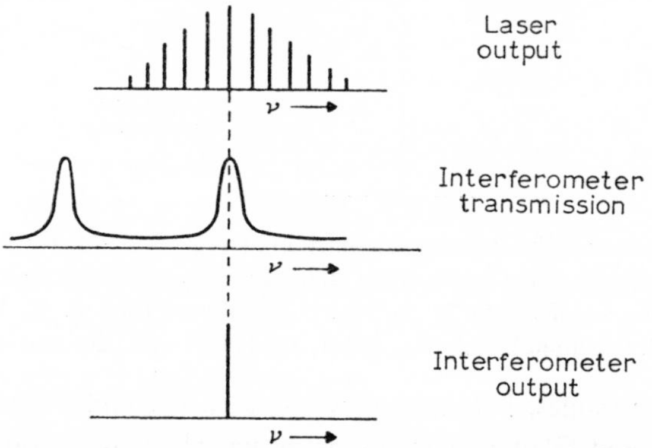

Fig. 11.23. Laser output, interferometer transmission, and interferometer output as a function of frequency. The interferometer is shown tuned to transmit a single frequency near the centre of the laser gain curve (reproduced from P. W. Smith, I.E.E.E. J. Quantum Electronics, **QE–1**, 343 (1965) with permission).

separated by a frequency larger than the oscillation bandwidth of the laser. The mode selection is summarized in fig. 11.23. In effect we are still using the $c/2d$ condition, but in this case the cavity is now the passive Fabry-Perot and is tuned so that it will transmit only one mode of the multimode output of the laser and reflect the rest. To prevent interaction of the reflected modes

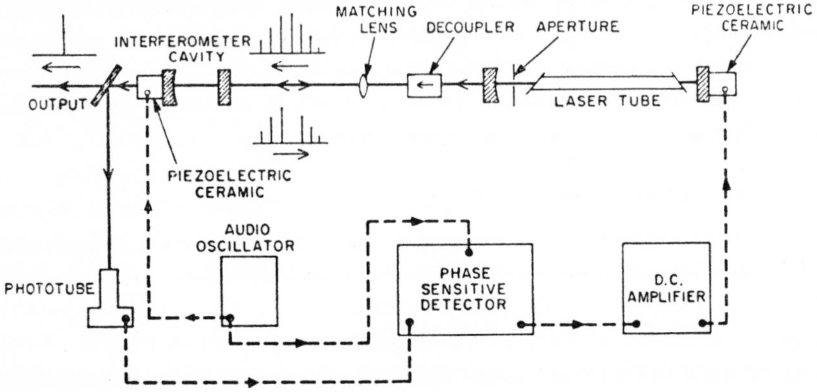

Fig. 11.24. Block diagram of the "External Mode Selector" method of obtaining single-frequency output (reproduced from P. W. SMITH, I.E.E.E. J. Quantum Electronics, **QE–1**, 343 (1965) with permission).

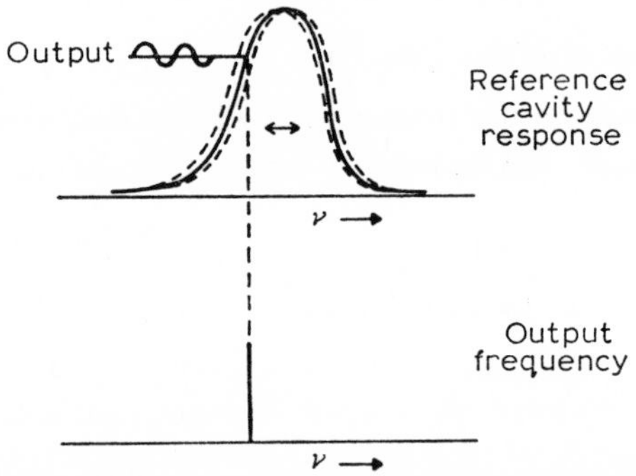

Fig. 11.25. Reference cavity response as a function of frequency. With the output frequency as shown, modulation of the length of the reference cavity modulates the amplitude of the output as shown schematically in the figure (reproduced from P. W. SMITH, I.E.E.E. J. Quantum Electronics, **QE–1**, 343 (1965) with permission).

with the laser modes, it is necessary to use a decoupler which may be a $\frac{1}{4}$-wave plate and Glan-Thompson prism, fig. 11.24. It is also necessary to match the Gaussian beam output of the laser to the boundary conditions imposed by the Fabry-Perot cavity as described in ch. 7 § 5. This is the purpose of the matching lens shown in fig. 11.24.

The feedback system shown in fig. 11.24 stabilizes the laser output. The resonant frequency of the reference cavity is modulated by altering the cavity length by means of the piezoceramic driven by the audio oscillator. The effect of this on the output is monitored by the phototube. As the cavity mirror oscillates, so the output amplitude also oscillates and passes through a peak intensity as shown in fig. 11.25. When the laser output drifts to one side of the peak of the reference cavity response, the output is amplitude modulated at the frequency of the audio oscillator. At resonance, there is no signal at the audio frequency (the signal is at 2× the frequency). On the other side of the reference cavity resonance, a signal of opposite phase is produced. Thus, a phase sensitive detector tuned to this audio frequency, may be used to derive a signal from the laser output whose amplitude is proportional to the deviation from the centre of the reference cavity response, and whose sign depends on the direction of the deviation.

8.3.2. *Internal mode selection* (*Smith and Fox*)

A very elegant method of internal mode selection has been described by SMITH [1965]. The scheme was proposed by Fox and involves replacing one of the mirrors of the laser cavity by three mirrors M_2, M_3, M_4 as shown in fig. 11.26. The three mirrors together form a secondary cavity which is

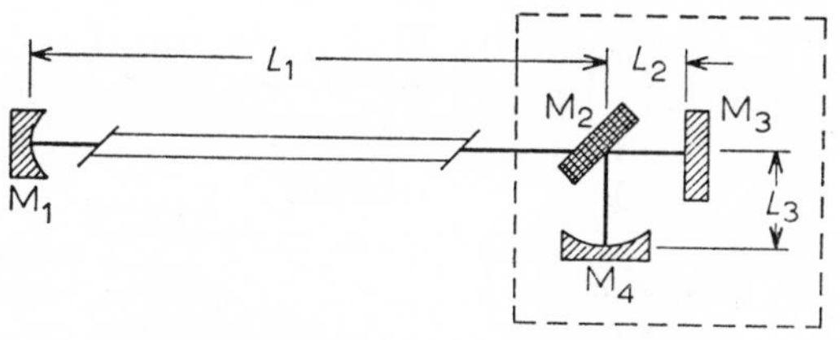

Fig. 11.26. Longitudinal mode selector. The three mirrors M_2, M_3, M_4, form a tunable reflector for the laser cavity (reproduced from P. W. SMITH, I.E.E.E. J. Quantum Electronics, **QE–1**, 343 (1965) with permission).

tunable. From the point of view of the laser cavity, the arrangement behaves like a single mirror whose reflectivity at a given frequency can be selected accurately. The secondary cavity is tuned to reflect the mode desired in the main laser cavity.

The mirrors M_3 and M_4 are totally reflecting so the width of the resonances is determined by the reflectivity of M_2 (cf. width of resonances of a Fabry-Perot depend on the reflectivities of the mirrors). The reflection characteristic of the three-mirror end reflector is shown in fig. 11.27. It should be noted

References on page 338

that the principle of selection is effectively the same as with mode selection by means of an external cavity.

The length of the secondary cavity is L_2+L_3 and this determines the separation between the resonances of this cavity. Thus, L_2+L_3 must be short so that the frequency separation between the resonances of the three

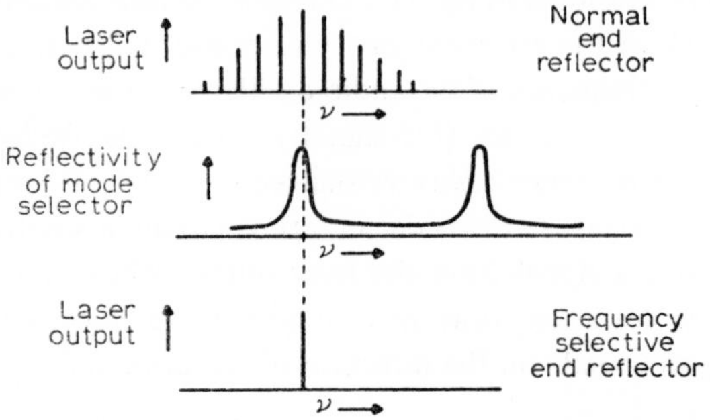

Fig. 11.27. Laser output with a normal end reflector, effective reflectivity of the three-mirror mode selector, and laser output with the three-mirror end reflector as a function of frequency. The mode selector is shown tuned so that the laser oscillates in a single mode near the centre of the laser gain curve (reproduced from P. W. Smith, I.E.E.E. J. Quantum Electronics, **QE–1,** 343 (1965) with permission).

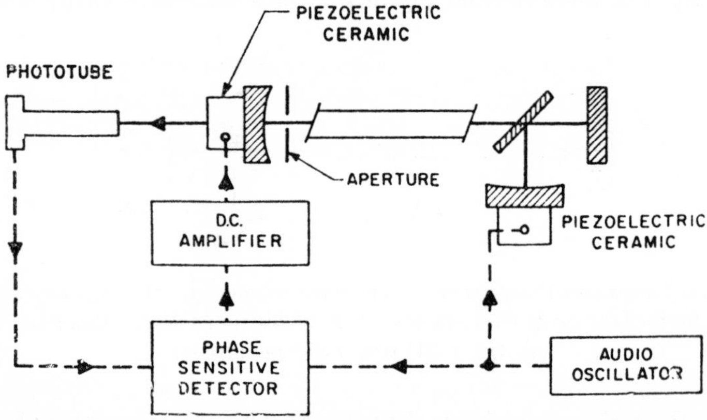

Fig. 11.28. Block diagram of the "Internal Mode Selector" method of obtaining single-frequency output (reproduced from P. W. Smith, I.E.E.E. J. Quantum Electronics, **QE–1,** 343 (1965) with permission).

mirror cavity is greater than the oscillation bandwidth of the laser. As with the external cavity, it is also necessary that the secondary and laser cavities are matched (see ch. 7 § 5 for matching requirements and calculations).

Smith investigated mode selection in a He-Ne laser. The mirror spacings were $L_1+L_2 = 150$ cm, $L_2+L_3 = 7.5$ cm and the mirror curvatures were $M_1 = 2$m, $M_2 = \infty$, $M_3 = \infty$, $M_4 = 10$ m. The laser was operated at 1 torr. The pressure broadened linewidth was estimated to be 150 MHz; the longitudinal mode spacing ($c/2L$) was 100 MHz. This situation is quite common, and in it, modes compete for atoms of the inverted population; selection of a single mode should generally result in an increase power output in that mode by removal of the competing modes. Another consideration is that of noise, for when modes compete for the same atoms, the individual modes tend to fluctuate in amplitude as the atoms supports first one mode, and then another. Clearly, these considerations indicate that the internal method of mode selection is preferable to the external method.

Figure 11.28 shows a block diagram of the apparatus used for obtaining a stabilized single frequency output.

8.4 *Mode selection by prism*

The angular divergence of the output from many high gain gas lasers and solid state lasers is one or two orders of magnitude greater than the diffraction minimum because of poor discrimination against off-axis modes. Diffraction losses in low gain systems help to confine oscillation to the lowest order modes. High gain systems may have mirrors of low reflectivity in order to optimize power output and so the diffraction losses of off-axis modes may be considerably less than the mirror losses. In such systems, some means of mode selection other than diffraction must be used. GIORDMAINE and KAISER [1964] introduced a resonator having one or both end reflectors in the form of a prism in which multiple internal reflections can occur. The principle of mode selection is based upon the rapid variation of reflectivity with angle of incidence near the critical angle. The prism has high reflectivity for only a narrow range ($\sim 1'$) about the axial direction desired.

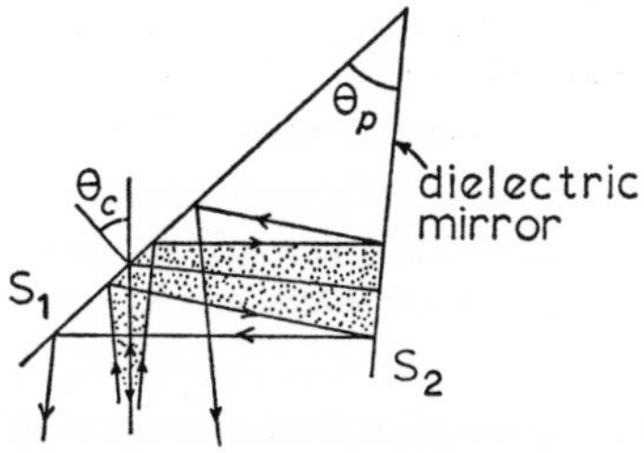

Fig. 11.29. Principle of discrimination against off-axis rays in a mode selector prism (after GIORDMAINE and KAISER [1964]).

References on page 338

Figure 11.29 shows the central and extreme rays of a divergent beam incident upon the prism mode selector. The prism angle, θ_p, is equal to or slightly greater than the critical angle θ_c. It may be seen that the off-axis rays have a strong tendency to walk out. They also incur greater reflection losses by meeting the surface S_1, at angles different from the critical angle. The axial rays, however, are reflected at the critical angle at S_1, at normal incidence at S_2 (which is a dielectric mirror), and then again at the critical angle at S_1 to return along their original path.

The reflected amplitudes $R_{\parallel}$ and $R_{\perp}$ for unit incident amplitude having polarization parallel and perpendicular, respectively, to the incident plane are given by

$$\begin{aligned} R_{\parallel} &= 1-\frac{2\mu(1-\mu^2\sin^2\theta_i)^{\frac{1}{2}}}{\cos\theta_i+\mu(1-\mu^2\sin^2\theta_i)^{\frac{1}{2}}}, \qquad \theta_i<\theta_c; \\ R_{\perp} &= 1-\frac{2(1-\mu^2\sin^2\theta_i)^{\frac{1}{2}}}{\mu\cos\theta_i+(1-\mu^2\sin^2\theta_i)^{\frac{1}{2}}}, \end{aligned} \tag{11.13}$$

where μ is the refractive index of the prism material relative to the air of the laboratory and θ_i is the angle of incidence.

In terms of the angle $\delta \equiv \theta_c - \theta_i \ll 1$, we use the approximation

$$\begin{aligned} R_{\parallel} &= 1-[2^{\frac{3}{2}}\mu^{\frac{3}{2}}/\{1-(1/\mu^2)\}^{\frac{1}{4}}]\delta^{\frac{1}{2}}, \\ R_{\perp} &= 1-[2^{\frac{3}{2}}/\mu^{\frac{1}{2}}\{1-(1/\mu^2)\}^{\frac{1}{4}}]\delta^{\frac{1}{2}}. \end{aligned} \tag{11.14}$$

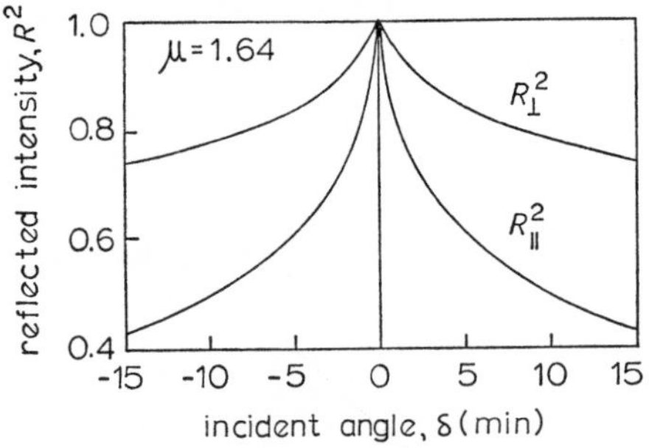

Fig. 11.30. Fractional reflected intensity of a mode selector prism, $\mu = 1.64$, with polarization parallel and perpendicular to the plane of incidence of the internal reflections; $\theta_p = \theta_c$ (after GIORDMAINE and KAISER [1964]).

Figure 11.30 shows the fractional reflected intensity calculated from (11.13) for a mode selector prism with $\mu = 1.64$, and polarization parallel and perpendicular to the plane of incidence of the internal reflections for $\theta_p = \theta_c$.

If θ_p is slightly greater than θ_c, the reflectivity of the prism is about 1.0 over a range of angles of width $2(\theta_p-\theta_c)$. Adjustment of $\theta_p-\theta_c$ can be made by varying the prism temperature. (E.g. a rise of temperature of 50 °C produces an increase $\sim 1''$ in $\theta_p-\theta_c$ for flint glasses.)

The prism used as described above discriminates only against rays off-axis in the plane of the diagram. To obtain selectivity in the vertical plane, a second prism must be added as shown in fig. 11.31. The half-wave plate is desirable so that the polarizations for both reflections can be parallel to the plane of incidence where the selectivity is higher.

GIORDMAINE and KAISER [1964] used the prism to obtain substantial reduction in beam width and an increase in emission near the beam centre. The improvement was greatest for parallel polarization.

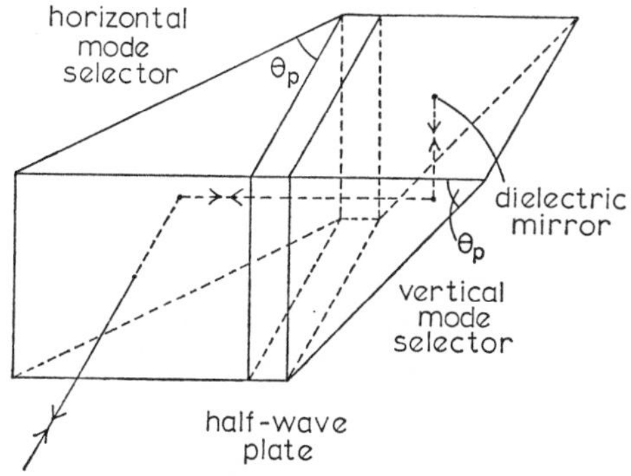

Fig. 11.31. Composite mode selector prism providing discrimination against off-axis rays in two dimensions (after GIORDMAINE and KAISER [1964]).

9. Stabilization of frequency of single mode lasers

A method of stabilization of gas discharge lasers which does not involve an external cavity (§ 8) has been successfully demonstrated by WHITE et al. [1964]. The laser transition itself is used as the basis for stabilization. The laser beam is passed through a cell containing the active laser gas in which a discharge is maintained and thus is either amplified or absorbed. When a longitudinal magnetic field is applied to the discharge, Zeeman splitting occurs and the medium becomes dichroic for circularly polarized light, fig. 11.32. If light propagating along the field is right hand circularly polarized (RHCP), it interacts with those atoms contributing to the low frequency absorption (amplification) profile; if the light is LHCP it interacts with those of the high frequency profile. To utilize this effect, the plane polarized output from the

References on page 338

laser is converted to CP light by a $\frac{1}{4}$-wave plate. The frequency of the laser output is undisturbed by the magnetic field and remains around ν_c; the objective is to stabilize the frequency as closely as possible to ν_c. Thus, the response of a detector beyond the absorption (amplifier) cell depends upon the frequency shift due to the Zeeman splitting. For a fixed magnetic field,

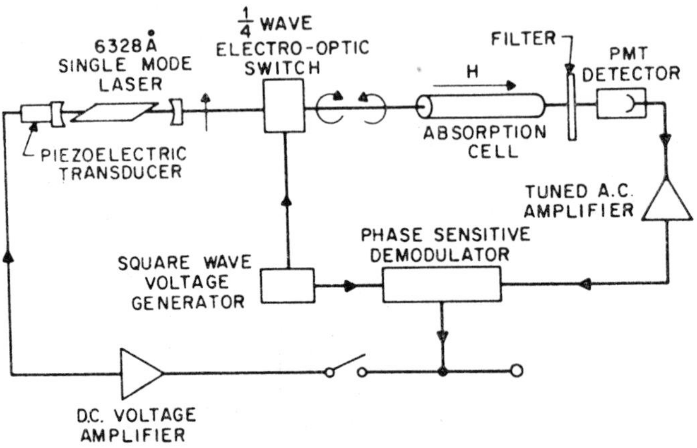

Fig. 11.32. Experimental feedback control system for frequency stabilization of 6328 Å gas laser (reproduced from A. D. WHITE, E. I. GORDON and E. F. LABUDA, Appl. Phys. Letters **5**, 97 (1964) with permission).

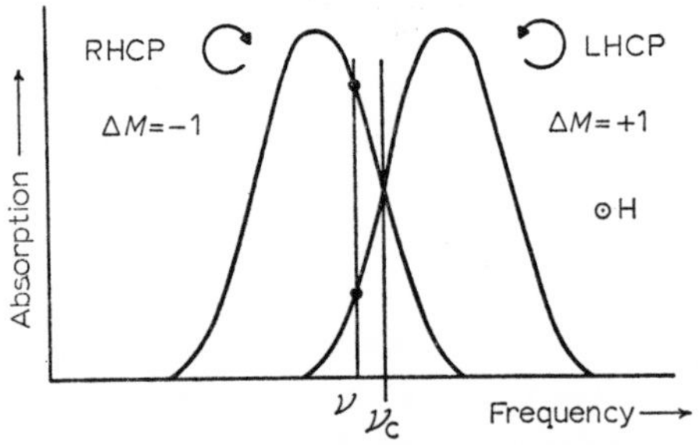

Fig. 11.33. Polarization sensitive double absorption profile produced by magnetic field (reproduced from A. D. WHITE, E. I. GORDON and E. F. LABUDA, Appl. Phys. Letters **5**, 97 (1964) with permission).

if the laser beam is switched from RHCP to LHCP there is no change in response at the detector if the laser frequency is ν_c; but if the laser frequency is ν, fig. 11.33 shows that the switching will produce a change in response.

Figure 11.32 shows the apparatus used by WHITE et al. The electro-optic switch was KDP. The absorption cell was filled with neon 20 (to avoid isotope shifts) to a pressure of 5 torr and operated at a discharge current of 20 – 50 mA. The axial magnetic field was ~350 gauss at which the absorption profiles are separated by 1.2 GHz. A narrow band filter (~10 Å) protected the S20 photomultiplier. The time-averaged frequency variation over a 5 min period was estimated to be ~± 1 MHz/sec.

10. Multilayer films

Multilayer films with periodic structure are used to obtain the reflectivities greater than 90 % generally necessary for the mirrors of a laser cavity and are basic to much of the technology of laser optics since they can be used to produce almost any required reflection or transmission characteristics of a surface for a given wavelength. BERNING [1963] has given a useful survey of the theory and calculations of optical thin films.

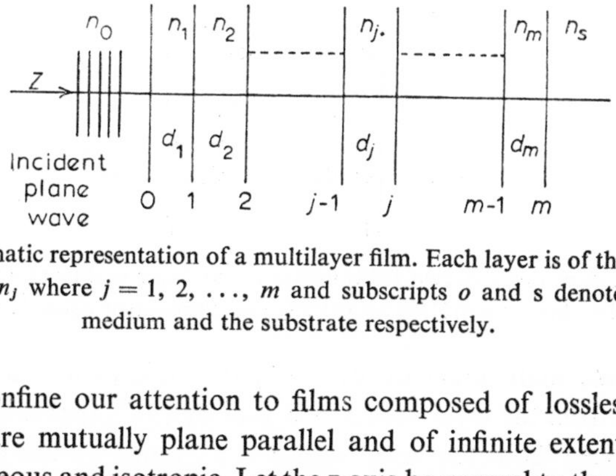

Fig. 11.34. Schematic representation of a multilayer film. Each layer is of thickness d_j and refractive index n_j where $j = 1, 2, \ldots, m$ and subscripts o and s denote the ambient medium and the substrate respectively.

We shall confine our attention to films composed of lossless dielectric layers which are mutually plane parallel and of infinite extent with each layer homogeneous and isotropic. Let the z-axis be normal to the plane of the layers and consider normally incident electromagnetic waves, monochromatic, plane polarized and travelling in the z-direction. Figure 11.34 shows a schematic representation of a multilayer film. At each interface some reflection occurs so that waves travelling in opposite directions exist in each layer. In the jth layer, the amplitudes of the electric and magnetic fields are given by

$$E(z)\exp(\mathrm{i}\omega t) = a_j \exp\{\mathrm{i}(\omega t - kn_j z + \alpha_j)\} + b_j \exp\{\mathrm{i}(\omega t + kn_j z + \beta_j)\}; \qquad (11.15)$$

$$H(z)\exp(\mathrm{i}\omega t) = n_j[a_j \exp\{\mathrm{i}(\omega t - kn_j z + \alpha_j)\} - b_j \exp\{\mathrm{i}(\omega t + kn_j z + \beta_j)\}], \qquad (11.16)$$

References on page 338

where a_j, b_j, α_j and β_j are constants given by the conditions that across any boundary $E(z)$ and $H(z)$ are continuous. Applying this condition to the boundary between the j and $(j+1)$ layers, we get

$$a_j \exp\{i(\alpha_j - kn_j z_j)\} + b_j \exp\{i(\beta_j + kn_j z_j)\} =$$
$$= a_{j+1} \exp\{i(\alpha_{j+1} - kn_{j+1} z_j)\} + b_{j+1} \exp\{i(\beta_{j+1} + kn_{j+1} z_j)\} \tag{11.17}$$

and

$$n_j[a_j \exp\{i(\alpha_j - kn_j z_j)\} - b_j \exp\{i(\beta_j + kn_j z_j)\}] =$$
$$= n_{j+1}[a_{j+1} \exp\{i(\alpha_{j+1} - kn_{j+1} z_j)\} - b_{j+1} \exp\{i(\beta_{j+1} + kn_{j+1} z_j)\}]. \tag{11.18}$$

These equations may be simplified by adopting the notation shown in fig. 11.35 in which the reflected wave is primed and the transmitted wave is

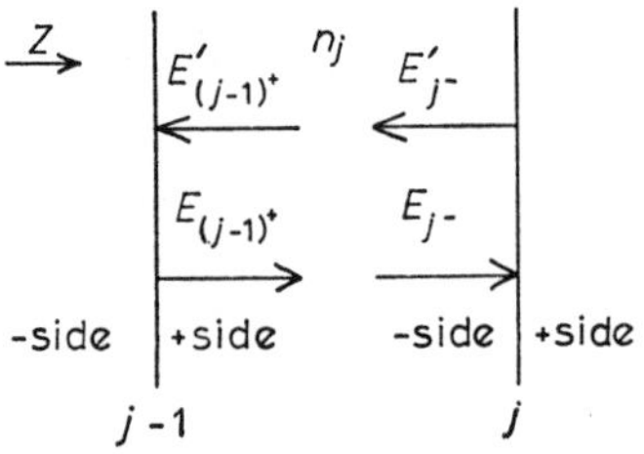

Fig. 11.35. Incident and reflected fields within the jth layer to illustrate the notation used in obtaining (11.19) to (11.22) and subsequent calculations.

unprimed. We shall designate by "+", that side of a boundary which is approached from the layer with $z > z_j$ and by "−", that side of the boundary approached from the layer with $z < z_j$. Since we are concerned only with time averaged and relative quantities we may omit the time dependency factors and put the amplitude at the mth boundary equal to unity. Thus we get

$$E_{j^-} = a_j \exp[i(\alpha_j - kn_j z_j)]; \tag{11.19}$$
$$E'_{j^-} = b_j \exp[i(\beta_j + kn_j z_j)]; \tag{11.20}$$
$$E_{(j-1)^+} = a_j \exp[i(\alpha_j - kn_j z_{j-1})]; \tag{11.21}$$
$$E'_{(j-1)^+} = b_j \exp[i(\beta_j + kn_j z_{j-1})]. \tag{11.22}$$

From (11.19) and (11.21) we get

$$E_{(j-1)^+} = a_j \exp[i\{\alpha_j - kn_j(z_j - d_j)\}] = E_{j^-} \exp[ikn_j d_j] =$$
$$= E_{j^-} \exp(i\Phi_j). \tag{11.23}$$

Similarly, from (11.20) and (11.22) we get

$$E'_{(j-1)^+} = E'_{j^-} \exp(-i\Phi_j), \tag{11.24}$$

where $d_j = (z_j - z_{j-1})$ is the geometric thickness of the jth layer and Φ_j is the phase thickness.

Applying this notation to (11.17) and (11.18) we get

$$E_{j^-} + E'_{j^-} = E_{j^+} + E'_{j^+}; \tag{11.25}$$

$$n_j[E_{j^-} - E'_{j^-}] = n_{j+1}[E_{j^+} - E'_{j^+}]. \tag{11.26}$$

For boundary $(j-1)$ we put $(j-1)$ for j in (11.25) and (11.26) and get

$$E_{(j-1)^-} + E'_{(j-1)^-} = E_{(j-1)^+} + E'_{(j-1)^+}; \tag{11.27}$$

$$n_{(j-1)}[E_{(j-1)^-} - E'_{(j-1)^-}] = n_j[E_{(j-1)^+} - E'_{(j-1)^+}]. \tag{11.28}$$

Combining (11.27) and (11.28) with (11.23) and (11.24) we get

$$E_{(j-1)^-} + E'_{(j-1)^-} = E_{j^-} \exp(i\Phi_j) + E'_{j^-} \exp(-i\Phi_j); \tag{11.29}$$

$$E_{(j-1)^-} - E'_{(j-1)^-} = \frac{n_j}{n_{j-1}} [E_{j^-} \exp(i\Phi_j) - E'_{j^-} \exp(-i\Phi_j)]. \tag{11.30}$$

Solving these equations for $E_{(j-1)^-}$ and $E'_{(j-1)^-}$ we get

$$E_{(j-1)^-} = \frac{1}{2}\left(1 + \frac{n_j}{n_{j-1}}\right) E_{j^-} \exp(i\Phi_j) + \frac{1}{2}\left(1 - \frac{n_j}{n_{j-1}}\right) E'_{j^-} \exp(-i\Phi_j); \tag{11.31}$$

$$E'_{(j-1)^-} = \frac{1}{2}\left(1 - \frac{n_j}{n_{j-1}}\right) E_{j^-} \exp(i\Phi_j) + \frac{1}{2}\left(1 + \frac{n_j}{n_{j-1}}\right) E'_{j^-} \exp(-i\Phi_j). \tag{11.32}$$

The Fresnel equations for normal incidence at the interface $(j-1)$ are given by

$$r_{j-1} = \frac{n_{j-1} - n_j}{n_{j-1} + n_j} \quad \text{and} \quad t_{j-1} = \frac{2n_{j-1}}{n_{j-1} + n_j}. \tag{11.33}$$

These equations will be more useful for present purposes in the forms

$$\frac{1}{t_{j-1}} = \frac{1}{2}\left(1 + \frac{n_j}{n_{j-1}}\right) \tag{11.34}$$

References on page **338**

and

$$\frac{r_{j-1}}{t_{j-1}} = \frac{1}{2}\left(1 - \frac{n_j}{n_{j-1}}\right). \tag{11.35}$$

The recursion equations (11.31) and (11.32) may be combined with (11.34) and (11.35) and written in matrix form thus,

$$\begin{bmatrix} E_{(j-1)^-} \\ E'_{(j-1)^-} \end{bmatrix} = \begin{bmatrix} \dfrac{\exp(\mathrm{i}\Phi_j)}{t_{j-1}} & \dfrac{r_{j-1}\exp(-\mathrm{i}\Phi_j)}{t_{j-1}} \\ \dfrac{r_{j-1}\exp(\mathrm{i}\Phi_j)}{t_{j-1}} & \dfrac{\exp(-\mathrm{i}\Phi_j)}{t_{j-1}} \end{bmatrix} \begin{bmatrix} E_{j^-} \\ E'_{j^-} \end{bmatrix}. \tag{11.36}$$

If we let E_j and H_j be the amplitudes of the resultant fields at boundary j we get

$$E_j = E_{j^-} + E'_{j^-}$$

and

$$H_j = n_j E_{j^-} - n_j E'_{j^-}.$$

In matrix form these equations become

$$\begin{bmatrix} E_j \\ H_j \end{bmatrix} = \begin{bmatrix} 1 & 1 \\ n_j & -n_j \end{bmatrix} \begin{bmatrix} E_{j^-} \\ E'_{j^-} \end{bmatrix}. \tag{11.37}$$

Hence, we obtain

$$\begin{bmatrix} E_{j^-} \\ E'_{j^-} \end{bmatrix} = \begin{bmatrix} 1 & 1 \\ n_j & -n_j \end{bmatrix}^{-1} \begin{bmatrix} E_j \\ H_j \end{bmatrix} = \begin{bmatrix} \dfrac{1}{2} & \dfrac{1}{2n_j} \\ \dfrac{1}{2} & -\dfrac{1}{2n_j} \end{bmatrix} \begin{bmatrix} E_j \\ H_j \end{bmatrix}. \tag{11.38}$$

Equation (11.37) written for boundary $(j-1)$ becomes

$$\begin{bmatrix} E_{j-1} \\ H_{j-1} \end{bmatrix} = \begin{bmatrix} 1 & 1 \\ n_{j-1} & -n_{j-1} \end{bmatrix} \begin{bmatrix} E_{(j-1)^-} \\ E'_{(j-1)^-} \end{bmatrix} \tag{11.39}$$

Combining (11.39), (11.36) and (11.38) we obtain

$$\begin{bmatrix} E_{j-1} \\ H_{j-1} \end{bmatrix} = \begin{bmatrix} 1 & 1 \\ n_{j-1} & -n_{j-1} \end{bmatrix} \cdot$$

$$\cdot \frac{1}{t_{j-1}} \begin{bmatrix} \exp(\mathrm{i}\Phi_j) & r_{j-1}\exp(-\mathrm{i}\Phi_j) \\ r_{j-1}\exp(\mathrm{i}\Phi_j) & \exp(-\mathrm{i}\Phi_j) \end{bmatrix} \begin{bmatrix} \dfrac{1}{2} & \dfrac{1}{2n_j} \\ \dfrac{1}{2} & -\dfrac{1}{2n_j} \end{bmatrix} \begin{bmatrix} E_j \\ H_j \end{bmatrix},$$

which, on multiplying out, gives

$$\begin{bmatrix} E_{j-1} \\ H_{j-1} \end{bmatrix} = \begin{bmatrix} \cos \Phi_j & (\mathrm{i}/n_j) \sin \Phi_j \\ \mathrm{i}n_j \sin \Phi_j & \cos \Phi_j \end{bmatrix} \begin{bmatrix} E_j \\ H_j \end{bmatrix}. \tag{11.40}$$

The determinant of the 2×2 matrix in (11.40) is unity. Some of the properties of these matrices relevant to the discussion to follow are given in appendix E. The 2×2 matrix in (11.40) is the transfer matrix of the layer and will be denoted by $\boldsymbol{M}_j$.

We can obtain the resultant fields at boundary $(j-2)$ from (11.40) by writing

$$\begin{bmatrix} E_{j-2} \\ H_{j-2} \end{bmatrix} = \boldsymbol{M}_{j-1} \begin{bmatrix} E_{j-1} \\ H_{j-1} \end{bmatrix} = \boldsymbol{M}_{j-1} \boldsymbol{M}_j \begin{bmatrix} E_j \\ H_j \end{bmatrix}.$$

Continuing this process, we see that the relationship between the fields at the first and last boundaries is given by the equation

$$\begin{bmatrix} E_0 \\ H_0 \end{bmatrix} = \boldsymbol{M}_1 \boldsymbol{M}_2 \cdots \boldsymbol{M}_m \begin{bmatrix} E_m \\ H_m \end{bmatrix}. \tag{11.41}$$

We may normalize by equating the amplitude E_m to unity, thus

$$E_m = E_{m^+} = 1 \tag{11.42}$$

and, therefore,

$$H_m = n_s E_{m^+} = n_s. \tag{11.43}$$

Combining (11.41), (11.42) and (11.43), we get

$$\begin{bmatrix} E_0 \\ H_0 \end{bmatrix} = \boldsymbol{M}_1 \boldsymbol{M}_2 \cdots \boldsymbol{M}_m \begin{bmatrix} 1 \\ n_s \end{bmatrix}. \tag{11.44}$$

Note that, by (11.40), the matrix $\boldsymbol{M}_j$ depends only on the thickness and refractive index of the jth layer.

The treatment given above applies to films made up of multiple layers with no special relationship between the layers, and in solving (11.44) there appears to be no alternative to forming the matrix product step-by-step. However, the films used for filters and laser mirrors are often of a simple periodic structure, which may consist of alternative quarter-wave layers of low and high refractive index, thus, LHLH . . . HLH. For films of periodic multilayers MIELENZ [1959] has shown very elegantly that the computation can be simplified considerably by using Chebyshev polynomials as described below.

References on page 338

If two multilayer films can be represented by the same matrix at a given wavelength, they are said to be equivalent at that wavelength. HERPIN [1947] has shown that any multilayer is equivalent, at one wavelength, to a fictitious bilayer. If the layer combination is of the form abc . . . cba, i.e. if radiation "sees" the same sequence of thicknesses and indices from either side of the film it is a symmetrical multilayer and its equivalent is a fictitious monolayer. Even if the multilayers are purely dielectric (real reffractive index), the equivalent fictitious monolayer or bilayer may have a complex index. Considering a periodic multilayer, this may be applied to the fundamental period of layers. Let the bilayer for the fundamental be represented by $\mathbf{A}_a\mathbf{A}_b$ and let the period occur m times. From (11.44) we get

$$\begin{bmatrix} E_0 \\ H_0 \end{bmatrix} = [\mathbf{A}_a\mathbf{A}_b]^m \begin{bmatrix} 1 \\ n_s \end{bmatrix}. \tag{11.45}$$

If it is a "symmetrical" multilayer, then, we have

$$\begin{bmatrix} E_0 \\ H_0 \end{bmatrix} = \mathbf{A}_\alpha\mathbf{A}_{\alpha-1} \cdots \mathbf{A}_2\mathbf{A}_1\mathbf{A}_2 \cdots \mathbf{A}_{\alpha-1}\mathbf{A}_\alpha \begin{bmatrix} 1 \\ n_s \end{bmatrix}.$$

Such a layer is called a "periodic symmetrical" multilayer if it consists of $(m+\frac{1}{2})$ fundamental periods, and, for this case, we have

$$\begin{bmatrix} E_0 \\ H_0 \end{bmatrix} = [\mathbf{A}_a\mathbf{A}_b]^m\mathbf{A}_a \begin{bmatrix} 1 \\ n_s \end{bmatrix}. \tag{11.46}$$

Equations (11.45) and (11.46) include the power of a 2×2 matrix which has been shown in appendix E to be given in terms of its elements and Chebyshev polynomials of the second kind by (E.36). Let the elements of the fundamental matrix be given by

$$\mathbf{A}_a\mathbf{A}_b = \begin{bmatrix} a_{11} & a_{12} \\ a_{21} & a_{22} \end{bmatrix}. \tag{11.47}$$

The argument of the Chebyshev polynomials is then

$$X = a_{11}+a_{22}. \tag{11.48}$$

The product $\mathbf{A}_a\mathbf{A}_b$ for the two fictitious (equivalent) layers a and b is

$$\begin{bmatrix} \cos\Phi_a & (\mathrm{i}/n_a)\sin\Phi_a \\ \mathrm{i}n_a\sin\Phi_a & \cos\Phi_a \end{bmatrix}\begin{bmatrix} \cos\Phi_b & (\mathrm{i}/n_b)\sin\Phi_b \\ \mathrm{i}n_b\sin\Phi_b & \cos\Phi_b \end{bmatrix}.$$

Forming this product and using (11.47) and (11.48) we obtain

$$X = 2\cos\Phi_a\cos\Phi_b - \frac{n_a^2+n_b^2}{n_a n_b}\sin\Phi_a\sin\Phi_b. \tag{11.49}$$

If the fictitious layers are each of the same optical thickness, then we have

$$\Phi_a = \Phi_b = \Phi \tag{11.50}$$

and (11.49) becomes

$$X = 2 - \frac{(n_a+n_b)^2}{n_a n_b}\sin^2\Phi. \tag{11.51}$$

For the periodic symmetrical multilayer, the matrix of the fictitious trilayer is

$$\mathbf{A}_a\mathbf{A}_b\mathbf{A}_a = \begin{bmatrix} \cos\Phi & (\mathrm{i}/n_a)\sin\Phi \\ \mathrm{i}n_a\sin\Phi & \cos\Phi \end{bmatrix}\begin{bmatrix} \cos\Phi & (\mathrm{i}/n_b)\sin\Phi \\ \mathrm{i}n_b\sin\Phi & \cos\Phi \end{bmatrix}\begin{bmatrix} \cos\Phi & (\mathrm{i}/n_a)\sin\Phi \\ \mathrm{i}n_a\sin\Phi & \cos\Phi \end{bmatrix},$$

which, when multiplied out, gives

$$\mathbf{A}_a\mathbf{A}_b\mathbf{A}_a = X\mathbf{A}_a - \mathbf{A}_b^{-1}, \tag{11.52}$$

where

$$\mathbf{A}_b^{-1} = \begin{bmatrix} \cos\Phi & -(\mathrm{i}/n_b)\sin\Phi \\ -\mathrm{i}n_b\sin\Phi & \cos\Phi \end{bmatrix}. \tag{11.53}$$

From (E.25) and (E.36) we get

$$[\mathbf{A}_a\mathbf{A}_b]^m\mathbf{A}_a = S_{m-1}(X)\mathbf{A}_a\mathbf{A}_b\mathbf{A}_a - S_{m-2}(X)\mathbf{I}\mathbf{A}_a,$$

which, using (E.31) and (11.52) gives

$$[\mathbf{A}_a\mathbf{A}_b]^m\mathbf{A}_a = S_m(X)\mathbf{A}_a - S_{m-1}(X)\mathbf{A}_b^{-1}. \tag{11.54}$$

This shows the important result that the two basic matrices $\mathbf{A}_a$ and $\mathbf{A}_b$ of the fictitious layers, together with the appropriate Chebyshev polynomials (appendix I), are all that is required to calculate the transfer matrix for the periodic-symmetrical multilayer if the fictitious layers have the same optical thickness. The multiplication necessary to calculate $[\mathbf{A}_a\mathbf{A}_b]^m\mathbf{A}_a$ without using the RHS of (11.54) is quite tedious and expensive in computer time.

The periodic symmetrical multilayer has an equivalent monolayer of

References on page 338

thickness Φ_1 and refractive index n_1. Hence, we can write

$$[\mathbf{A}_a\mathbf{A}_b]^m\mathbf{A}_a = \begin{bmatrix} \cos\Phi_1 & (i/n_1)\sin\Phi_1 \\ in_1\sin\Phi_1 & \cos\Phi_1 \end{bmatrix}. \tag{11.55}$$

From (11.53) and (11.54) we get

$$[\mathbf{A}_a\mathbf{A}_b]^m\mathbf{A}_a = S_m(X)\begin{bmatrix} \cos\Phi & (i/n_a)\sin\Phi \\ in_a\sin\Phi & \cos\Phi \end{bmatrix} - S_{m-1}(X) \begin{bmatrix} \cos\Phi & -(i/n_b)\sin\Phi \\ -in_b\sin\Phi & \cos\Phi \end{bmatrix}. \tag{11.56}$$

Comparison of elements in (11.55) and (11.56) gives

$$\cos\Phi_1 = [S_m(X) - S_{m-1}(X)]\cos\Phi \tag{11.57}$$

and

$$n_1^2 = \frac{n_a S_m(X) + n_b S_{m-1}(X)}{\dfrac{1}{n_a}S_m(X) + \dfrac{1}{n_b}S_{m-1}(X)}. \tag{11.58}$$

The argument X is a real number if the multilayers are purely dielectric, and if the individual matrices $\mathbf{A}_a$ and $\mathbf{A}_b$ represent individual layers. This is the case for a film made of alternate quarter-wave layers (at λ_0) of low and high refractive index. We then have

$$\Phi_a = \Phi_b = \phi = \frac{\pi}{2}\frac{\lambda_0}{\lambda} \tag{11.59}$$

and

$$X = x = 2 - \frac{n_a + n_b}{n_a n_b}\sin^2\phi \tag{11.60}$$

is real.

In general, the intensity reflection coefficient (reflectance) R may be defined by the equation

$$R = \left|\frac{E'_{0-}}{E_{0-}}\right|^2, \tag{11.61}$$

and the intensity transmission coefficient (transmittance) T, by

$$T = \frac{\mathscr{Re}(n_s)}{n_0}\left|\frac{E_{m+}}{E_{0-}}\right|^2, \tag{11.62}$$

where $\mathscr{Re}$ denotes the "real part of".

Substituting $j = 0$ into (11.38) and using the resulting equation in (11.61) and (11.62) (remembering that E_{m+} is taken to be unity), we get

$$R = \left| \frac{\mathsf{E}_0 - H_0/n_0}{\mathsf{E}_0 + H_0/n_0} \right|^2, \tag{11.63}$$

and

$$T = \frac{4n_s}{n_0|\mathsf{E}_0 + (H_0/n_0)|^2}. \tag{11.64}$$

Example. To show how the above analysis may be applied we shall consider a laser mirror suitable for the argon II system and shall calculate the reflection coefficient at 4880 Å of a multilayer system consisting of layers which are each a quarter wave at 5145 Å. Let the system have 11 alternating layers of zinc sulphide and magnesium fluoride on glass. We have

$$n_0 = 1 \text{ (air)}, \qquad n_b = 1.38 \text{ (MgF}_2\text{)},$$

$$n_a = 2.3 \text{ (ZnS)}, \qquad n_s = 1.52 \text{ (glass)}.$$

From (11.59) we get $\phi = 1.65631$ and, therefore, $\cos\phi = -0.08540$, $\sin\phi = 0.99634$. Then we use (11.46) and (11.56) to obtain

$$\begin{bmatrix} \mathsf{E}_0 \\ H_0 \end{bmatrix} = \left[S_5(X) \begin{pmatrix} -0.08540 & 0.99634\mathrm{i}/2.3 \\ 2.3 \times 0.99634\mathrm{i} & -0.08540 \end{pmatrix} - \right.$$

$$\left. - S_4(X) \begin{pmatrix} -0.08540 & -0.99634\mathrm{i}/1.38 \\ -1.38 \times 0.99634\mathrm{i} & -0.08540 \end{pmatrix} \right] \begin{bmatrix} 1 \\ 1.52 \end{bmatrix}.$$

The argument (X) of the Chebyshev polynomials $S(X)$ is given by (11.60), $x = -2.23549$. Using (E.33), we then calculate $S_5(X)$ and $S_4(X)$ to be

$$S_5 = -17.84927, \qquad S_4 = +10.98192,$$

and, finally, we get

$$\begin{bmatrix} \mathsf{E}_0 \\ H_0 \end{bmatrix} = \begin{bmatrix} 2.46219 + 0.29897\mathrm{i} \\ -25.80359\mathrm{i} + 3.74253 \end{bmatrix},$$

which gives

$$\mathsf{E}_0 = 2.46219 + 0.29897\mathrm{i},$$

$$H_0 = 3.74253 - 25.80359\mathrm{i}.$$

Using these values in (11.61), (11.62) we get the reflectance and transmittance of the given multilayer at 4880 Å to be

$$R = 99.12\% \quad \text{and} \quad T = 0.88\%.$$

References on page 338

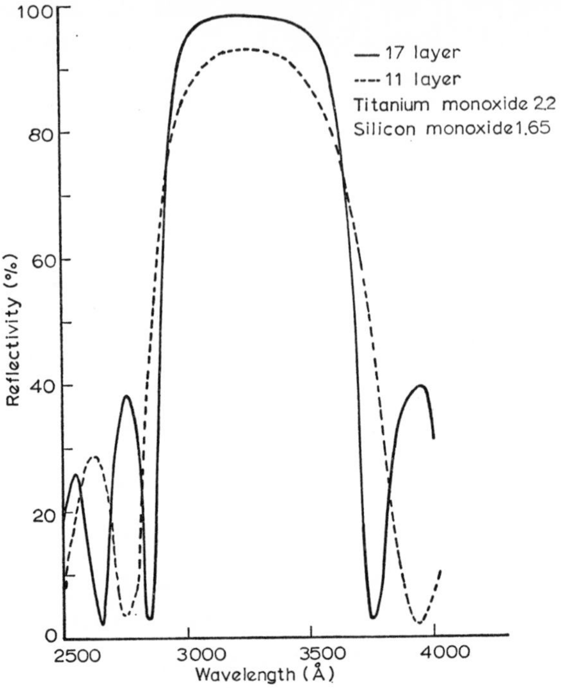

Fig. 11.36. Reflectivity as a function of wavelength for multilayer dielectric layers centred on 3250 Å.

Figure 11.36 shows the reflectivity as a function of wavelength for mirrors with $\lambda_0 = 3250$ Å. The mirror substrates are fused quartz and a total of 17 layers and 11 layers respectively of titanium monoxide and silicon monoxide constitute the mirrors. The refractive indices used for the calculation are 2.2 (TiO) and 1.65 (SiO). Since the refractive indices vary with details of the evaporation process, average values have been chosen. Absorption has been neglected.

References

Berning, P. H., 1963, Theory and Calculations of Optical Thin Film, in: Hass, G., ed., Physics of Thin Films, Vol. I (Academic Press, New York) pp. 69–121.

Bloom, A. L., 1963(a), Appl. Phys. Letters **2**, 101.

Bloom, A. L., 1963(b), Spectra-Physics, Inc., Mountain View, Calif., Laser Technical Bulletin No. 2, Properties of Laser Resonators Giving Uniphase Wave Fronts.

Bobroff, D. L., 1964, Appl. Optics **3**, 1485.

Boyd, G. D. and H. Kogelnik, 1962, Bell System Tech. J. **41**, 1347.

Fox, A. G. and T. Li, 1961, Bell System Tech. J. **40**, 453.

Fox, A. G. and T. Li, 1963, Proc. I.E.E.E. **51**, 80.

Goldsborough, J. P., 1964, Appl. Optics **3**, 267.

Giordmaine, J. A. and W. Kaiser, 1964, J. Appl. Phys. **35**, 3446.

Herpin, A., 1947, Compt. Rend. Acad. Sci. **225**, 182.

Javan, A., W. R. Bennett, Jr., and D. R. Herriott, 1961, Phys. Rev. Letters **6**, 106.

Kolomnikov, Yu. D., Yu. V. Troitskiy and V. P. Chebotayev, 1965, Radio Engng. Electronic Phys. (USA) **10**, 312.

Li, T., 1963, Bell Syst. Tech. J. **42**, 2609.

Mielenz, K. D., 1959, J. Research Natl. Bur. Standards (A) **63** (A), 297.

Mielenz, K. D., K. F. Nefflen and K. E. Gillilland, 1964, Appl. Optics **3**, 785.

Peck, E. R., 1962, J. Opt. Soc. Am. **52**, 253.

Rigrod, W. W., H. Kogelnik, D. J. Brangaccio and D. R. Herriott, 1962, J. Appl. Phys. **33**, 743.

Rigrod, W. W., 1964, Isolation of Axi-Symmetrical Optical Resonator Modes, In: Grivet, P. and N. Bloembergen, eds., Proceedings of the Third International Congress on Quantum Electronics (Columbia University Press, New York) pp. 1285–1290.

Sinclair, D. C., 1964, Appl. Optics **3**, 1067.

Smith, P. W., 1965, I.E.E.E. J. Quantum Electronics, **QE–1**, 343.

Statz, H. and C. L. Tang, 1965, J. Appl. Phys. **36**, 1816.

White, A. D., 1964, Appl. Optics **3**, 431.

White, A. D., E. I. Gordon and E. F. Labuda, 1964, Appl. Phys. Letters **5**, 97.

White, A. D., 1965, I.E.E.E. J. Quantum Electronics, **QE–1**, 349.

Yariv, A. and J. P. Gordon, 1963, Proc. I.E.E.E. **51**, 4.

APPENDIX A

COMPLEX INTEGRATION

To evaluate the type of integral appearing in dispersion relations and Hilbert transforms, contour integration in the complex plane is used. Cauchy's integral theorem is to the effect that the integral of an analytic function along any closed curve (contour) lying completely within the domain of analyticity is zero. Consider the function $F(z) = f(z)/p(z)$ where $f(z)$ is analytic and $p(z)$ is a polynomial of order n. If the roots of the polynomial are $a_1, \ldots, a_n$ we have $f(z)/(z-a_1) \ldots (z-a_n)$ and $F(z)$ is not analytic at the points $z = a_1, \ldots, a_n$. The function $F(z)$ is said to have poles at $z = a_1, \ldots, a_n$. Since the roots a_i need not be real, they can be points anywhere in the complex plane. The residue of the function at a pole is important in evaluating an integral. If the pole is at $z = a_i$, the residue is obtained by multiplying the function $f(z)/p(z)$ by $(z-a_i)$ and then putting $z = a_i$; thus, the residue at the pole a_i is given by

$$\frac{f(a_i)}{(a_i-a_1)(a_i-a_2)\cdots(a_i-a_{i-1})(a_i-a_{i+1})\cdots(a_i-a_a)}. \tag{A.1}$$

Dispersion relations

The starting point in deriving the dispersion relations is the Cauchy integral formula. If function $f(z)$ of the complex variable z is analytic everywhere within and along a closed contour C, the value of $f(z)$ at any point a is completely determined by its value along any closed contour enclosing a. Thus we have,

$$\begin{aligned} \frac{1}{2\pi i}\oint_C \frac{f(z)dz}{z-a} &= f(a) \quad \text{if } a \text{ is inside } C, \\ &= 0 \quad \text{if } a \text{ is outside } C; \\ \frac{1}{\pi i} P\oint_C \frac{f(z)dz}{z-a} &= f(a) \quad \text{if } a \text{ lies on } C, \end{aligned} \tag{A.2}$$

where P is the principal value of the integral (see below). The contour C_1 fig. A.1 may be chosen to be from $-\infty$ to $+\infty$ along the real axis, and a

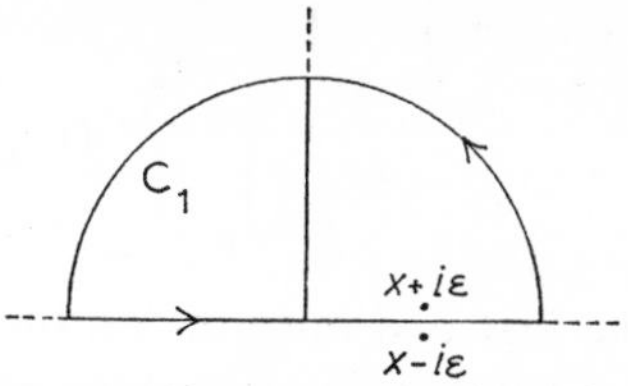

Fig. A.1.

semicircle with infinite radius in the upper half of the complex plane returning to $-\infty$ (i.e. the integration is taken anticlockwise). Let a be very close to the real axis fig. A.1 so that

$$a = x_a \pm \mathrm{i}\varepsilon \qquad (\varepsilon > 0). \tag{A.3}$$

Then, since $x_a + \mathrm{i}\varepsilon$ lies inside the contour of integration but $x_a - \mathrm{i}\varepsilon$ lies outside it, we have

$$\frac{1}{2\pi \mathrm{i}} \oint_{C_1} \frac{f(z)\mathrm{d}z}{z-a} = f(x_a + \mathrm{i}\varepsilon) \quad (+\text{sign}), \qquad = 0 \quad (-\text{sign}). \tag{A.4}$$

If the contribution to this integral from the semicircle vanishes as the radius tends to ∞, we can neglect the semicircle and get

$$\frac{1}{2\pi \mathrm{i}} \int_{-\infty}^{\infty} \frac{f(x)\mathrm{d}x}{x-(x_a \pm \mathrm{i}\varepsilon)} = f(x_a + \mathrm{i}\varepsilon) \quad (+\text{sign}), \qquad = 0 \quad (-\text{sign}). \tag{A.5}$$

We now take the limit $\varepsilon \to 0$ corresponding to z being a real quantity. This is straightforward as far as the right hand side of this equation is concerned and gives $f(x_a)$. To obtain the limit $\varepsilon \to 0$ for the left hand side of the equation we take a contour C_2 from $-\infty$ to $+\infty$ as before but avoid the point

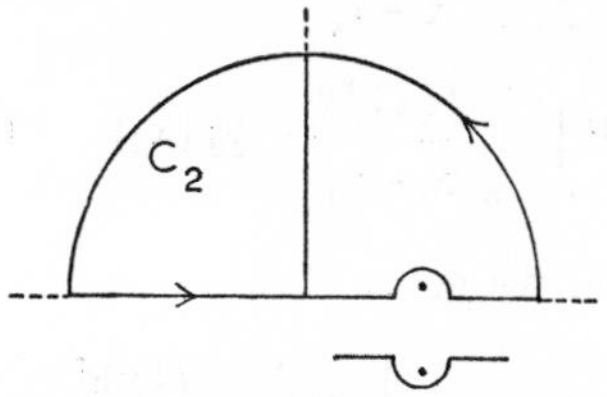

Fig. A.2.

x_a ($\varepsilon = 0$) by introducing a small semicircle of radius r which is either above or below x_a as shown in fig. A.2. We require the limit of the contour integral as $r \to 0$.

We have

$$\frac{1}{2\pi i}\oint_{C_2}\frac{f(z)dz}{z-a} = \frac{1}{2\pi i}\left(\int_{-\infty}^{x_a-r}+\int_{x_a+r}^{\infty}\right)\frac{f(x)dx}{x-x_a}+ + \frac{1}{2\pi i}\int_{\pi}^{0 \text{ or } 2\pi}\frac{f(x_a+re^{i\theta})ire^{i\theta}d\theta}{(x_a+re^{i\theta})-x_a}. \tag{A.6}$$

In the limit $r \to 0$, the first two integrals on the right make an integral called Cauchy's principal value which is designated by P, thus

$$P\int_{-\infty}^{\infty}\frac{f(x)dx}{x-x_a} = \lim_{r\to 0}\left(\int_{-\infty}^{x_a-r}\frac{f(x)dx}{x-x_a}+\int_{x_a+r}^{\infty}\frac{f(x)dx}{x-x_a}\right). \tag{A.7}$$

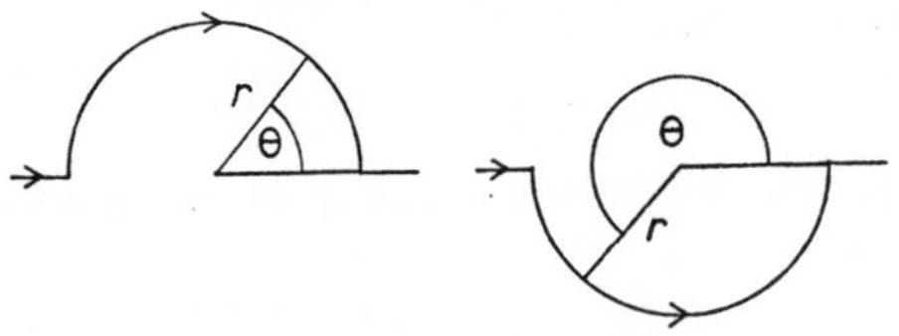

Fig. A.3.

The third integral on the right hand side may be obtained by considering fig. A.3 from which it readily follows that

$$z-a = r(\cos\theta+i\sin\theta) = re^{i\theta}, \qquad dz = ire^{i\theta}d\theta. \tag{A.8}$$

The upper limit 0 corresponds to the contour above x_a, and the limit 2π corresponds to the contour below x_a.

In the limits as $\varepsilon \to 0$ and $r \to 0$ from (A.5), (A.6) and (A.7) we get

$$f(x_a) = \frac{1}{2\pi i}P\int_{-\infty}^{\infty}\frac{f(x)dx}{x-x_a}+\tfrac{1}{2}f(x_a) \qquad (+\text{sign}), \tag{A.9}$$

$$0 = \frac{1}{2\pi i}P\int_{-\infty}^{\infty}\frac{f(x)dx}{x-x_a}-\tfrac{1}{2}f(x_a) \qquad (-\text{sign}). \tag{A.10}$$

Adding (A.9) and (A.10) we get

$$f(x_a) = \frac{1}{\pi i}P\int_{-\infty}^{\infty}\frac{f(x)dx}{x-x_a}. \tag{A.11}$$

Let $f(x)$ be a complex function given by

$$f(x) = \mathcal{R}(x)+i\mathcal{I}(x). \tag{A.12}$$

From (A.11) and (A.12) we get

$$\mathscr{R}(x_a)+\mathrm{i}\mathscr{I}(x_a) = \frac{1}{\pi \mathrm{i}} P\int_{-\infty}^{\infty} \frac{\mathscr{R}(x)\mathrm{d}x}{x-x_a} + \frac{1}{\pi \mathrm{i}} P\int_{-\infty}^{\infty} \frac{\mathrm{i}\mathscr{I}(x)\mathrm{d}x}{x-x_a}.$$

Equating real and imaginary parts we get

$$\mathscr{R}(x_a) = \frac{1}{\pi} P\int_{-\infty}^{\infty} \frac{\mathscr{I}(x)\mathrm{d}x}{x-x_a}; \tag{A.13a}$$

$$\mathscr{I}(x_a) = -\frac{1}{\pi} P\int_{-\infty}^{\infty} \frac{\mathscr{R}(x)\mathrm{d}x}{x-x_a}. \tag{A.14}$$

These relations between the real and imaginary parts of a function are the dispersion relations. They are valid when there are no poles in the upper half-plane.

If the function $f(x)$ is such that it satisfies the crossing symmetry relationship

$$f(-x) = f^*(x),$$

where the asterisk denotes the complex conjugate, then from (A.12) we have

$$f(-x) = \mathscr{R}(-x)+\mathrm{i}\mathscr{I}(-x) \quad \text{and} \quad f^*(x) = \mathscr{R}(x)-\mathrm{i}\mathscr{I}(x).$$

Equating real and imaginary parts we get

$$\mathscr{R}(-x) = \mathscr{R}(x) \quad \text{and} \quad \mathscr{I}(-x) = -\mathscr{I}(x),$$

which show that $\mathscr{R}$ and $\mathscr{I}$ are even and odd functions, respectively.

The integral in (A.13) can now be written thus

$$\int_{-\infty}^{\infty} \frac{\mathscr{I}(x)\mathrm{d}x}{x-x_a} = \int_{-\infty}^{0} \frac{\mathscr{I}(x)\mathrm{d}x}{x-x_a} + \int_{0}^{\infty} \frac{\mathscr{I}(x)\mathrm{d}x}{x-x_a},$$

and we can use the odd character of $\mathscr{I}$ to obtain

$$\mathscr{R}(x_a) = \frac{2}{\pi} P\int_{0}^{\infty} \frac{x\mathscr{I}(x)\mathrm{d}x}{x^2-x_a^2}. \tag{A.13b}$$

This form of a dispersion relation is more directly applicable to physical quantities which cannot be negative, e.g. frequency.

Hilbert transforms

The Hilbert transform of a function $f(x)$ may be designated by $H[f(x)]$ and is given by

$$H[f(x)] = \frac{1}{\pi}\int_{-\infty}^{\infty} \frac{f(x)\mathrm{d}x}{x-a} = g(a), \tag{A.15}$$

where the principal value of the integral is taken if x and a are both real. Inversion gives

$$f(a) = -H[g(x)] = -\frac{1}{\pi}\int_{-\infty}^{\infty} \frac{g(x)\mathrm{d}x}{x-a}. \tag{A.16}$$

Equations (A.13) and (A.14) show that the real and imaginary part of $f(x)$ are Hilbert transforms of each other. An integral which is useful in the theory of coherence is of the form

$$P\int_{-\infty}^{\infty} \frac{\mathrm{e}^{-\mathrm{i}x\omega}\mathrm{d}x}{x-a}. \tag{A.17}$$

This integral has the values given by

$$\begin{aligned} P\int_{-\infty}^{\infty} \frac{\mathrm{e}^{-\mathrm{i}x\omega}\mathrm{d}x}{x-a} &= -\pi \mathrm{i}\mathrm{e}^{-\mathrm{i}a\omega} \quad \text{for} \quad \omega > 0, \\ &= \pi \mathrm{i}\mathrm{e}^{-\mathrm{i}a\omega} \quad \text{for} \quad \omega < 0. \end{aligned} \tag{A.18}$$

Expansions of complex functions

If a function $f(z)$ is analytic everywhere within a circle centred at a, then it may be expanded into a Taylor series. The radius of the circle about a within which $f(z)$ is analytic is the circle of convergence of the Taylor series. The expansion is given by

$$f(z) = \sum_{n=0}^{\infty} c_n(z-a)^n, \tag{A.19}$$

where the coefficients c_n are given by

$$c_n = \frac{f^{(n)}(a)}{n!}. \tag{A.20}$$

If $f(z)$ is not analytic at a but is analytic along and between two concentric circles C_1 and C_2 about the centre a, the more general Laurent expansion must be used. This expansion is given by

$$f(z) = \sum_{n=-\infty}^{\infty} c_n(z-a)^n, \tag{A.21}$$

where the coefficients c_n are given by

$$c_n = \frac{1}{2\pi \mathrm{i}}\int_{\mathrm{C}} \frac{f(z)\mathrm{d}z}{(z-a)^{n+1}} \tag{A.22}$$

and C is any closed path taken anticlockwise about a within the annular region between C_1 and C_2. The radius of the inner circle about a can be infinitesimally small.

Integral representation of Bessel functions

The representation of a Bessel function of 1st kind and nth order as an integral is required to establish equation (6.34a). It may be derived as follows. The generating function (I.28) for the Bessel function $J_n(x)$ is given by

$$g(x, t) = \exp\left[\frac{x}{2}\left(t - \frac{1}{t}\right)\right] = \sum_{n=-\infty}^{\infty} J_n(x)t^n. \tag{A.23}$$

where $J_n(x)$ is defined in the region $0 \leqslant x < \infty$. The expansion given by (A.23) is a Laurent series and the coefficients $J_n(x)$ may be determined by applying (A.22), from which we get

$$J_n(x) = \frac{1}{2\pi \mathrm{i}} \int_C \frac{g(x, t)}{t^{n+1}} \mathrm{d}t$$

where C encloses the point $t = 0$ and will be taken as a circle of unit radius in the t-plane where $t = \exp(\mathrm{i}\theta)$. We then obtain

$$J_n(x) = \frac{1}{2\pi \mathrm{i}} \int_0^{2\pi} \frac{\exp[x(\mathrm{e}^{\mathrm{i}\theta} - \mathrm{e}^{-\mathrm{i}\theta})/2]\mathrm{i}\exp(\mathrm{i}\theta)\mathrm{d}\theta}{[\exp(\mathrm{i}\theta)]^{n+1}}$$

and, finally,

$$J_n(x) = \frac{1}{2\pi} \int_0^{2\pi} \exp \mathrm{i}(x \sin\theta - n\theta)\mathrm{d}\theta. \tag{A.24}$$

References on page 398

HAMILTONIAN FOR CHARGED PARTICLE IN ELECTROMAGNETIC FIELD

The force acting on a particle of charge q and mass m moving with velocity $\boldsymbol{v}$ in an electromagnetic field is given in Gaussian units by

$$\boldsymbol{F} = q\left[\boldsymbol{E} + \frac{1}{c}\boldsymbol{v}\times\boldsymbol{B}\right]. \qquad \text{(B.1)}$$

This force is called the Lorentz force.

Maxwell's equations are:

$$\nabla\times\boldsymbol{E} + \frac{1}{c}\frac{\partial\boldsymbol{B}}{\partial t} = 0; \qquad \nabla\cdot\boldsymbol{D} = 4\pi\rho; \qquad \text{(B.2a, b)}$$

$$\nabla\times\boldsymbol{H} - \frac{1}{c}\frac{\partial\boldsymbol{D}}{\partial t} = \frac{4\pi\boldsymbol{j}}{c}; \qquad \nabla\cdot\boldsymbol{B} = 0. \qquad \text{(B.2c, d)}$$

It is seen that when a time varying magnetic field is present, $\boldsymbol{E}$ is not the gradient of a scalar function since $\nabla\times\boldsymbol{E} \neq 0$. However, $\nabla\cdot\boldsymbol{B} = 0$ so $\boldsymbol{B}$ may be expressed as the curl of a vector, $\boldsymbol{A}$, the magnetic vector potential, thus

$$\boldsymbol{B} = \nabla\times\boldsymbol{A}. \qquad \text{(B.3)}$$

Hence, from (B.2a) we get,

$$\nabla\times\boldsymbol{E} + \frac{1}{c}\frac{\partial}{\partial t}(\nabla\times\boldsymbol{A}) = \nabla\times\left(\boldsymbol{E} + \frac{1}{c}\frac{\partial\boldsymbol{A}}{\partial t}\right) = 0. \qquad \text{(B.4)}$$

Equation (B.4) shows that we may introduce a scalar potential, ϕ, thus

$$\boldsymbol{E} + \frac{1}{c}\frac{\partial\boldsymbol{A}}{\partial t} = -\nabla\phi$$

giving

$$\boldsymbol{E} = -\nabla\phi - \frac{1}{c}\frac{\partial\boldsymbol{A}}{\partial t}. \qquad \text{(B.5)}$$

Equations (B.1), (B.3) and (B.5) may now be combined to give

$$\boldsymbol{F} = q\left[-\nabla\phi - \frac{1}{c}\frac{\partial \boldsymbol{A}}{\partial t} + \frac{1}{c}\boldsymbol{v}\times(\nabla\times\boldsymbol{A})\right]. \tag{B.6}$$

To derive the Lagrangian from (B.6) we must express the vector potential terms rather differently. When a particle is moving in a time varying electromagnetic field, the value of $\boldsymbol{A}$ at any given point in the field changes with time and the particle changes its location with time. The relevant value of $\boldsymbol{A}$ is that at the location of the particle. If the particle were stationary we should only be concerned with the changes of $\boldsymbol{A}$ due to the variation of $\boldsymbol{A}$ with time, i.e. $\partial \boldsymbol{A}/\partial t$. However, since $\boldsymbol{A}$ varies spatially, the motion of the particle also contributes to the total change in $\boldsymbol{A}$ with time. Hence, the total time derivative of $\boldsymbol{A}$ is

$$\frac{\mathrm{d}\boldsymbol{A}}{\mathrm{d}t} = \frac{\partial \boldsymbol{A}}{\partial t} + (\boldsymbol{v}\cdot\nabla)\boldsymbol{A}. \tag{B.7}$$

Elementary vector algebra shows that the other vector potential term of (B.6) is

$$\boldsymbol{v}\times\nabla\times\boldsymbol{A} = \nabla(\boldsymbol{v}\cdot\boldsymbol{A}) - (\boldsymbol{v}\cdot\nabla)\boldsymbol{A}. \tag{B.8}$$

Equations (B.7) and (B.8) may now be used in (B.6) to give

$$\boldsymbol{F} = q\left[-\nabla\left(\phi - \frac{1}{c}\boldsymbol{v}\cdot\boldsymbol{A}\right) - \frac{1}{c}\frac{\mathrm{d}\boldsymbol{A}}{\mathrm{d}t}\right].$$

This may be written

$$\boldsymbol{F} = q\left[-\nabla\left(\phi - \frac{1}{c}\boldsymbol{v}\cdot\boldsymbol{A}\right) - \frac{1}{c}\frac{\mathrm{d}}{\mathrm{d}t}\{\nabla_v(\boldsymbol{v}\cdot\boldsymbol{A})\}\right], \tag{B.9}$$

where (in Cartesian coordinates),

$$\nabla_v = \left(\boldsymbol{i}\frac{\partial}{\partial v_x}\right) + \left(\boldsymbol{j}\frac{\partial}{\partial v_y}\right) + \left(\boldsymbol{k}\frac{\partial}{\partial v_z}\right).$$

The information necessary to construct the Lagrangian may now be obtained from (B.9) by putting

$$U = q\phi - (q/c)\boldsymbol{v}\cdot\boldsymbol{A}. \tag{B.10}$$

Writing the x-component of the combination of (B.9) and (B.10), we have

$$F_x = -\frac{\partial U}{\partial x} + \frac{\mathrm{d}}{\mathrm{d}t}\frac{\partial U}{\partial v_x}.$$

References on page 398

It will now be recognized that U is a generalized potential (see GOLDSTEIN [1950] for a discussion of this), so the Lagrangian for a charged particle in an electromagnetic field is

$$L = T - U = T - q\phi + \frac{q}{c}\boldsymbol{v}\cdot\boldsymbol{A}. \tag{B.11}$$

Hence, for an electron, we get

$$L = \left[\frac{mv^2}{2} + e\phi - \frac{e}{c}\boldsymbol{v}\cdot\boldsymbol{A}\right]. \tag{B.12}$$

In terms of generalized coordinates p, q we can write, for components $i = x, y, z$,

$$v_i = \dot{q}_i \quad \text{and} \quad p_i = \frac{\partial L}{\partial \dot{q}_i} = mv_i - \frac{e}{c}A_i. \tag{B.13}$$

By definition, the Hamiltonian is given by

$$H \equiv \sum_i p_i \dot{q}_i - L. \tag{B.14}$$

Hence, using (B.12), we get

$$H = \left(mv^2 - \frac{e}{c}\boldsymbol{v}\cdot\boldsymbol{A}\right) - \left(\frac{mv^2}{2} + e\phi - \frac{e}{c}\boldsymbol{v}\cdot\boldsymbol{A}\right) = \frac{mv^2}{2} - e\phi.$$

Equation (B.13) then gives

$$H = \frac{1}{2m}\left(\boldsymbol{p} + \frac{e}{c}\boldsymbol{A}\right)^2 - e\phi. \tag{B.15}$$

RELATIONSHIP BETWEEN VECTOR POTENTIAL AND ENERGY DENSITY OF ELECTROMAGNETIC FIELD

Let a plane electromagnetic field be described by the equation

$$A = A_0 \cos 2\pi\nu t. \tag{C.1}$$

From (B.5) we have, since $\phi = 0$,

$$E = -\frac{1}{c}\frac{\partial A}{\partial t} = \frac{2\pi\nu A_0}{c}\sin(2\pi\nu t).$$

Hence,

$$E^2 = \frac{4\pi^2\nu^2 A_0^2}{c^2}\sin^2(2\pi\nu t).$$

In the case where the energy in the radiation field is distributed over a finite frequency range, $|E|^2$ is re-defined as being associated with unit frequency interval of the radiation field so that $|E(\nu)|^2 d\nu$ is the field intensity associated with frequency interval $d\nu$ about ν. Hence, we have,

$$\rho(\nu) = E(\nu)^2/4\pi, \tag{C.2}$$

where $\rho(\nu)$ is the spectral energy density per unit frequency interval.

References on page 398

OBLIQUE INCIDENCE OF ELECTROMAGNETIC WAVES ON A DIELECTRIC SURFACE

Consider a beam of light incident on the surface of a transparent material. The beam is moving from a medium of lower, to one of higher refractive index. At normal incidence to a plane surface some light is reflected (about 4 % for $\mu = 1.5$) and the rest is transmitted into the medium. The electric vector of the radiation is parallel to the surface and considerations of symmetry lead us to conclude that the reflected and refracted beams will have the same polarization characteristics as the incident beam. At oblique incidence, the symmetry is lost and the orientation of the plane of the electric vector relative to the plane of the surface becomes important in determining the polarized state of the reflected and refracted beams (i.e. these beams have different polarization characteristics from the incident beam).

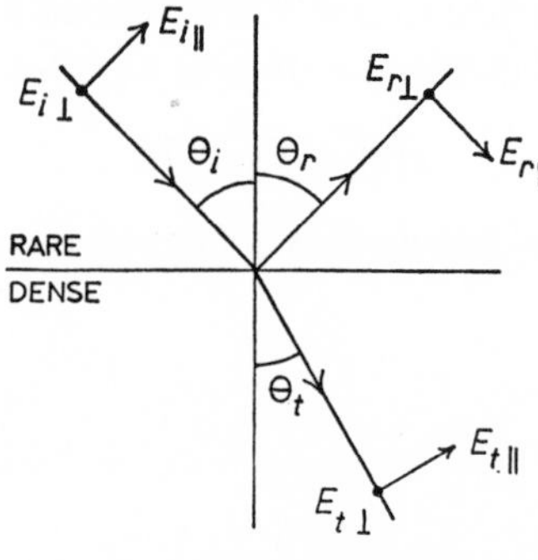

Fig. D.1.

Let $E_{\parallel}$ and $E_{\perp}$ be the components of the electric vector of a plane wave incident on a plane surface as shown in fig. D.1 in which suffixes i, r, and t refer to the incident, reflected and transmitted (refracted) rays.

The Fresnel equations for reflection and refraction are

$$\frac{E_{r\perp}}{E_{i\perp}} = -\frac{\sin(\theta_i-\theta_t)}{\sin(\theta_i+\theta_t)}, \qquad \frac{E_{t\perp}}{E_{i\perp}} = \frac{2\cos\theta_i \sin\theta_t}{\sin(\theta_i+\theta_t)},$$
$$\frac{E_{r\parallel}}{E_{i\parallel}} = -\frac{\tan(\theta_i-\theta_t)}{\tan(\theta_i+\theta_t)}, \qquad \frac{E_{t\parallel}}{E_{i\parallel}} = \frac{2\cos\theta_i \sin\theta_t}{\sin(\theta_i+\theta_t)\cos(\theta_i-\theta_t)}. \tag{D.1}$$

Intensity is the square of the amplitude, and the reflection coefficient, R, is defined as the ratio of the reflected to the incident intensity. Hence,

$$R_{\parallel} = \frac{\tan^2(\theta_i-\theta_t)}{\tan^2(\theta_i+\theta_t)} \quad \text{and} \quad R_{\perp} = \frac{\sin^2(\theta_i-\theta_t)}{\sin^2(\theta_i+\theta_t)}. \tag{D.2}$$

The intensity of the refracted beam is proportional to μ as well as to E^2 and there is also a change in beam cross section.

The relationship between the reflectivities $R_{\parallel}$ and $R_{\perp}$ and the reflectivity R for the electric vector from which the components were derived is

$$R = R_{\parallel}\cos^2\alpha + R_{\perp}\sin^2\alpha, \tag{D.3}$$

where α is the angle between the electric vector and the plane of incidence.

At normal incidence (i.e. small angles), the equations of (D.2) reduce to

$$R_{\perp} = R_{\parallel} = \left(\frac{\mu-1}{\mu+1}\right)^2,$$

since $\theta_i \approx \mu\theta_t$ for small angles.

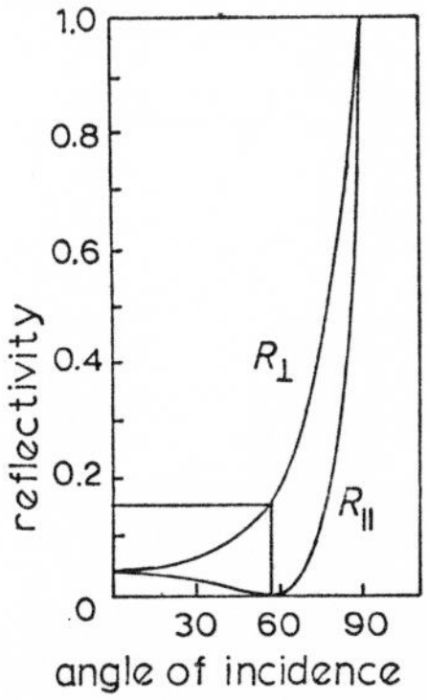

Fig. D.2. Reflectivity versus angle of incidence for two planes of polarization. Reflectivity is zero at the Brewster angle for the $R_{\parallel}$ component.

References on page 398

Equation (D.2) shows that $R_{\parallel}$ is zero when $\theta_i = \theta_t = 90°$ since $\tan(\theta_i + \theta_t) = \infty$. Hence, we get

$$\mu = \tan \theta_{iB} \tag{D.4}$$

where θ_{iB} is the angle at which the parallel component $E_{\parallel}$ is not reflected. This is the Brewster angle.

The curves for $R_{\perp}$ and $R_{\parallel}$ obtained from (D.2) are given in fig. D.2 and represent upper and lower limits of the reflectivities between which the reflectivities corresponding to all other orientations of the electric vector must lie. At the Brewster angle, the $\parallel$-component is not reflected at all while about 15 % of the $\perp$-component is reflected.

Inspection of fig. D.2 shows that radiation polarized with the electric vector in a plane at a small angle to the incident plane could have quite low reflection losses.

At normal incidence, the parallel and perpendicular components are equal. At grazing incidence, both components are totally reflected.

When a beam is travelling from a medium of higher to one of lower refractive index the fresnel equations, (D.1) still apply with angles θ_i and θ_t interchanged. The previous angles θ_{iB} and θ_{tB} are unchanged, that is, the beam incident at the previous angle θ_{tB} but travelling from the optically denser to the rarer medium is refracted at θ_{iB}. At the critical angle, the reflectivity is 100 %. At angles of incidence greater than the critical angle, Fresnel's equations contain imaginary quantities and we shall not consider these. At normal incidence, the reflectivity is again about 4 % for $\mu = 1.5$.

MATRIX THEORY

Physics and engineering courses now generally include matrix theory, but the details tend to be forgotten rather easily. The reader who has not made a formal study of matrix theory should consult one of the many texts available (e.g. GANTMACHER [1959], BOWMAN [1962], SCHWARTZ [1961]. The matrix theory we review is just that relevant to the topics considered in this book.

The representation of a set of equations by a matrix and the rules for the multiplication of matrices may be deduced from the following examples.

1. $$\begin{aligned} x' &= ax+by \\ y' &= cx+dy \end{aligned} \rightarrow \begin{bmatrix} x' \\ y' \end{bmatrix} = \begin{bmatrix} a & b \\ c & d \end{bmatrix} \begin{bmatrix} x \\ y \end{bmatrix}.$$

2. $$\begin{bmatrix} x'' \\ y'' \end{bmatrix} = \begin{bmatrix} e & f \\ g & h \end{bmatrix} \begin{bmatrix} a & b \\ c & d \end{bmatrix} \begin{bmatrix} x \\ y \end{bmatrix} =$$

$$= \begin{bmatrix} (ea+fc) & (eb+fd) \\ (ga+hc) & (gb+hd) \end{bmatrix} \begin{bmatrix} x \\ y \end{bmatrix} =$$

$$= \begin{bmatrix} (ea+fc)x & (eb+fd)y \\ (ga+hc)x & (gb+hd)y \end{bmatrix};$$

$$x'' = (ea+fc)x+(eb+fd)y,$$

$$y'' = (ga+hc)x+(gb+hd)y.$$

To multiply a matrix by a scalar, each element of the matrix is multiplied by the scalar. A factor common to each element of a matrix may be removed. For example

$$m \begin{bmatrix} a & b \\ c & d \end{bmatrix} = \begin{bmatrix} ma & mb \\ mc & md \end{bmatrix}.$$

Addition and subtraction of matrices is demonstrated by the following example

References on page 398

$$\begin{bmatrix} a & b \\ c & d \end{bmatrix} \pm \begin{bmatrix} k & l \\ m & n \end{bmatrix} = \begin{bmatrix} a \pm k & b \pm l \\ c \pm m & d \pm n \end{bmatrix}.$$

If the elements of a matrix are time-dependent, the derivative of the matrix is obtained by replacing each element of the matrix with its time derivative; thus

$$\frac{\mathrm{d}}{\mathrm{d}t}\begin{bmatrix} a & b \\ c & d \end{bmatrix} = \begin{bmatrix} \dot{a} & \dot{b} \\ \dot{c} & \dot{d} \end{bmatrix},$$

where the dot represents differentiation with respect to time.

Equation (9.41) provides a useful example of the above; writing it out fully we have

$$\mathrm{i}\frac{\partial}{\partial t}\begin{bmatrix} \rho_{aa} & \rho_{ab} \\ \rho_{ba} & \rho_{bb} \end{bmatrix} =$$

$$= \mathrm{i}\begin{bmatrix} \lambda_a & 0 \\ 0 & \lambda_b \end{bmatrix} + \frac{1}{\hbar}\begin{bmatrix} E_a & V \\ V & E_b \end{bmatrix}\begin{bmatrix} \rho_{aa} & \rho_{ab} \\ \rho_{ba} & \rho_{bb} \end{bmatrix} - \frac{1}{\hbar}\begin{bmatrix} \rho_{aa} & \rho_{ab} \\ \rho_{ba} & \rho_{bb} \end{bmatrix}\begin{bmatrix} E_a & V \\ V & E_b \end{bmatrix} -$$

$$- \frac{\mathrm{i}}{2}\begin{bmatrix} \gamma_a & 0 \\ 0 & \gamma_b \end{bmatrix}\begin{bmatrix} \rho_{aa} & \rho_{ab} \\ \rho_{ba} & \rho_{bb} \end{bmatrix} - \frac{\mathrm{i}}{2}\begin{bmatrix} \rho_{aa} & \rho_{ab} \\ \rho_{ba} & \rho_{bb} \end{bmatrix}\begin{bmatrix} \gamma_a & 0 \\ 0 & \gamma_b \end{bmatrix}.$$

A separate equation can be written for each element of this matrix equation. To derive (9.42) we multiply the matrices to obtain the elements corresponding to ρ_{ab} and write the appropriate equation, thus

$$\hbar\mathrm{i}\dot{\rho}_{ba} = V\rho_{aa} + E_b\rho_{ba} - \rho_{ba}E_a - \rho_{bb}V - \frac{\mathrm{i}}{2}\gamma_b\hbar\rho_{ba} - \frac{\mathrm{i}}{2}\rho_{ba}\gamma_a\hbar.$$

This gives

$$\dot{\rho}_{ba} = -\mathrm{i}V(\rho_{aa} - \rho_{bb})/\hbar + (\mathrm{i}\omega_{ab} - \gamma)\rho_{ba},$$

where

$$\omega_{ab} = (E_a - E_b)/\hbar, \quad \text{and} \quad \gamma = (\gamma_a + \gamma_b)/2.$$

The element of a matrix situated in the ith row and jth column may be designated a_{ij}. The *transpose* of a matrix $\mathbf{A}$ is designated $\tilde{\mathbf{A}}$ and is obtained by systematically changing columns to rows and rows to columns without changing the order in which they occur; i.e. $\mathbf{A} = [a_{ij}]$ when transposed becomes $\tilde{\mathbf{A}} = [a_{ji}]$. To obtain the *cofactor* of a_{ij}, delete the row and column passing through this element, form the determinant of the resulting $(n-1)$th order matrix, and multiply the determinant by $(-1)^{i+j}$. To obtain the *adjoint* of a matrix $\mathbf{A}$ we replace each element in $\mathbf{A}$ by its cofactor and then transpose the resulting matrix. The *reciprocal* of a square matrix $\mathbf{A}$ is $\mathbf{A}^{-1}$ and is defined as (adjoint $\mathbf{A}$/determinant $\mathbf{A}$), e.g.

$$\begin{bmatrix} a & b \\ c & d \end{bmatrix}^{-1} = \begin{bmatrix} \dfrac{d}{ad-bc} & \dfrac{-b}{ad-bc} \\ \dfrac{-c}{ad-bc} & \dfrac{a}{ad-bc} \end{bmatrix}.$$

A square matrix $\mathbf{A}$ of order n is given. If this is applied to any column matrix $\mathbf{x}$ with n elements, another column matrix is formed. It may be possible to choose $\mathbf{x}$ such that each element of the resulting column matrix is proportional to each element of $\mathbf{x}$, thus

$$\mathbf{A}\mathbf{x} = \lambda \mathbf{x}, \tag{E.1}$$

where λ is the constant of proportionality. When this equation is satisfied, $\mathbf{x}$ is an *eigenvector* of $\mathbf{A}$ and λ is an *eigenvalue*† of $\mathbf{A}$. Let $\mathbf{A}$ be a 2×2 matrix. Writing eq. (E.1) in full for this case we get

$$\begin{aligned} a_{11}x_1 + a_{12}x_2 &= \lambda x_1, \\ a_{21}x_1 + a_{22}x_2 &= \lambda x_2. \end{aligned} \tag{E.2}$$

By transferring terms from the right hand side to the left hand side, these equations become,

$$\begin{aligned} (a_{11}-\lambda)x_1 + a_{12}x_2 &= 0, \\ a_{21}x_1 + (a_{22}-\lambda)x_2 &= 0. \end{aligned} \tag{E.3}$$

This homogeneous system of linear equations has a non-trivial (i.e. non-zero) solution if, and only if, the determinant of the coefficients is zero; that is

$$\begin{vmatrix} (a_{11}-\lambda) & a_{12} \\ a_{21} & (a_{22}-\lambda) \end{vmatrix} = 0. \tag{E.4}$$

By expanding the determinant in (E.4) we obtain the characteristic equation of the matrix $\mathbf{A}$ (in general this is a polynomial). The eigenvalues of $\mathbf{A}$ are

† Other terms used for eigenvector are: proper vector, latent vector and characteristic vector.

Other terms used for eigenvalue are: proper value, latent value, latent root, latent number and characteristic number.

References on page 398

the roots of the characteristic equation,

$$\lambda^2-(a_{11}-a_{22})\lambda+(a_{11}a_{22}-a_{21}a_{12}) = 0.$$

This equation may be written †

$$\lambda^2-(\text{Trace } \mathbf{A})\lambda+(\text{Determinant } \mathbf{A}) = 0. \tag{E.5}$$

If the roots are λ_1 and λ_2, elementary theory of quadratic equations shows that

$$\lambda_1+\lambda_2 = \text{Trace } \mathbf{A} \quad \text{and} \quad \lambda_1\lambda_2 = \text{Det } \mathbf{A}. \tag{E.6}$$

It is sometimes convenient to transform a matrix into another matrix. Such a transform may be described as follows. Let a column matrix $\mathbf{y}$ be obtained when a square matrix $\mathbf{A}$ is applied to a column matrix $\mathbf{x}$ thus

$$\mathbf{y} = \mathbf{Ax}. \tag{E.7}$$

Let us decide to obtain $\mathbf{x}$ from some other column matrix $\mathbf{x}'$ by applying the square matrix $\mathbf{B}$ to $\mathbf{x}'$, and let us obtain $\mathbf{y}$ from $\mathbf{y}'$ by using the same matrix $\mathbf{B}$, thus

$$\mathbf{x} = \mathbf{Bx}', \qquad \mathbf{y} = \mathbf{By}'. \tag{E.8}$$

Then, from eqs. (E.7) and (E.8) we have

$$\mathbf{By}' = \mathbf{ABx}'.$$

Hence

$$\mathbf{y}' = \mathbf{B}^{-1}\mathbf{ABx}'. \tag{E.9}$$

The matrix $\mathbf{B}^{-1}\mathbf{AB}$ is the transform of the matrix $\mathbf{A}$ by the matrix $\mathbf{B}$; we shall call this matrix $\mathbf{C}$, thus

$$\mathbf{C} = \mathbf{B}^{-1}\mathbf{AB}.$$

It may be shown [HEADING, 1958] that the eigenvalues of $\mathbf{C}$ are identical with the eigenvalues of $\mathbf{A}$ and that the trace and determinant of a matrix are invariants, i.e. do not change as a result of a transformation.

There are several types of square matrix; we shall give some terminology and describe some of them. The main diagonal (or leading diagonal) is from upper left to lower right. A *unit matrix* (or identity matrix), $\mathbf{I}$, has all its main diagonal elements equal to unity and all its off-diagonal elements zero, thus $a_{ij} = \delta_{ij}$ where δ_{ij} is the Kronecker symbol. A *diagonal matrix* has leading diagonal elements which may be different from one another, and off-diagonal elements which are all zero; i.e. $a_{ij} = a_{ij}\delta_{ij}$. A *symmetric matrix* is such that $a_{ij}=a_{ji}$, i.e. $\tilde{\mathbf{A}}=\mathbf{A}$. A matrix is *skew symmetric* if $a_{ij} = -a_{ji}$,

† The *trace* of a square, finite matrix is the sum of the diagonal elements.

i.e. $\tilde{\boldsymbol{A}} = -\boldsymbol{A}$. The diagonal elements of such a matrix are all zero. An *orthogonal matrix* is one for which $\boldsymbol{A}\tilde{\boldsymbol{A}} = \tilde{\boldsymbol{A}}\boldsymbol{A} = \boldsymbol{I}$, that is, $\boldsymbol{A}^{-1} = \tilde{\boldsymbol{A}}$. If the columns of an orthogonal matrix are denoted by $\boldsymbol{a}_i$, then we have $\tilde{\boldsymbol{a}}_i\boldsymbol{a}_j = \delta_{ij}$. The rows have the same property.

e.g. The following matrices are each orthogonal

$$\boldsymbol{A} = \tfrac{1}{4}\begin{bmatrix} \sqrt{2} & \sqrt{2} & -2\sqrt{3} \\ \sqrt{6} & \sqrt{6} & 2 \\ 2\sqrt{2} & -2\sqrt{2} & 0 \end{bmatrix}, \quad \boldsymbol{B} = \tfrac{1}{5}\begin{bmatrix} 4 & 3 & 0 \\ -3 & 4 & 0 \\ 0 & 0 & 1 \end{bmatrix}.$$

Orthogonality can be verified for each matrix by noting that for each column (row), the scalar product with itself is unity when the appropriate factor ($\frac{1}{4}$ or $\frac{1}{5}$) is used, and the scalar product of any column (row) with a different column (row) is zero.

The elements of a matrix may be complex quantities. A matrix, $\boldsymbol{A}$, is *Hermitian* if $a_{ij} = a_{ji}^*$, i.e. if $(\tilde{\boldsymbol{A}})^* = \boldsymbol{A}$.

The following matrix is Hermitian,

$$\boldsymbol{A} = \begin{bmatrix} 3 & 1+2\mathrm{i} & 4+3\mathrm{i} \\ 1-2\mathrm{i} & 4 & 3-\mathrm{i} \\ 4-3\mathrm{i} & 3+\mathrm{i} & 5 \end{bmatrix}.$$

The complex conjugate of the transpose of a matrix $\boldsymbol{A}$ is sometimes denoted by $\boldsymbol{A}^\dagger$ rather than $(\tilde{\boldsymbol{A}})^*$. A *unitary matrix*, $\boldsymbol{u}$, is a matrix such that $(\tilde{\boldsymbol{u}})^* = \boldsymbol{u}^{-1}$. Hence, we get $\boldsymbol{I} = \boldsymbol{u}\boldsymbol{u}^{-1} = \boldsymbol{u}\tilde{\boldsymbol{u}}^*$ and $\boldsymbol{I} = \boldsymbol{u}^{-1}\boldsymbol{u} = \tilde{\boldsymbol{u}}^*\boldsymbol{u}$, or $\boldsymbol{u}\boldsymbol{u}^\dagger = \boldsymbol{u}^\dagger\boldsymbol{u} = \boldsymbol{I}$. If the elements of $\boldsymbol{u}$ are u_{ij}, then $\boldsymbol{u}^\dagger$ has elements u_{ji}^* and we have $\Sigma_k u_{ik}u_{jk}^* = \delta_{ij}$ and $\Sigma_k u_{ki}^* u_{kj} = \delta_{ij}$. In this case the rows form a mutually orthogonal set of unit vectors and so do the columns. A unitary transformation describes a coordinate transformation resulting from a rotation of the coordinate axes. Both Hermitian and unitary matrices are extremely important in quantum mechanics.

Matrix types may be summarized as follows.

Matrix	Relation between elements	Types of matrix
$\boldsymbol{I}$	$a_{ij} = \delta_{ij}$	identity matrix, unit matrix
	$a_{ij} = a_{ij}\delta_{ij}$	diagonal matrix
$\tilde{\boldsymbol{A}} = \boldsymbol{A}$	$a_{ij} = a_{ji}$	symmetric matrix
$\tilde{\boldsymbol{A}} = -\boldsymbol{A}$	$a_{ij} = -a_{ji}$	skew-symmetric matrix
$\boldsymbol{A}^{-1} = \tilde{\boldsymbol{A}}$	$\tilde{\boldsymbol{a}}_i\boldsymbol{a}_j = \delta_{ij}$	orthogonal matrix
$(\tilde{\boldsymbol{A}})^* = \boldsymbol{A}^\dagger = \boldsymbol{A}$	$a_{ij} = a_{ji}^*$	Hermitian matrix
$\boldsymbol{u}\boldsymbol{u}^\dagger = \boldsymbol{u}^\dagger\boldsymbol{u} = \boldsymbol{I}$	$\Sigma_k a_{ik}a_{jk}^* = \delta_{ij}$	unitary matrix

References on page 398

Polynomials involving matrices

Matric polynomials and their representation by Chebychev polynomials have been considered in the treatment of multilayer dielectric films in ch. 11 § 10.

If the elements of a matrix $\mathbf{A}$ are polynomials in a scalar, x, then we may expand $\mathbf{A}$ as shown in the following example,

$$\mathbf{A} = \begin{bmatrix} x^2-2x-1 & x^2+4 \\ x^4-x & x^3+2x+3 \end{bmatrix} =$$

$$= \begin{bmatrix} 0 & 0 \\ 1 & 0 \end{bmatrix} x^4 + \begin{bmatrix} 0 & 0 \\ 0 & 1 \end{bmatrix} x^3 + \begin{bmatrix} 1 & 1 \\ 0 & 0 \end{bmatrix} x^2 + \begin{bmatrix} -2 & 0 \\ -1 & 2 \end{bmatrix} x + \begin{bmatrix} -1 & 4 \\ 0 & 3 \end{bmatrix}.$$

In general, we have

$$\mathbf{A} \equiv \mathbf{A}_m x^m + \mathbf{A}_{m-1} x^{m-1} + \cdots + \mathbf{A}_1 x + \mathbf{A}_0 .$$

Polynomials in the square matrix $\mathbf{A}$ of order n may be described as follows,

$$f(\mathbf{A}) \equiv c_m \mathbf{A}^m + c_{m-1} \mathbf{A}^{m-1} + \cdots + c_1 \mathbf{A} + c_0 \mathbf{I}, \tag{E.10}$$

where $f(\mathbf{A})$ denotes the function, the coefficients $c_m, \cdots, c_0$ are scalars, and $\mathbf{I}$ is the unit matrix of order n. The algebra of polynomials in the square matrix $\mathbf{A}$ with scalar coefficients is the same as the familiar algebra of scalar polynomials.

Any square matrix $\mathbf{A}$ of order n satisfies a polynomial equation of the form

$$\mathbf{A}^k + c_{k-1} \mathbf{A}^{k-1} + \cdots + c_0 \mathbf{I} = \mathbf{0} \tag{E.11}$$

where $k \leqslant n$ and $c_{k-1}, \ldots, c_0$ are scalars. Thus a 3×3 matrix always satisfies a cubic, quadratic, or linear equation.

Example

If the matrix $\mathbf{A}$ is given by

$$\mathbf{A} = \begin{bmatrix} 1 & -2 \\ 3 & 4 \end{bmatrix},$$

then it is easy to show that for $c_1 = -5$ and $c_0 = 10$, $(\mathbf{A}^2 + c_1 \mathbf{A} + c_0 \mathbf{I})$ reduces to the zero matrix to satisfy (E.11). The working is as follows:

$$\begin{bmatrix} -5 & -10 \\ 15 & 10 \end{bmatrix} + c_1 \begin{bmatrix} 1 & -2 \\ 3 & 4 \end{bmatrix} + c_0 \begin{bmatrix} 1 & 0 \\ 0 & 1 \end{bmatrix} = \mathbf{A}^2 + c_1 \mathbf{A} + c_0 \mathbf{I} = \begin{bmatrix} 0 & 0 \\ 0 & 0 \end{bmatrix}.$$

By considering corresponding elements to obtain the zero matrix we obtain the above values for c_1 and c_0.

Associated with the matrix $\boldsymbol{A} = [a_{ij}]$ is the characteristic matrix of $\boldsymbol{A}$ given by

$$\boldsymbol{A}-\lambda\boldsymbol{I} = \begin{bmatrix} a_{11}-\lambda & a_{12} & \cdots & a_{1n} \\ a_{21} & a_{22}-\lambda & \cdots & a_{2n} \\ \cdots & \cdots & \cdots & \cdots \\ a_{n1} & a_{n2} & \cdots & a_{nn}-\lambda \end{bmatrix}, \qquad \text{(E.12)}$$

where λ is an arbitrary scalar. The determinant $|\boldsymbol{A}-\lambda\boldsymbol{I}|$ is the characteristic function of $\boldsymbol{A}$ and the equation $|\boldsymbol{A}-\lambda\boldsymbol{I}| = 0$ is the characteristic equation of $\boldsymbol{A}$ whose roots $\lambda_1, \lambda_2, \ldots, \lambda_n$ are the eigenvalues of $\boldsymbol{A}$. Expanding $|\boldsymbol{A}-\lambda\boldsymbol{I}|$, we get

$$|\boldsymbol{A}-\lambda\boldsymbol{I}| \equiv (-1)^n(\lambda^n+p_1\lambda^{n-1}+p_2\lambda^{n-2}+\cdots+p_n). \qquad \text{(E.13)}$$

If λ is replaced by $\boldsymbol{A}$ in the characteristic function (E.13), the polynomial obtained reduces to the zero matrix

$$\boldsymbol{A}^n+p_1\boldsymbol{A}^{n-1}+p_2\boldsymbol{A}^{n-2}+\cdots+p_n\boldsymbol{I} = \boldsymbol{0}. \qquad \text{(E.14)}$$

Example

$$\boldsymbol{A} = \begin{bmatrix} 1 & 2 \\ -3 & 4 \end{bmatrix},$$

$$|\boldsymbol{A}-\lambda\boldsymbol{I}| = \begin{vmatrix} 1-\lambda & 2 \\ -3 & 4-\lambda \end{vmatrix} = 10-5\lambda+\lambda^2.$$

$$\therefore\ \boldsymbol{A}^2-5\boldsymbol{A}+10 = \begin{bmatrix} -5 & 10 \\ -15 & 10 \end{bmatrix} -5\begin{bmatrix} 1 & 2 \\ -3 & 4 \end{bmatrix} +10\begin{bmatrix} 1 & 0 \\ 0 & 1 \end{bmatrix} = $$
$$= \begin{bmatrix} 0 & 0 \\ 0 & 0 \end{bmatrix}.$$

In fact, every square matrix $\boldsymbol{A}$ satisfies its characteristic equation. This is the Cayley-Hamilton theorem.

It follows from the Cayley-Hamilton theorem that a rational function of a square matrix $\boldsymbol{A}$ of order n can be expressed as a polynomial in $\boldsymbol{A}$ whose highest degree is $n-1$. For, if $f(\boldsymbol{A})$ is a polynomial of degree $k \geqslant n$, we may use (E.14) and get

$$\boldsymbol{A}^k = \boldsymbol{A}^{k-n}\boldsymbol{A}^n = \boldsymbol{A}^{k-n}(-p_1\boldsymbol{A}^{n-1}-p_2\boldsymbol{A}^{n-2}-\cdots) = -p_1\boldsymbol{A}^{k-1}-p_2\boldsymbol{A}^{k-2}.$$

Thus $\boldsymbol{A}^k$ can be replaced by a polynomial of degree $k-1$, which corresponds to the reduction of the degree of $f(\boldsymbol{A})$ from k to $k-1$. In the same way, we can reduce the degree to $k-2$, and so on until $k = n-1$ where we have $\boldsymbol{A}^{n-1} = \boldsymbol{A}^{(n-1)-n}\boldsymbol{A}^n$. Hence, if $f(\boldsymbol{A})$ is a polynomial of any degree in the square matrix $\boldsymbol{A}$, we can always write

$$f(\boldsymbol{A}) = C_1\boldsymbol{A}^{n-1}+C_2\boldsymbol{A}^{n-2}+\cdots+C_n\boldsymbol{I}. \qquad \text{(E.15)}$$

References on page 398

We can now derive the Lagrange-Sylvester interpolation polynomial. Since each eigenvalue, λ_s, of $\mathbf{A}$ satisfies the identity $|\mathbf{A}-\lambda\mathbf{I}| \equiv 0$, we obtain from (E.13)

$$\lambda_s^n \equiv -p_1\lambda_s^{n-1}-p_2\lambda_s^{n-2}-\cdots-p_n. \tag{E.16}$$

We can, therefore, reduce the polynomial $f(\lambda_s)$ as above to the form

$$f(\lambda_s) = C_1\lambda_s^{n-1}+C_2\lambda_s^{n-2}+\cdots+C_n, \tag{E.17}$$

where the C's are the same as in (E.15) for $s = 1, 2, \ldots, n$.

The n equations of (E.17) together with (E.15) form a system of $(n+1)$ equations (assuming each λ_s is a single distinct value). Eliminating C_1, $C_2, \ldots C_n$, from this system we get

$$\begin{vmatrix} f(\mathbf{A}) & \mathbf{A}^{n-1} & \mathbf{A}^{n-2} & \cdots & \mathbf{A} & \mathbf{I} \\ f(\lambda_1) & \lambda_1^{n-1} & \lambda_1^{n-2} & \cdots & \lambda_1 & 1 \\ f(\lambda_2) & \lambda_2^{n-1} & \lambda_2^{n-2} & \cdots & \lambda_2 & 1 \\ \cdot & \cdot & \cdot & \cdot & \cdot & \cdot \\ f(\lambda_n) & \lambda_n^{n-1} & \lambda_n^{n-2} & \cdots & \lambda_n & 1 \end{vmatrix} = \mathbf{0}. \tag{E.18}$$

Expanding the determinant from the first column we get

$$f(\mathbf{A})D_0 = f(\lambda_1)\mathbf{D}_1+f(\lambda_2)\mathbf{D}_2+\cdots+f(\lambda_n)\mathbf{D}_n, \tag{E.19}$$

where D_0 is the determinant (of a type called an alternant) given by

$$D_0 = \begin{vmatrix} \lambda_1^{n-1} & \lambda_1^{n-2} & \cdots & \lambda_1 & 1 \\ \lambda_2^{n-1} & \lambda_2^{n-2} & \cdots & \lambda_2 & 1 \\ \cdot & \cdot & \cdot & \cdot & \cdot \\ \lambda_n^{n-1} & \lambda_n^{n-2} & \cdots & \lambda_n & 1 \end{vmatrix} = \begin{matrix} (\lambda_1-\lambda_2)(\lambda_1-\lambda_3)\cdots(\lambda_1-\lambda_n) \\ (\lambda_2-\lambda_3)\cdots(\lambda_2-\lambda_n) \\ \cdots\cdots\cdots\cdots \\ (\lambda_{n-1}-\lambda_n). \end{matrix}$$

To prove this equality we put $\lambda_1 = \lambda_2$ and the first and second rows become identical which means that $(\lambda_1-\lambda_2)$ is a factor of D_0. The other factors follow similarly.

Inspection of (E.18) shows that the determinant D_1 is given by

$$D_1 = \begin{vmatrix} \mathbf{A}^{n-1} & \mathbf{A}^{n-2} & \cdots & \mathbf{A} & \mathbf{I} \\ \lambda_2^{n-1} & \lambda_2^{n-2} & \cdots & \lambda_2 & 1 \\ \cdot & \cdot & \cdot & \cdot & \cdot \\ \lambda_n^{n-1} & \lambda_n^{n-2} & \cdots & \lambda_n & 1 \end{vmatrix} = \begin{matrix} (\mathbf{A}-\lambda_2\mathbf{I})(\mathbf{A}-\lambda_3\mathbf{I})\cdots(\mathbf{A}-\lambda_n\mathbf{I}) \\ (\lambda_2-\lambda_3)\cdots(\lambda_2-\lambda_n) \\ \cdots\cdots\cdots\cdots \\ (\lambda_{n-1}-\lambda_n). \end{matrix}$$

It will be noticed that this may be obtained from D_0 by replacing λ_1 by $\mathbf{A}$. The determinants $\mathbf{D}_2, \ldots, \mathbf{D}_n$ may be obtained by replacing $\lambda_2, \ldots, \lambda_n$, by $\mathbf{A}$ in turn. Dividing (E.19) by D_0, cancelling common factors, and putting $\mathbf{L}_s = \mathbf{D}_s/D_0$, we get

$$f(\mathbf{A}) = f(\lambda_1)\mathbf{L}_1+f(\lambda_2)\mathbf{L}_2+\cdots+f(\lambda_n)\mathbf{L}_n, \tag{E.20}$$

where

$$\boldsymbol{L}_1 = \frac{(\boldsymbol{A}-\lambda_2\boldsymbol{I})(\boldsymbol{A}-\lambda_3\boldsymbol{I})\cdots(\boldsymbol{A}-\lambda_n\boldsymbol{I})}{(\lambda_1-\lambda_2)(\lambda_1-\lambda_3)\cdots(\lambda_1-\lambda_n)},$$

$$\boldsymbol{L}_2 = \frac{(\boldsymbol{A}-\lambda_1\boldsymbol{I})(\boldsymbol{A}-\lambda_3\boldsymbol{I})\cdots(\boldsymbol{A}-\lambda_n\boldsymbol{I})}{(\lambda_2-\lambda_1)(\lambda_2-\lambda_3)\cdots(\lambda_2-\lambda_n)}. \tag{E.21}$$

Each of the matrices $\boldsymbol{L}_1, \boldsymbol{L}_2, \ldots \boldsymbol{L}_n$, when multiplied out, forms a polynomial in $\boldsymbol{A}$ of degree $n-1$. We have seen in ch. 11 § 10 that the mth power of a matrix occurs in the theory of the optical properties of dielectric multilayers. If $f(\boldsymbol{A}) = \boldsymbol{A}^m$ where m is a positive integer, we have

$$\boldsymbol{A}^m = \lambda_1^m\boldsymbol{L}_1+\lambda_2^m\boldsymbol{L}_2+\cdots+\lambda_n^m\boldsymbol{L}_n. \tag{E.22}$$

Using (E.21) we may write (E.22) in the form

$$\boldsymbol{A}^m = \sum_{r=1}^{n}(\lambda_r)^m \frac{\prod\limits_{s\neq r}(\boldsymbol{A}-\lambda_s\boldsymbol{I})}{\prod\limits_{s\neq r}(\lambda_s-\lambda_r)}. \tag{E.23}$$

The matrices which occur in the study of dielectric multilayers are 2×2 matrices so for these (E.23) gives

$$\boldsymbol{A}^m = \lambda_1^m\frac{\boldsymbol{A}-\lambda_2\boldsymbol{I}}{\lambda_1-\lambda_2}+\lambda_2^m\frac{\boldsymbol{A}-\lambda_1\boldsymbol{I}}{\lambda_2-\lambda_1}.$$

Hence, we get

$$\boldsymbol{A}^m = \frac{\lambda_1^m-\lambda_2^m}{\lambda_1-\lambda_2}\boldsymbol{A}-\lambda_1\lambda_2\frac{\lambda_1^{m-1}-\lambda_2^{m-1}}{\lambda_1-\lambda_2}\boldsymbol{I}. \tag{E.24}$$

The matrices occurring in the study of the optics of multilayer dielectrics and of lens sequences (ch. 5 § 5.2) have determinants of unity. For this case, since $\lambda_1\lambda_2 = 1$, (E.24) shows that the coefficient of $\boldsymbol{A}$ is a polynomial of degree $m-1$ and that of $\boldsymbol{I}$ is a polynomial of degree $m-2$. Thus, for this case, we can write

$$\boldsymbol{A}^m = S_{m-1}\boldsymbol{A}-S_{m-2}\boldsymbol{I}, \tag{E.25}$$

where S_{m-1} and S_{m-2} are polynomials of degree $m-1$ and $m-2$, respectively.

Abelès [1950] and Mielenz [1959] have shown that when $\boldsymbol{A}$ is a 2×2 matrix with determinant unity, the powers of $\boldsymbol{A}$ can be expressed as Chebyshev polynomials (see appendix I for some details of these polynomials).

References on page 398

Following Mielenz, we expand **A** as follows†:

$$\mathbf{A} = \begin{bmatrix} a_{11} & a_{12} \\ a_{21} & a_{22} \end{bmatrix} = a\sigma_0 + b\sigma_1 + c\sigma_2 + d\sigma_3, \tag{E.26}$$

where

$$\sigma_0 = \begin{bmatrix} 1 & 0 \\ 0 & 1 \end{bmatrix}, \quad \sigma_1 = \begin{bmatrix} 0 & 1 \\ 1 & 0 \end{bmatrix}, \quad \sigma_2 = \begin{bmatrix} 0 & -\mathrm{i} \\ \mathrm{i} & 0 \end{bmatrix}, \quad \sigma_3 = \begin{bmatrix} 1 & 0 \\ 0 & -1 \end{bmatrix}.$$

It is easy to show that

$$\mathbf{A} = \begin{bmatrix} a+d & b-\mathrm{i}c \\ b+\mathrm{i}c & a-d \end{bmatrix},$$

from which we obtain

$$a_{11} + a_{22} = 2a = X \qquad \text{(say)}, \tag{E.27}$$

and

$$a^2 - (b^2 + c^2 + d^2) = 1.$$

Squaring **A** and noting that

$$\sigma_0^2 = \sigma_1^2 = \sigma_2^2 = \sigma_3^2 = \sigma_0;$$

$$\sigma_0\sigma_1 = \sigma_1\sigma_0 = \sigma_1,$$

$$\sigma_0\sigma_2 = \sigma_2\sigma_0 = \sigma_2,$$

$$\sigma_0\sigma_3 = \sigma_3\sigma_0 = \sigma_3;$$

$$\sigma_1\sigma_2 = \mathrm{i}\sigma_3, \qquad \sigma_2\sigma_3 = \mathrm{i}\sigma_1, \qquad \sigma_3\sigma_1 = \mathrm{i}\sigma_2,$$

$$\sigma_2\sigma_1 = -\mathrm{i}\sigma_3, \qquad \sigma_3\sigma_2 = -\mathrm{i}\sigma_1, \qquad \sigma_1\sigma_3 = -\mathrm{i}\sigma_2,$$

we get, finally

$$\mathbf{A}^2 = 2a\mathbf{A} - \sigma_0. \tag{E.28}$$

Multiplying (E.25) by **A** and using (E.28) we get

$$\begin{aligned} \mathbf{A}^m\mathbf{A} &= S_{m-1}\mathbf{A}^2 - S_{m-2}\mathbf{A}, \\ \mathbf{A}^{m+1} &= 2aS_{m-1}\mathbf{A} - S_{m-1}\sigma_0 - S_{m-2}\mathbf{A}. \end{aligned} \tag{E.29}$$

We may also obtain the equation for $\mathbf{A}^{m+1}$ directly from (E.25), thus

$$\mathbf{A}^{m+1} = S_m\mathbf{A} - S_{m-1}\sigma_0. \tag{E.30}$$

By comparing coefficients in (E.29) and (E.30) and using (E.27), we obtain

$$S_m(X) = XS_{m-1}(X) - S_{m-2}(X). \tag{E.31}$$

† Any 2×2 matrix can be expressed as a linear combination of the unit matrix, σ_0, and the Pauli matrices, σ_1, σ_2, σ_3.

Putting $m = 2$ in (E.25) and equating coefficients with (E.28), we get

$$S_0 = 1 \quad \text{and} \quad S_1(X) = X. \tag{E.32}$$

Equations (E.31) and (E.32) give the following values for the polynomials $S_m(X)$:

$$\begin{aligned} S_0 &= 1, & S_4 &= X^4-3X^2+1, \\ S_1 &= X, & S_5 &= X^5-4X^3+3X, \\ S_2 &= X^2-1, & S_6 &= X^6-5X^4+6X^2-1, \\ S_3 &= X^3-2X, & &\text{etc.} \end{aligned} \tag{E.33}$$

A general expression for these polynomials is given by

$$S_m(X) = \sum_{r=0}^{(m/2)} (-1)^r \binom{m}{r} X^{m-2r},$$

where $(m/2)$ denotes the largest integer contained in $(m/2)$, e.g. $(5/2)$ denotes 2.

Comparing (E.33) with (I.43) and using (I.44) and (I.45) we see that if $X = 2x$, the S_m's defined by (E.33) are the Chebyshev polynomials of the second kind. Hence, we have

$$S_m(X) = \frac{\sin(m+1)\theta}{\sin\theta}, \qquad X = 2\cos\theta, \tag{E.34a}$$

and

$$S_m(X) = \frac{\sinh(m+1)\phi}{\sinh\phi}, \qquad X = 2\cosh\phi. \tag{E.34b}$$

Equations (E.34a, b) may also be written in a form analogous to (I.40), viz.

$$S_m(X) = \frac{[X+(X^2-4)^{\frac{1}{2}}]^{m+1}-[X-(X^2-4)^{\frac{1}{2}}]^{m+1}}{2^{m+1}(X^2-4)^{\frac{1}{2}}}. \tag{E.35}$$

For real X, these polynomials are also real.

If θ and ϕ are to be real, (E.34a) must be used for $|X| \leqslant 2$, and (E.34b) for $|X| \geqslant 2$.

The matrix $\mathbf{A}^m$, in terms of its elements and Chebyshev polynomials is given by

$$\mathbf{A}^m = S_{m-1}\begin{bmatrix} a_{11} & a_{12} \\ a_{21} & a_{22} \end{bmatrix} - S_{m-1}\begin{bmatrix} 1 & 0 \\ 0 & 1 \end{bmatrix}.$$

Hence, we get

$$\mathbf{A}^m = \begin{bmatrix} S_{m-1}a_{11}-S_{m-2} & S_{m-1}a_{12} \\ S_{m-1}a_{21} & S_{m-1}a_{22}-S_{m-2} \end{bmatrix}. \tag{E.36}$$

References on page 398

APPENDIX F

SCALAR THEORY OF DIFFRACTION OF LIGHT

Light, being electromagnetic, consists of two coupled vector fields. Calculation of diffraction patterns by solving Maxwell's equations is very difficult and has been successful only in a few special cases. Diffraction patterns may be calculated much more easily and with adequate accuracy by adopting the simplification of representing the light wave by a scalar field† and using Kirchhoff's formulation of Fresnel's theory of diffraction. The questions of charge density and current density at the boundaries of the obstacle causing diffraction are avoided by using the scalar approximation. We represent both the electric field and the magnetic field by a single scalar function, u. This is equivalent to considering a single component of the vector field in a region where there are no currents and charges. The wave equation to be considered for the propagation of the scalar is

$$\nabla^2 u - (1/c^2)\partial^2 u/\partial t^2 = 0. \tag{F.1}$$

As in electromagnetic theory, the principle of superposition holds in scalar theory, for if u_1 and u_2 are any two waves satisfying (F.1) the linearity of the equation shows that $u_1 + u_2$ is also a solution.

A scalar wave which satisfies (F.1) is

$$u(r, t) = (A/r) \exp[i(kr - \omega t)], \tag{F.2}$$

where $A = A_0 \exp(i\theta)$, $k = 2\pi/\lambda = \omega/c$ and the wave is emitted by a point source. At a point P which does not coincide with a source, at time, t, the value of $u(P, t)$ due to many sources may be calculated from the positions, amplitudes, and phases of the sources by superimposing the waves from each source. However, the necessary detailed information about the sources is not readily available so we use the Huygens-Fresnel principle which regards

† Since the physical dimensions of laser cavities are generally much greater than the wavelength of the radiation oscillating, this simplification is quite accurate.

every point of a wavefront as the source of spherical wavelets, together with the integral theorem of Kirchhoff to calculate $u(P, t)$. This theorem expresses the solution of the homogeneous wave equation at P, in terms of the solution and its spatial derivative in the normal direction at all points on an arbitrary closed surface S surrounding P but excluding any sources. The theorem is derived as follows.

Let us consider those solutions of (F.1) that may be split into spatial and time dependent parts, thus

$$u = u_p \exp(-i\omega t). \tag{F.3}$$

Substitution of the value of u given by (F.3) into (F.1) gives

$$\nabla^2 u_p + k^2 u_p = 0. \tag{F.4}$$

This is the Helmholtz equation.

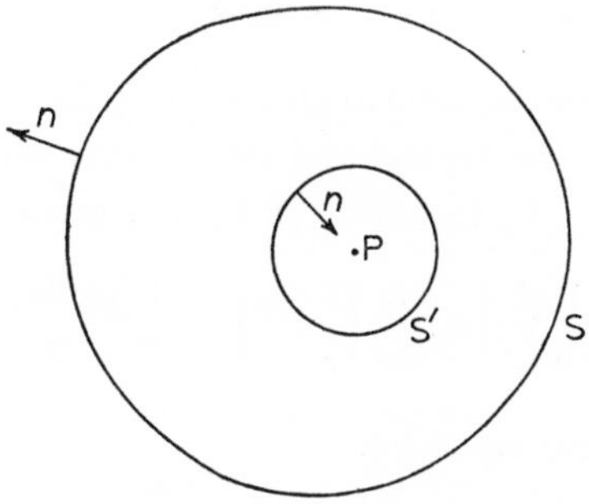

Fig. F.1.

Consider a volume bounded by a closed surface S and let P be any point within it surrounded by a sphere S' of radius ε, fig. F.1. Green's theorem applied to the volume v between S and S' gives for any arbitrary scalar fields U, V

$$\int_{SS'} \left(V \frac{\partial U}{\partial n} - U \frac{\partial V}{\partial n} \right) \mathrm{d}S = \int_v (V\nabla^2 U - U\nabla^2 V) \mathrm{d}v, \tag{F.5}$$

where $\partial/\partial n$ is in the direction of the outward normal to S at $\mathrm{d}S$. We shall use Green's theorem to calculate the effect at P due to waves arriving at P from sources outside the surface S. Let V be the function u_p and let U be the function $U = (1/r) \exp(ikr)$, where r is the distance from P. The function U has a singularity at P where r is zero. The sphere S' around P is taken to avoid this violation of the requirements of Green's theorem.

References on page 398

Over the surface of the sphere S', the outward normal from the volume v is in the opposite direction to the radius so $\partial/\partial n = -\partial/\partial r$. We shall evaluate the integrals of (F.5) in the limit $\varepsilon \to 0$. Substituting the values of U and V in (F.5) shows that the volume integrand is zero throughout v (since U and V satisfy (F.4)) and thus, we get

$$\int_S \left[u_p \frac{\partial}{\partial n}\left(\frac{1}{r}\,\mathrm{e}^{\mathrm{i}kr}\right) - \frac{1}{r}\,\mathrm{e}^{\mathrm{i}kr}\frac{\partial u_p}{\partial n}\right] \mathrm{d}S + \\ + \int_{S'} \left[-u_p \frac{\partial}{\partial r}\left(\frac{1}{r}\,\mathrm{e}^{\mathrm{i}kr}\right) + \frac{1}{r}\,\mathrm{e}^{\mathrm{i}kr}\frac{\partial u_p}{\partial r}\right] \mathrm{d}S' = 0, \qquad \text{(F.6)}$$

where r is the distance of the surface element ($\mathrm{d}S$, $\mathrm{d}S'$) from P. The integral over S' becomes

$$\int_{S'} \left[\frac{u_p}{r^2} - \frac{\mathrm{i}ku_p}{r} + \frac{1}{r}\frac{\partial u_p}{\partial r}\right] \mathrm{e}^{\mathrm{i}kr}\mathrm{d}S'.$$

We may express this in terms of solid angles by using $\mathrm{d}S' = r^2\mathrm{d}\Omega$. If we also take the limit $r = \varepsilon = 0$, the second and third terms vanish and $\mathrm{e}^{\mathrm{i}kr} \approx 1$. Integrating we obtain $4\pi u_p(P)$. Equation (F.6) then gives

$$u_p(P) = -\frac{1}{4\pi}\int_S \left[u_p \frac{\partial}{\partial n}\left(\frac{1}{r}\,\mathrm{e}^{\mathrm{i}kr}\right) - \frac{1}{r}\,\mathrm{e}^{\mathrm{i}kr}\frac{\partial u_p}{\partial n}\right] \mathrm{d}S. \qquad \text{(F.7)}$$

This is Kirchhoff's integral equation.

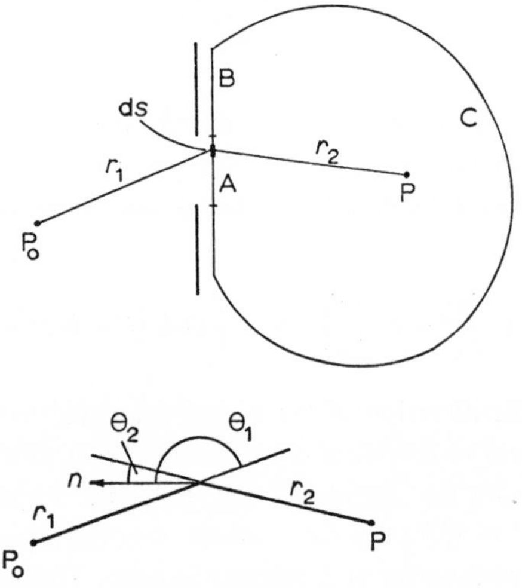

Fig. F.2.

We shall now apply the Kirchhoff integral equation to determine the amplitude at a point P due to radiation from a source P_0 which reaches P by diffraction through an aperture in an opaque screen between P and P_0, fig. F.2. Assume that the linear dimensions of the aperture are very much greater than a wavelength and that the aperture dimensions are themselves very much smaller than the distances of P_0 and P respectively from the screen. The closed surface over which we take Kirchhoff's integral is divided into three regions A, B, C for convenience, see fig. F.2. Region A is the aperture itself; region B is part of the opaque plane on the shadow side; region C is centred on P and is of very large radius. Equation (F.7) then becomes

$$u_p(P) = -\frac{1}{4\pi}\left[\int_A + \int_B + \int_C\right]\left[u_p \frac{\partial}{\partial n}\left(\frac{1}{r}\,e^{ikr}\right) - \frac{1}{r}\,e^{ikr}\frac{\partial u_p}{\partial n}\right] dS. \qquad (F.8)$$

The values of u_p and $\partial u_p/\partial n$ on A, B, C are not known exactly and it is at this point that the boundary conditions which form the basis of Kirchhoff's diffraction theory are introduced. Kirchhoff assumed that u_p and $\partial u_p/\partial n$ in the region A are the same with the screen as without it, and in region B they are zero. These assumptions are quite accurate for the conditions described. In region C, if the radius of C is large, the integral over C vanishes. (The vanishing of the integral over C is discussed by STONE [1963] and by BORN and WOLF [1964].) The source emits spherical waves whose amplitude at dS is $u_p = (A/r_1)\exp(ikr_1)$.

Since $\partial r_1/\partial n = \cos\theta_1$ and $\partial r_2/\partial n = \cos\theta_2$, we can write

$$\frac{\partial u_p}{\partial n} = \frac{\partial u_p}{\partial r_1}\cos\theta_1 = A\cos\theta_1\left(-\frac{1}{r_1^2} + \frac{ik}{r_1}\right)e^{ikr_1}$$

and

$$\frac{\partial}{\partial n}\left(\frac{1}{r_2}\,e^{ikr_2}\right) = \cos\theta_2\left(-\frac{1}{r_2^2} + \frac{ik}{r_2}\right)e^{ikr_2}.$$

Using these expressions in (F.8), and neglecting terms $1/r_2$ and $1/r_1$ which are small compared with k (this is the high frequency approximation) we get

$$u_p(P) = -\frac{iAk}{4\pi}\int_A \frac{e^{ik(r_1+r_2)}}{r_1 r_2}(\cos\theta_2 - \cos\theta_1)\,dS_A. \qquad (F.9)$$

This is the Fresnel-Kirchhoff diffraction equation. The factor $(\cos\theta_2 - \cos\theta_1)$ is the obliquity factor. The field at the aperture due to waves from source P_0 is $u_{pa} = (A/r_1)\exp(ikr_1)$. Hence, the field at P can be expressed in terms of the aperture field u_{pa} thus

$$u_p(P) = -\frac{ik}{4\pi}\int_A u_{pa}\frac{e^{ikr_2}}{r_2}(\cos\theta_2 - \cos\theta_1)\,dS_A. \qquad (F.10)$$

References on page 398

INTEGRAL EQUATIONS

Much of the theory of laser cavities is devoted to methods for solving the integral equations obtained. The purpose of this appendix is to provide a rudimentary background to these equations and their relevance to laser cavities.

An integral equation is an equation containing an integral whose integrand includes an unknown function $f(x)$ which is to be determined. Of course, the equation may contain other terms both under the integral and outside it. Let $F(x)$ and $K(x, y)$ be known functions. For the purposes of discussion we shall consider only the following two types of integral equation

$$F(x) = \int_a^b K(x, y)f(y)\mathrm{d}y. \tag{G.1}$$

$$f(x) = \lambda\int_a^b K(x, y)f(y)\mathrm{d}y, \tag{G.2}$$

where λ is a numerical factor (generally complex).

The known function $K(x, y)$ under the integral is the *kernel* of the integral equation and must be defined in the rectangle $a \leqslant x \leqslant b$, $a \leqslant y \leqslant b$. The functions $F(x)$ and $f(x)$ must be defined in the interval $a \leqslant x \leqslant b$. Integral equations are *linear* if the unknown function appears linearly in the equation. Both the above equations are *linear integral equations.* When both limits of integration are constants, the equation is a *Fredholm* integral equation. An equation in which the unknown function occurs only under the integral is called an *integral equation of the first kind*; (G.1) is an example of this type. An *integral equation of the second kind* is one in which the unknown function occurs both under the integral and elsewhere; (G.2) is an example of this type. If an equation is such that every term contains the unknown function, it is a *homogeneous* integral equation. If the equation contains one term which does not contain the unknown function, it is *inhomogeneous*. Thus (G.2) is an example of a homogeneous Fredholm equation of the second kind. Solution of an integral equation is concerned with inverting the linear integral transformation given by (G.1) to determine $f(y)$, or with determining $f(x)$, as in (G.2).

An important integral equation of the first kind is the Fourier integral equation

$$F(x) = \frac{1}{\sqrt{(2\pi)}} \int_{-\infty}^{+\infty} f(y) \exp(-ixy) \mathrm{d}y. \tag{G.3}$$

In this equation $(1/\sqrt{2\pi}) \exp(-ixy)$ is the kernel.

Equation (G.2) will be recognized as the type which arises in the theory of laser cavities given in ch. 6. Every value of λ (possibly complex) for which the integral equation has continuous non-zero solutions is an *eigenvalue* of the homogeneous integral equation, or of the kernel, $K(x, y)$. The corresponding solutions are called the *eigenfunctions* of the integral equation or of the kernel for the eigenvalue λ. Rigorous proof of the existence of eigenvalues and eigenfunctions of some equations with optical resonator kernels has been given by COCHRAN [1965], NEWMAN and MORGAN [1964] and HOCHSTADT [1966].

The eigenfunctions give the field distribution at the mirrors for each of the possible modes of oscillation and the eigenvalue corresponding to a mode gives the factor (generally complex) by which the field of the mode is diminished as a result of diffraction losses in a single transit. The phase of the eigenvalue determines the mirror spacings which will support oscillation at a given frequency.

The kernels which arise in laser cavity theory are generally complex-symmetric but not Hermitian, that is

$$K(x, y) = K(y, x), \qquad K(x, y) \neq K^*(y, x). \tag{G.4}$$

Integral equations of the second kind with symmetric kernels may be solved by the expansion method of Schmidt and Hilbert. This technique has been used by STREIFER and GAMO [1964]. Integral equations of the second kind can also be solved by the method of successive approximations (due largely to Neumann, Liouville and Voltera) as described by HILDEBRAND [1952]. The method gives $f(x)$ in the form of a power series in λ, the coefficients being functions of x. When the resulting series converges rapidly, the method is particularly useful and is analogous to the iterative process used by FOX and LI [1961] and described in ch. 4 § 2.

The eigenfunctions of a homogeneous integral equation with a symmetric

References on page 398

kernel are orthogonal. This means that

$$\int_a^b f_n(x) f_m(x) \mathrm{d}x = 0, \qquad n \neq m, \tag{G.5}$$

where

$$f_n(x) = \lambda_n \int_a^b K(x, y) f_n(y) \mathrm{d}y; \tag{G.6}$$

$$f_m(x) = \lambda_m \int_a^b K(x, y) f_m(y) \mathrm{d}y. \tag{G.7}$$

To show this we assume that $\lambda_m \neq \lambda_n$, and multiply (G.6) by $\lambda_m f_m(x)$ and (G.7) by $\lambda_n f_n(x)$, subtract the resulting equations, and integrate to get

$$(\lambda_m - \lambda_n) \int_a^b f_n(x) f_m(x) \mathrm{d}x =$$

$$= \lambda_n \lambda_m \int_a^b \int_a^b [K(x, y) f_n(y) f_m(x) - K(x, y) f_m(y) f_n(x)] \mathrm{d}y \mathrm{d}x.$$

The right hand side of this equation is zero, as may be seen by interchange of the variables of integration in the second half of the double integral, and $K(x, y) = K(y, x)$. Since $\lambda_m \neq \lambda_n$, it follows that the eigenfunctions are orthogonal.

HERMITE POLYNOMIALS

The first few Hermite polynomials are:

$$
\begin{aligned}
H_0(x) &= 1, \\
H_1(x) &= 2x, \\
H_2(x) &= 4x^2-2, \\
H_3(x) &= 8x^3-12x, \\
H_4(x) &= 16x^4-48x^2+12, \\
H_5(x) &= 32x^5-160x^3+120x, \\
H_6(x) &= 64x^6-480x^4+720x^2-120.
\end{aligned}
$$

References on page 398

ORTHOGONAL FUNCTIONS

Orthogonal functions and polynomials are extremely important in pure and applied mathematics and have extensive literature. We shall sketch some of the background and introduce only those aspects which are useful in the treatments of laser cavities and quantum theory which we consider.

Two real functions $f(x)$ and $g(x)$ are said to be orthogonal to each other in the interval $a \leqslant x \leqslant b$ if

$$\int_a^b f(x)g(x)\mathrm{d}x = 0. \tag{I.1}$$

The concept of orthogonality arises by analogy with vector theory where (in 3 dimensions) two vectors $\boldsymbol{A}$ and $\boldsymbol{B}$ are said to be orthogonal if

$$\boldsymbol{A} \cdot \boldsymbol{B} = A_1 B_1 + A_2 B_2 + A_3 B_3 = 0.$$

In a space of N dimensions, the vectors are orthogonal if $\Sigma_{i=1}^{N} A_i B_i = 0$. In a space with an infinite number of dimensions in which A_i and B_i are distributed continuously, i becomes a continuous variable (x) and $\Sigma_i A_i B_i$ becomes $\int A(x)B(x)\mathrm{d}x$. In effect we have translated the concept of a function into that of a vector in an infinite-dimensional space. If the integral is zero, the functions $A(x)$ and $B(x)$ are orthogonal. Thus, we can see that

$$\int_a^b f(x)g(x)\mathrm{d}x = f \cdot g. \tag{I.2}$$

defines a scalar product in function space. In such a space, a scalar product (inner product) can be defined in the same way as in finite vector spaces and hence orthogonality may be so defined. If the functions considered are complex, then we have, by definition

$$(f \cdot g^*) = (f^* \cdot g) = \int_a^b f^*(x)g(x)\mathrm{d}x. \tag{I.3}$$

The function $f^*(x)$ is orthogonal to $g(x)$ if $(f^* \cdot g) = 0$.

The norm of a function $f(x)$ (which may be complex) is defined as

$$N(f) = (f \cdot f) = \int_a^b |f(x)|^2 \mathrm{d}x, \tag{I.4}$$

and a function is said to be normalized if the norm is unity. A set of functions $f_1, f_2, \ldots, f_n$ is orthogonal if

$$\int_a^b f_i(x) f_j(x) \mathrm{d}x = \lambda_i \delta_{ij}, \tag{I.5}$$

where $\delta_{ij} = 1$ when $i = j$ and $\delta_{ij} = 0$ when $i \neq j$.

We may define a new function such that

$$\phi_i = f_i / \sqrt{\lambda_i} \tag{I.6}$$

and then obtain

$$\int_a^b \phi_i(x) \phi_j(x) \mathrm{d}x = \delta_{ij}. \tag{I.7}$$

The functions $\phi_i(x)$ are orthonormal and the process of obtaining ϕ_i from f_i is normalization. An orthogonal system can always be made orthonormal. If the functions are complex, we have,

$$\int_a^b \phi_i^*(x) \phi_j(x) \mathrm{d}x = \delta_{ij}. \tag{I.8}$$

An example of an important orthonormal set defined over any interval of length 2π is given by the functions

$$\frac{1}{\sqrt{2\pi}}, \quad \frac{\cos x}{\sqrt{\pi}}, \quad \frac{\sin x}{\sqrt{\pi}}, \ldots, \frac{\cos nx}{\sqrt{\pi}}, \quad \frac{\sin nx}{\sqrt{\pi}}, \ldots.$$

An example of a complex orthonormal set in the interval $0 \leqslant x \leqslant 2\pi$ is that given by

$$\frac{1}{\sqrt{2\pi}}, \quad \frac{\mathrm{e}^{\pm ix}}{\sqrt{2\pi}}, \ldots, \frac{\mathrm{e}^{\pm inx}}{\sqrt{2\pi}}, \ldots.$$

The n functions $f_1, f_2, \ldots, f_n$ are linearly dependent if the linear relation $\Sigma_{i=1}^n c_i f_i = 0$ is true for all x when c_i are non-zero constants. The functions of an orthogonal system are always linearly independent, since multiplying $\Sigma_{i=1}^n c_i f_i = 0$ by f_j and integrating over the domain of definition gives $c_j = 0$.

A set of orthonormal functions (polynomials) can be used to approximate an arbitrary function. Orthonormal functions rather than more general

References on page 398

linearly independent functions are particularly suitable for this purpose because the coefficient matrix is diagonal and unity for orthonormal functions (see, for example, RICE [1964]), i.e. the coefficients are easy to calculate. The suggestion that the arbitrary function be represented in terms of a set of orthonormal functions is given by Weierstrass's approximation theorem which states that it is always possible to approximate an arbitrary continuous function $f(x)$ defined in a finite interval $a \leqslant x \leqslant b$ over the whole interval (a, b) as closely as we please by a power polynomial of sufficiently high degree. (For proof, see for example, COURANT and HILBERT [1953].) Let $f(x)$ be the system to be approximated and let the orthonormal set be $\phi_0(x), \phi(x), \ldots, \phi_n(x)$ defined over a common interval $a \leqslant x \leqslant b$. We shall use the least squares approximation and show that

$$c_0\phi_0(x)+c_1\phi_1(x)+c_2\phi_2(x)+\cdots+c_n\phi_n(x)$$

may represent $f(x)$ with an accuracy such that the mean square error is given by

$$M = \int_a^b |f(x)-\sum_{i=0}^n c_i\phi_i(x)|^2\mathrm{d}x. \tag{I.9}$$

To find the value of the expansion coefficients c_i which minimize M, we proceed as follows. Integrating (I.9), we get

$$M = \int_a^b \left[f(x)-\sum_{i=0}^n c_i\phi_i(x)\right]\left[f^*(x)-\sum_{i=0}^n c_i^*\phi_i^*(x)\right]\mathrm{d}x =$$

$$= \int_a^b |f(x)|^2\mathrm{d}x - \sum_{i=0}^n (c_i a_i^* + c_i^* a_i) + \sum_{i=0}^n |c_i|^2,$$

where

$$a_i = \int_a^b f(x)\phi_i^*(x)\mathrm{d}x.$$

This may be written

$$M = \int_a^b |f(x)|^2\mathrm{d}x + \sum_{i=0}^n |c_i - a_i|^2 - \sum_{i=0}^n |a_i|^2, \tag{I.10}$$

which shows that M is a minimum when $c_i = a_i$ and we then have

$$\sum_{i=0}^n |a_i|^2 \leqslant \int_a^b |f(x)|^2\mathrm{d}x. \tag{I.11}$$

This is Bessel's inequality which holds for all orthonormal systems and, since it is true for all n, it shows that the sum of the squares of the expansion coefficients always converges.

The orthonormal set $\phi_1, \phi_2, \ldots, \phi_n$, is a complete set if

$$\lim_{n\to\infty} \int_a^b |f(x) - \sum_{i=0}^{n} a_i \phi_i|^2 \mathrm{d}x = 0. \tag{I.12}$$

which, using (I.10), leads to

$$\int_a^b |f(x)|^2 \mathrm{d}x = \sum_{i=0}^{\infty} |a_i|^2; \tag{I.13}$$

this is called Parseval's equality and is the completeness relation. Equation (I.13) shows that the orthonormal set ϕ_i is complete when

$$a_i = \int_a^b f(x)\phi_i^* \mathrm{d}x. \tag{I.14}$$

The coefficients a_i are the Fourier coefficients of $f(x)$ with respect to the set $\{\phi_i(x)\}$.

Summarizing, then, any function $f(x)$ for which $\int_a^b |f(x)|^2 \mathrm{d}x$ exists can be represented by the complete orthonormal set ϕ_i as follows

$$f(x) = \sum_{i=0}^{n} a_i \phi_i, \tag{I.15}$$

where

$$a_i = \int_a^b f(x)\phi_i^* \mathrm{d}x. \tag{I.16}$$

If we assume from the start that $f(x)$ can be accurately represented by an expansion in terms of a complete set of orthonormal functions $\phi_i(x)$ we arrive very quickly at (I.15) and (I.16), thus

$$f(x) = a_0\phi_0(x) + a_1\phi_1(x) + a_2\phi_2(x) + \cdots + a_i\phi_i(x) + \cdots. \tag{I.15}$$

Multiplying by $\phi_j^*(x)$ and integrating (remembering that terms for which $j \neq i$ are zero) we get

$$\int_a^b f(x)\phi_i^*(x)\mathrm{d}x = a_i \int_a^b \phi_i(x)\phi_i^*(x)\mathrm{d}x = a_i. \tag{I.16}$$

Some sets of functions F_0, F_1, are not orthogonal but for some important non-orthogonal functions we can sometimes find a function $w(x)$ (see,

References on page 398

for example, LANCZOS [1957]) such that, on the interval (a, b) we have

$$\int_a^b w(x)F_i(x)F_j(x)\mathrm{d}x = 0 \qquad (i \neq j). \tag{I.17}$$

The function $w(x)$ is the weight function and the set $\{F_i\}$ is orthogonal with respect to $w(x)$. If one, or both limits are infinite, we may introduce a weight function to prevent the integrals (I.17) from diverging. The norm of F_i is given by

$$N(F_i) = \int_a^b w(x)F_i^2(x)\mathrm{d}x. \tag{I.18}$$

If the norm of each F_i is 1, the set is orthonormal with respect to $w(x)$. Clearly, we can define a set of functions g_i by the equation $g_i = \sqrt{w}\, F_i$ and then we have

$$\int_a^b g_i(x)g_j(x)\mathrm{d}x = 0, \qquad (i \neq j). \tag{I.19}$$

If the functions g_i are to be real, then $w(x) \geqslant 0$.

Orthogonalization

If we are given an arbitrary function $f(x)$ continuous in the interval (a, b) and wish to expand it in terms of a set of preassigned linearly independent functions $f_i(x)$ for which $\int_a^b |f_i(x)|^2\mathrm{d}x$ exists, we could approximate it in the mean by the linear aggregate $\Sigma_{i=1}^n c_i f_i(x)$, but as noted above, the coefficients c_i are more difficult to determine than is the case if our set is orthonormal. We proceed by obtaining a set of orthonormal functions from the given set $\{f_i(x)\}$, i.e. we orthogonalize it. In this process we replace the functions $f_1(x), f_2(x), \ldots, f_n(x), \ldots$ by the same number of new functions $\phi_1(x), \phi_2(x), \ldots, \phi_n(x), \ldots$, where each new function is a linear combination of the old, thus

$$\phi_n(x) = c_1^{(n)}f_1(x) + c_2^{(n)}f_2(x) + \cdots + c_n^{(n)}f_n(x). \tag{I.20}$$

The procedure is known as Gramm-Schmidt orthogonalization.

The step-by-step procedure may be described as follows. We choose ϕ_1 and then calculate a normalized function ϕ_2 which is orthogonal to ϕ_1. We next obtain a normalized function ϕ_3 which is orthogonal to ϕ_1 and ϕ_2, and so on until we have constructed the required orthonormal set.

The process of orthogonalization can be applied to functions given in the interval (a, b) as described below or it can be applied to functions given on the contour C, in which case we merely take $\int_c$ in place of $\int_a^b$ everywhere.

Let us now obtain an expression to give $\phi_1(x), \ldots, \phi_n(x)$ in succession. We start with $\phi_1(x)$ which will have the form $af_1(x)$ where the constant a is obtained from the condition $\int_a^b \phi_1^2 dx = 1$ and get

$$\phi_1(x) = f_1(x) \Big/ \left[\int_a^b |f_1(x)|^2 dx\right]^{\frac{1}{2}}. \qquad (I.21)$$

After determining the first n functions $\phi_1(x), \phi_2(x), \ldots, \phi_n(x)$ we come to $\phi_{n+1}(x)$ which must be a linear combination of these and the function $f_{n+1}(x)$, therefore,

$$\phi_{n+1}(x) = a_1\phi_1(x) + a_2\phi_2(x) + \cdots + a_n\phi_n(x) + af_{n+1}(x). \qquad (I.22)$$

To determine the constants a_i we multiply (I.22) by $\phi_i(x)$ and integrate; remembering that the functions $\phi_1(x), \ldots, \phi_n(x)$ are orthonormal, we get

$$a_i + a\int_a^b f_{n+1}(x)\phi_i(x)dx = 0. \qquad (I.23)$$

We may obtain a_i from this and substitute in (I.22); the condition $\int_a^b \phi_{n+1}^2 dx = 1$ gives the constant a; and then we finally obtain

$$\phi_{n+1}(x) = \frac{f_{n+1}(x) - \sum_{i=1}^{n} \left[\int_a^b f_{n+1}\phi_i dx\right]\phi_i(x)}{\left[\int_a^b \left\{f_{n+1}(x) - \sum_{i=1}^{n}\left(\int_a^b f_{n+1}\phi_i dx\right)\phi_i(x)\right\}^2 dx\right]^{\frac{1}{2}}}. \qquad (I.24)$$

This equation enables us to obtain $\phi_1(x), \ldots, \phi_n(x)$ in succession.

A neater expression of the above is given in terms of determinants as follows

$$\phi_n(x) = \frac{\begin{vmatrix} (f_1 \cdot f_1) & (f_1 \cdot f_2) & \cdots & (f_1 \cdot f_{n-1}) & f_1(x) \\ (f_2 \cdot f_1) & (f_2 \cdot f_2) & \cdots & (f_2 \cdot f_{n-1}) & f_2(x) \\ \cdots & \cdots & \cdots & \cdots & \cdots \\ (f_n \cdot f_1) & (f_n \cdot f_2) & \cdots & (f_n \cdot f_{n-1}) & f_n(x) \end{vmatrix}}{(\Delta_{n-1} \cdot \Delta_n)^{\frac{1}{2}}}, \qquad (I.25)$$

where Δ_n is the Gramm determinant of the functions $f_1, f_2, \ldots, f_n$, given by

$$\Delta_n = \begin{vmatrix} (f_1 \cdot f_1) & (f_1 \cdot f_2) & \cdots & (f_1 \cdot f_n) \\ (f_2 \cdot f_1) & (f_2 \cdot f_2) & \cdots & (f_2 \cdot f_n) \\ \cdots & \cdots & \cdots & \cdots \\ (f_n \cdot f_1) & (f_n \cdot f_2) & \cdots & (f_n \cdot f_n) \end{vmatrix}. \qquad (I.26)$$

The set of functions $f_1(x), f_2(x), \ldots, f_n(x)$ are linearly independent if Δ_n does not vanish (for proof, see SANSONE [1959]).

References on page 398

As an example, consider the functions $1, x, x^2, x^3$ in the interval $0 \leqslant x \leqslant 1$. The Gramm determinant is obtained from the definition of a scalar product (I.2) and putting $f_1 = 1, f_2 = x, f_3 = x^2, f_4 = x^3$ we get

$$\begin{vmatrix} 1 & \frac{1}{2} & \frac{1}{3} & \frac{1}{4} \\ \frac{1}{2} & \frac{1}{3} & \frac{1}{4} & \frac{1}{5} \\ \frac{1}{3} & \frac{1}{4} & \frac{1}{5} & \frac{1}{6} \\ \frac{1}{4} & \frac{1}{5} & \frac{1}{6} & \frac{1}{7} \end{vmatrix}.$$

Since this is not zero, the functions $1, x, x^2, x^3$ are linearly independent in the interval $0 \leqslant x \leqslant 1$. More generally, it can be shown that the set $\{x^n\}$ defined over any finite interval $a \leqslant x \leqslant b$ are linearly independent. If we multiply each member of the set $\{x^n\}$ by $\sqrt{w(x)}$ where $w(x)$ is a positive weight function we obtain $\sqrt{w(x)}, x\sqrt{w(x)}, x^2\sqrt{w(x)}, \ldots$ which are linearly independent.

By orthogonalizing the functions $\sqrt{w(x)}, x\sqrt{w(x)}, x^2\sqrt{w(x)}, \ldots$ in the interval $a \leqslant x \leqslant b$ we obtain the orthonormal set $\phi_0, \phi_1, \phi_2, \ldots, \phi_n, \ldots$ where

$$\phi_n(x) = \sqrt{w(x)}\mathscr{P}_n(x) \tag{I.27}$$

and $\mathscr{P}_n(x)$ is a polynomial of degree n. Some important sets of polynomials are those of Legendre, Chebyshev, Jacobi (hypergeometric), Laguerre, and Hermite; they are obtained (except for a constant factor) by orthogonalizing in the following conditions:

Polynomial		Weight function	Interval
Legendre	$P_n(x)$	1	$-1 \leqslant x \leqslant 1$
Chebyshev	$T_n(x)$	$\dfrac{1}{\sqrt{1-x^2}}$	$-1 \leqslant x \leqslant 1$
Jacobi		$x^{q-1}(1-x)^{p-q}$ $q > 0 \quad p-q > -1$	$0 \leqslant x \leqslant 1$
Laguerre	$L_n(x)$	e^{-x}	$0 \leqslant x < \infty$
Hermite	$H_n(x)$	e^{-x^2}	$-\infty < x < \infty$

Generating functions

The polynomials may also be obtained by means of a generating function. A generating function $g(x, t)$ is one which, when expanded as a power series in t, gives the complete set of functions required as the coefficients of successive powers of t, thus we have

$$g(x, t) = \sum_{n=0}^{\infty} a_n \phi_n(x) t^n, \tag{I.28}$$

where $\phi_n(x)$ is one of the required functions, and a_n is independent of x and t. As an example we consider the Hermite polynomials for which the generating function is given by

$$g(x, t) = \exp(-t^2+2tx) = \exp(x^2) \exp\{-(t-x)^2\}. \tag{I.29}$$

Therefore, we have

$$\exp(x^2) \exp\{-(t-x)^2\} = \sum_{n=0}^{\infty} \frac{H_n(x)}{n!} t^n. \tag{I.30}$$

By differentiating this n times with respect to t we get for the left-hand side:

$$\exp(x^2) \frac{\partial^n}{\partial t^n} \exp\{-(t-x)^2\} = \exp(x^2)(-1)^n \frac{\partial^n}{\partial x^n} \exp\{-(t-x)^2\}$$

and for the right-hand side, we get

$$H_n(x) + H_{n+1}(x)t + H_{n+2}(x)t^2 + \cdots.$$

On putting $t = 0$, we get

$$H_n(x) = (-1)^n \exp(x^2) \frac{d^n}{dx^n} \exp(-x^2), \tag{I.31}$$

since

$$\frac{\partial}{\partial t} \exp\{-(t-x)^2\} = -\frac{\partial}{\partial x} \exp\{-(t-x)^2\}.$$

Thus, to determine $H_n(x)$ we differentiate the function $\exp(-x^2)$, n times.

If we differentiate (I.29) with respect to x, we get,

$$\frac{\partial g(x, t)}{\partial x} = 2tg(x, t),$$

which gives

$$\sum_{n=0}^{\infty} \frac{H_n'(x)}{n!} t^n = 2t \sum_{n=0}^{\infty} \frac{H_n(x)}{n!} t^n, \tag{I.32}$$

where $H_n'(x)$ denotes the derivative of $H_n(x)$ with respect to x. Equating the

References on page 398

coefficients of t^n in (I.32) we get

$$H_n'(x) = 2nH_{n-1}(x). \tag{I.33}$$

Recurrence relations

A recurrence relation enables all polynomials of a given type to be calculated from the first two. We may obtain a recurrence relation for the Hermite polynomials as follows. Differentiating (I.29) with respect to t, we get

$$\frac{\partial g}{\partial t} = -2(t-x)g(x, t),$$

$$\sum_{n=1}^{\infty} \frac{H_n(x)}{(n-1)!} t^{n-1} + 2(t-x) \sum_{n=0}^{\infty} \frac{H_n(x)}{n!} t^n = 0,$$

$$\sum_{n=0}^{\infty} \left[\frac{H_{n+1}(x)}{n!} - 2x \frac{H_n(x)}{n!} + \frac{2H_{n-1}(x)}{(n-1)!} \right] t^n = 0,$$

and, finally,

$$H_{n+1}(x) - 2xH_n(x) + 2nH_{n-1}(x) = 0. \tag{I.34}$$

From (I.33) and (I.34) we can obtain the differential equation.

$$H_n''(x) - 2xH_n'(x) + 2nH_n(x) = 0, \qquad n = 0, 1, 2, \cdots. \tag{I.35}$$

It follows from (I.35) that the function $u = H_n(x)$ is a particular integral of the second order linear differential equation

$$u'' - 2xu' + 2nu = 0. \tag{I.36}$$

Hermite polynomials

We shall now use the generating function to show that the Hermite polynomials satisfy simple integral equations with symmetric kernels. Replacing x by y in the expansion (I.30), then multiplying the result by $\exp[\mathrm{i}xy - (y^2/2)]$ and integrating, we get

$$\int_{-\infty}^{\infty} \exp[2yt - t^2 + \mathrm{i}xy - (y^2/2)] \mathrm{d}y = \int_{-\infty}^{\infty} \exp[\mathrm{i}xy - (y^2/2)] \sum_{n=0}^{\infty} \frac{H_n(y)}{n!} t^n \, \mathrm{d}y.$$

Integrating the left-hand side and interchanging integration and summation on the right-side we get

$$\sqrt{2\pi} \exp(-x^2/2) \sum_{n=0}^{\infty} \frac{(\mathrm{i}t)^n}{n!} H_n(x) = \sum_{n=0}^{\infty} \frac{t^n}{n!} \int_{-\infty}^{\infty} \exp[\mathrm{i}xy - (y^2/2)] H_n(y) \mathrm{d}y.$$

Since this is true for all values of t, the coefficients of corresponding powers of t are equal, and so we get

$$\exp(-x^2/2)H_n(x) = \frac{1}{\mathrm{i}^n\sqrt{2\pi}}\int_{-\infty}^{\infty} \exp[\mathrm{i}xy-(y^2/2)]H_n(y)\mathrm{d}y. \qquad \text{(I.37)}$$

This is a homogeneous Fredholm integral equation of the second kind,

$$f(x) = \lambda\int_a^b K(x, y)f(y)\mathrm{d}y. \qquad \text{(G.2)}$$

This integral equation is obtained on analysing the propagation of radiation within a cavity, see ch. 6, and its solution is seen above to be given by the Hermite polynomials.

Chebyshev polynomials

The Chebyshev polynomials of the first kind are defined by

$$T_0 = 1, \qquad T_n(x) = \cos(n\cos^{-1} x) \qquad \text{(I.38a)}$$

where $n = 1, 2, 3, \cdots$. They form an orthogonal set with weight function $w(x) = 1/(1-x^2)^{\frac{1}{2}}$ in the interval $-1 \leqslant x \leqslant 1$, since, by substituting $\theta = \cos^{-1} x$, we get

$$\int_{-1}^{+1} T_n(x)T_m(x)\frac{\mathrm{d}x}{(1-x^2)^{\frac{1}{2}}} = \int_0^{\pi}\cos n\theta\cos m\theta\,\mathrm{d}\theta = 0 \qquad (m \neq n).$$

To show that the $T_n(x)$ are polynomials, we note that by de Moivre's theorem

$$\cos n\phi + \mathrm{i}\sin n\phi = (\cos\phi + \mathrm{i}\sin\phi)^n.$$

Expanding the binomial, taking the real parts of both sides, and replacing the even powers of $\sin\phi$ by

$$(\sin^2\phi)^k = (1-\cos^2\phi)^k,$$

we see that $\cos n\phi$ is a polynomial of degree n in $\cos\phi$. Therefore, since $\cos(\cos^{-1}x) = x$ we see that $T_n(x)$ $(= \cos(n\cos^{-1} x))$ is a polynomial of degree n in x.

Chebyshev polynomials are important because, of all polynomials of degree n, $T_n(x)$ has the smallest maximum value in the interval $-1 \leqslant x \leqslant 1$. To prove this we consider the points $x_k = \cos(k\pi/n)$ where $k = 0, 1, \ldots, n$ for which the departure of $T_n(x)$ from zero is greatest. For $\theta = 0, \pi/n, 2\pi/n, \ldots, \pi$ we have $T_n(x_k) = (-1)^k$. We now suppose that there is a polynomial $P_n(x) = x_n + a_{n-1}x^{n-1} + \ldots$ whose deviation from zero in the interval $-1 \leqslant x \leqslant 1$ is less than that of $T_n(x)$ and consider the rational function $T_n(x)-P_n(x)$ at the points x_k. We see that for this function we get

$$T_n(x_0)-P_n(x_0) > 0, \qquad T_n(x_1)-P_n(x_1) < 0,$$
$$T_n(x_2)-P_n(x_2) > 0, \qquad T_n(x_3)-P_n(x_3) < 0,$$

etc.,

and see that it is alternately positive and negative and so has at least n roots. But since $T_n(x)-P_n(x)$ is of degree $(n-1)$ at the most, it cannot have more than $(n-1)$ roots and so $P_n(x)$ cannot be less than $T_n(x)$. This property of having the smallest maximum value in the interval $-1 \leqslant x \leqslant 1$ is used in ch. 5 § 5 in considering a cavity as a lens sequence.

Chebyshev polynomials of the second kind are defined by

$$U_0 = 1, \qquad U_n(x) = \sin(n \cos^{-1} x), \tag{I.38b}$$

where $n = 1, 2, 3, \cdots$.

The Chebyshev polynomials may also be written in the forms

$$T_n(x) = \tfrac{1}{2}[\{x+i\sqrt{(1-x^2)}\}^n + \{x-i\sqrt{(1-x^2)}\}^n]; \tag{I.39}$$
$$U_n(x) = -\tfrac{1}{2}i[\{x+i\sqrt{(1-x^2)}\}^n - \{x-i\sqrt{(1-x^2)}\}^n]. \tag{I.40}$$

These may be proved by letting $x = \cos\theta$, converting the trigonometric functions to exponential form, and then applying de Moivre's theorem, after which we replace $\cos\theta$ by x and $\sin\theta$ by $(1-x^2)^{\frac{1}{2}}$.

By expanding (I.39) and (I.40) by the binomial theorem we obtain equations which may be used to calculate the first few Chebyshev polynomials; these equations are given by

$$T_n(x) = \sum_{r=0}^{(n/2)} (-1)^r \frac{n!}{(2r)!(n-2r)!} (1-x^2)^r x^{n-2r}; \tag{I.41}$$

$$U_n(x) = \sum_{r=0}^{(n-1)/2} (-1)^r \frac{n!}{(2r+1)!(n-2r-1)!} (1-x^2)^{r+\frac{1}{2}} x^{n-2r-1}. \tag{I.42}$$

The polynomials we obtain are as follows:

$$\begin{aligned}
T_0(x) &= 1, & U_0(x) &= 0,\\
T_1(x) &= x, & U_1(x) &= (1-x^2)^{\frac{1}{2}},\\
T_2(x) &= 2x^2-1, & U_2(x) &= (1-x^2)^{\frac{1}{2}}2x,\\
T_3(x) &= 4x^3-3x, & U_3(x) &= (1-x^2)^{\frac{1}{2}}(4x^2-1),\\
T_4(x) &= 8x^4-8x^2+1, & U_4(x) &= (1-x^2)^{\frac{1}{2}}(8x^3-4x),\\
T_5(x) &= 16x^5-20x^3+5x; & U_5(x) &= (1-x^2)^{\frac{1}{2}}(16x^4-12x^2+1).
\end{aligned} \tag{I.43}$$

To avoid the factor $(1-x^2)^{\frac{1}{2}}$, the Chebyshev polynomial of the second kind is sometimes defined by

$$\mathscr{U}_n(x) = \frac{\sin\{(n+1)\cos^{-1} x\}}{(1-x^2)^{\frac{1}{2}}} = \frac{U_{n+1}}{(1-x^2)^{\frac{1}{2}}}. \tag{I.44}$$

By letting $\cos^{-1} x = \theta$, (I.44) may also be written in the form

$$\mathscr{U}_n(x) = \frac{\sin(n+1)\theta}{\sin\theta}. \tag{I.45}$$

The Chebyshev polynomials may also be defined in terms of the hyperbolic functions, thus

$$T_n(x) = \cosh(n\cosh^{-1} x) \tag{I.46}$$

and

$$U_n(x) = \sinh(n\cosh^{-1} x). \tag{I.47}$$

Equations (I.46) and (I.47) may be shown to give equations (I.39) and (I.40) by letting $x = \cosh\phi$, converting the hyperbolic functions to exponential form, noting that $(\cosh\phi+\sinh\phi) = e^{\phi}$, and $(\cosh\phi-\sinh\phi) = e^{-\phi}$, after which we replace $\cosh\phi$ by x and $\sinh\phi$ by $(x^2-1)^{\frac{1}{2}}$.

In hyperbolic form, (I.45) may be written

$$\mathscr{U}_n(x) = \frac{\sinh(n+1)\phi}{\sinh\phi}, \tag{I.48}$$

where $\cosh^{-1} x = \phi$.

For negative arguments, find $\mathscr{U}_n(|x|)$ and use the relation

$$\mathscr{U}_n(-x) = (-1)^n\mathscr{U}_n(x). \tag{I.49}$$

References on page 398

DOPPLER AND NATURAL BROADENING

Doppler Broadening

Doppler broadening is one of the most important in gas discharge lasers, since at the low pressures involved, the gain profile is determined by the Doppler effect. Pressure effects only become a consideration when considering hole burning. At the low pressures generally involved, the collision period is greater than the radiative lifetime.

Consider a gas whose stationary atoms (molecules, ions) emit or absorb frequencies within a very narrow band peaked at ν_0. (This narrow band is due to natural broadening and will be neglected.) For moving atoms, the frequencies observed are shifted by the Doppler effect. Only velocities in the line of sight are important. Let the atom's velocity component in the line of sight by u, and let c be the velocity of light. The frequency observed is (to first order in u/c)

$$\nu = \nu_0 \{1 \pm (u/c)\}, \tag{J.1}$$

where the plus sign refers to an approaching atom and the minus to a receding atom. From (J.1) we get

$$u = (\nu - \nu_0)(c/\nu_0) \quad \text{and} \quad \mathrm{d}u = (c/\nu_0)\mathrm{d}\nu. \tag{J.2}$$

The probability that an atom of mass m, in a gas at temperature T has a velocity between u and $u + \mathrm{d}u$ is given by the Maxwellian velocity distribution which applies to components in one direction,

$$P(u)\mathrm{d}u = \left(\frac{m}{2\pi kT}\right)^{\frac{1}{2}} \exp\left(-\frac{mu^2}{2kT}\right)\mathrm{d}u. \tag{J.3}$$

Strictly speaking the Maxwellian distribution applies only to thermal equilibrium but the departure from this among the radiating species (atoms, molecules, or ions) of a gas discharge under the conditions of interest to lasers are generally negligible, even among the population of species excited to the upper laser level. The probability that the frequency emitted in the u-direction is in the range ν, $\nu+d\nu$ is, from (J.2) and (J.3),

$$p(\nu-\nu_0)d\nu = \frac{c}{\nu_0}\left(\frac{m}{2\pi kT}\right)^{\frac{1}{2}} \exp\left[-\frac{m}{2kT}\frac{c^2}{\nu_0^2}(\nu-\nu_0)^2\right] d\nu. \tag{J.4}$$

This is the fraction of the total number of atoms which emit or absorb in the line of sight of the observer. Since intensity is proportional to the number of emitters, it follows that

$$\mathscr{I} = \mathscr{I}_0 \exp\left[-\frac{m}{2kT}\frac{c^2}{\nu_0^2}(\nu-\nu_0)^2\right], \tag{J.5}$$

where $\mathscr{I}_0$ is the intensity at $\nu = \nu_0$. The shape of the Doppler broadened profile is Gaussian.

The line width $\Delta\nu_D$ is defined as the frequency separation of the points at which the intensity has half its maximum value (or where p is half maximum). We then have

$$\ln\frac{\mathscr{I}}{\mathscr{I}_0} = \ln\tfrac{1}{2} = -\frac{m}{2kT}\frac{c^2}{\nu_0^2}\left(\frac{\Delta\nu_D}{2}\right)^2,$$

$$\Delta\nu_D = 2\nu_0\left(\frac{2kT}{mc^2}\ln 2\right)^{\frac{1}{2}}. \tag{J.6}$$

In terms of wavelength (in Å) we have

$$\frac{\Delta\lambda_D}{\lambda_0} = 7.16\times10^{-7}\left(\frac{T}{M}\right)^{\frac{1}{2}}, \tag{J.7}$$

where M is the molecular weight. Example: for neon with $\lambda = 6328$ Å, $M = 20$, $T = 300$ °K, $\Delta\lambda_D \approx 1.6\times10^{-2}$ Å.

Equation (J.4) is often written in terms of the Doppler width, thus

$$p(\nu-\nu_0) = \frac{2(\ln 2)^{\frac{1}{2}}}{\pi^{\frac{1}{2}}\Delta\nu_D} \exp\left[-\frac{(4\ln 2)(\nu-\nu_0)^2}{(\Delta\nu_D)^2}\right]. \tag{J.8}$$†

The gain (absorption) coefficient is proportional to the population concerned. The fractional intensity gain per unit length is defined by

† In Ch. 1 § 2.1 we have used $\mathscr{D}(\nu'_{mn}, \nu_{mn})$ instead of $p(\nu-\nu_0)$, since a more precise nomenclature is required to handle the two linewidths considered.

References on page 398

$g(\nu) = (1/\mathscr{I})(\mathrm{d}\mathscr{I}/\mathrm{d}x)$ and is given by

$$g(\nu) = g_0 \exp\left[-\frac{(4 \ln 2)(\nu-\nu_0)^2}{(\Delta\nu_D)^2}\right] \approx$$

$$\approx g_0 \exp\left[-\left(\frac{\nu-\nu_0}{0.6\Delta\nu_D}\right)^2\right]. \tag{J.9}$$

Natural broadening

We have seen above that radiation from a system of moving particles is spread over a frequency range which may be calculated by considering the Doppler effect. We shall now consider the spontaneous radiation from a single isolated particle and show that it too is spread over a frequency range. The broadening process we are to consider is natural broadening and we shall use a classical approach as in ch. 2 § 4.4. From (2.45) we have

$$\ddot{x}+\gamma\dot{x}+\omega_0^2 x = 0, \tag{2.45}$$

where ω_0 is the angular frequency of the oscillating electron and γ is the classical radiation damping constant. A solution of (2.45) is given by

$$x = A \exp(-\gamma t/2) \exp(\mathrm{i}\omega_1 t), \tag{J.10}$$

where $\omega_1 = \omega_0^2-(\gamma/2)^2$ and A is an arbitrary amplitude. Since the intensity of the radiation emitted is proportional to the square of the amplitude, the intensity varies as $\exp(-\gamma t)$ and so falls to the fraction e^{-1} of its initial value in time τ given by

$$\tau = \gamma^{-1}. \tag{J.11}$$

(Quantum mechanically, τ is the mean lifetime of an excited state.) We wish to known the range of frequencies introduced into the radiation emitted by the classical oscillator by the damping as described by (J.10) for $t \geqslant 0$.

From Fourier analysis we can write

$$f(t) = \frac{1}{(2\pi)^{\frac{1}{2}}}\int_{-\infty}^{\infty} g(\omega) \exp(\mathrm{i}\omega t)\mathrm{d}\omega \tag{J.12}$$

and

$$g(\omega) = \frac{1}{(2\pi)^{\frac{1}{2}}}\int_{-\infty}^{\infty} f(t) \exp(-\mathrm{i}\omega t)\mathrm{d}t. \tag{J.13}$$

Letting $f(t) = A \exp(-\gamma t/2) \exp(i\omega_1 t)$ and using (J.13), we get

$$g(\omega) = \frac{A}{(2\pi)^{\frac{1}{2}}} \int_0^\infty \exp\left[-\frac{\gamma t}{2} + i(\omega_1 - \omega)t\right] dt =$$

$$= \frac{A}{(2\pi)^{\frac{1}{2}}} \left| \frac{\exp(-\gamma t/2) \exp[i(\omega_1 - \omega)t]}{i(\omega_1 - \omega + i\gamma/2)} \right|_0^\infty =$$

$$= \frac{iA}{(2\pi)^{\frac{1}{2}}(\omega_1 - \omega + i\gamma/2)}.$$

This refers to the amplitude. The intensity distribution is given as a function of frequency by the equation, $\mathscr{I} = g(\omega)g^*(\omega)$. Hence, we have,

$$\mathscr{I}(\omega) = \frac{A^2}{2\pi[(\omega_1 - \omega)^2 + \gamma^2/4]}.$$

The reader should plot the function $f(t)$ against t, and $|g(\omega)|^2$ against ω. The latter is a resonance curve with peak at ω_1 and falling to half the peak value at

$$\omega = \omega_1 \pm \gamma/2.$$

The full width at half maximum value is γ.

There are many other types of broadening and these have been described by a number of authors, e.g. BREENE [1957].

REAL CORRELATION FUNCTIONS AND THE MUTUAL COHERENCE FUNCTION

We shall follow MANDEL [1963] in deriving relations between real correlation functions and the mutual coherence function.

From the definition of the analytic signal, the mutual coherence function may be decomposed in the following way

$$\Gamma_{12}(\tau) = \overline{V_1^{(r)}(t+\tau)V_2^{(r)}(t)} + \overline{V_1^{(i)}(t+\tau)V_2^{(i)}(t)} + \\ + \mathrm{i}\overline{V_1^{(i)}(t+\tau)V_2^{(r)}(t)} - \mathrm{i}\overline{V_1^{(r)}(t+\tau)V_2^{(i)}(t)}. \tag{K.1}$$

We now require to show that

$$\overline{V_1(t+\tau)V_2(t)} = 0, \tag{K.2}$$

for if this is the case, then by decomposing the analytic signals in (K.2) into their real and imaginary parts, the following two identities are obtained

$$\begin{aligned} \overline{V_1^{(r)}(t+\tau)V_2^{(r)}(t)} &= \overline{V_1^{(i)}(t+\tau)V_2^{(i)}(t)}, \\ \overline{V_1^{(i)}(t+\tau)V_2^{(r)}(t)} &= -\overline{V_1^{(r)}(t+\tau)V_2^{(i)}(t)}. \end{aligned} \tag{K.3}$$

In which case, the required relationships follow by substituting (K.3) into (K.1) and separating real and imaginary parts, namely

$$\begin{aligned} \overline{V_1^{(r)}(t+\tau)V_2^{(r)}(t)} &= \overline{V_1^{(i)}(t+\tau)V_2^{(i)}(t)} = \tfrac{1}{2}\,\mathcal{Re}\,[\Gamma_{12}(\tau)], \\ \overline{V_1^{(i)}(t+\tau)V_2^{(r)}(t)} &= -\overline{V_1^{(r)}(t+\tau)V_2^{(i)}(t)} = \tfrac{1}{2}\mathcal{I}[\Gamma_{12}(\tau)]. \end{aligned} \tag{K.4}$$

In order to demonstrate (K.2), the ensemble average is replaced by a time average, on the assumption that the random processes are ergodic

$$\overline{V_1(t+\tau)V_2(t)} = \lim_{T\to\infty} \frac{1}{T}\int_{-T/2}^{+T/2} V_{1T}(t+\tau)V_{2T}(t)\mathrm{d}t. \tag{K.5}$$

Introducing the Fourier transforms $\tilde{V}_{1T}(\nu)$, $\tilde{V}_{2T}(\nu)$ of the analytic signals

$V_{1T}(t)$, $V_{2T}(t)$, (K.5) can be written

$$\overline{V_1(t+\tau)V_2(t)} = \lim_{T\to\infty} \frac{1}{T}\int_{-T/2}^{+T/2}\int_0^\infty\int_0^\infty \tilde{V}_{1T}(\nu)\tilde{V}_{2T}(\nu')\cdot$$

$$\cdot \exp\{2\pi \mathrm{i}[\nu\tau+(\nu+\nu')t]\}\mathrm{d}\nu\mathrm{d}\nu'\mathrm{d}t. \qquad \text{(K.6)}$$

Interchanging the order of integration, and integrating over t, (K.6) becomes

$$\overline{V_1(t+\tau)V_2(t)} = \lim_{T\to\infty} \int_0^\infty\int_0^\infty \tilde{V}_{1T}(\nu)\tilde{V}_{2T}(\nu''-\nu)\cdot$$

$$\cdot \exp(2\pi \mathrm{i}\nu\tau)\cdot\frac{\sin(\pi\nu''T)}{\pi\nu''T}\cdot \mathrm{d}\nu''\mathrm{d}\nu, \qquad \text{(K.7)}$$

where $\nu'' = (\nu+\nu')$.

As $T\to\infty$, then $\sin(\pi\nu''T)/(\pi\nu''T)\to 0$ unless $\nu'' = 0$. However, in this case $\tilde{V}_{2T}(\nu''-\nu)$ vanishes since by definition $\tilde{V}_{2T}(\nu)$ and $\tilde{V}_{1T}(\nu)$ are zero for negative or zero frequencies. It follows therefore that in the limit the above integral is zero, so that the required identity (K.2) holds. The relations (K.4) between the real correlation functions and the mutual coherence function are therefore valid for the case of random processes where the ergodic hypothesis applies.

References on page 398

SOME ELECTROMAGNETIC WAVE THEORY

Maxwell's equations in rationalized MKS units (SI units) for propagation of electromagnetic waves are as follows

$$\nabla \times \boldsymbol{E} = -\dot{\boldsymbol{B}}; \tag{L.1}$$

$$\nabla \times \boldsymbol{H} = \boldsymbol{J}+\dot{\boldsymbol{D}}; \tag{L.2}$$

$$\nabla \cdot \boldsymbol{D} = \rho; \tag{L.3}$$

$$\nabla \cdot \boldsymbol{B} = 0, \tag{L.4}$$

where ρ is the free charge density.

By Ohm's law, the current density, $\boldsymbol{J}$, and conductivity σ, are related by

$$\boldsymbol{J} = \sigma \boldsymbol{E}. \tag{L.5}$$

The constitutive relations between the field vectors $\boldsymbol{E}$, $\boldsymbol{D}$, $\boldsymbol{B}$ and $\boldsymbol{H}$ are given by

$$\boldsymbol{D} = \varepsilon_0 \boldsymbol{E}+\boldsymbol{P} \tag{L.6}$$

and

$$\boldsymbol{H} = \frac{\boldsymbol{B}}{\mu_0}+\boldsymbol{M}, \tag{L.7}$$

where $\boldsymbol{P}$ (polarization) and $\boldsymbol{M}$ (magnetization) are the net dipole moments per unit volume. Differentiating (L.2) with respect to time, multiplying by μ_0 and using (L.5) and (L.6), we get

$$\nabla \times \mu_0 \frac{\partial \boldsymbol{H}}{\partial t} = \mu_0 \sigma \frac{\partial \boldsymbol{E}}{\partial t} + \mu_0 \varepsilon_0 \frac{\partial^2 \boldsymbol{E}}{\partial t^2} + \mu_0 \frac{\partial^2 \boldsymbol{P}}{\partial t^2}\,. \tag{L.8}$$

Since $\boldsymbol{M}$ is negligible in media of interest to lasers, we use (L.1) and (L.7) to obtain

$$\nabla \times \nabla \times \boldsymbol{E} = -\left[\mu_0 \sigma \frac{\partial \boldsymbol{E}}{\partial t} + \mu_0 \varepsilon_0 \frac{\partial^2 \boldsymbol{E}}{\partial t^2} + \mu_0 \frac{\partial^2 \boldsymbol{P}}{\partial t^2}\right].$$

Using the identity

$$\nabla\times\nabla\times\boldsymbol{E} = \nabla(\nabla\cdot\boldsymbol{E})-\nabla^2\boldsymbol{E}$$

and noting that in free space, $\nabla\cdot\boldsymbol{E} = 0$, we get

$$\nabla^2\boldsymbol{E} = \mu_0\sigma\frac{\partial\boldsymbol{E}}{\partial t}+\mu_0\varepsilon_0\frac{\partial^2\boldsymbol{E}}{\partial t^2}+\mu_0\frac{\partial^2\boldsymbol{P}}{\partial t^2}\,. \tag{L.9}$$

This equation may be used to obtain equations (9.1), (9.7) and (9.13).

References on page 398

CYLINDRICAL REPRESENTATION OF PLANE WAVES

Consider plane electromagnetic waves propagated in a direction defined by the unit vector $\boldsymbol{n}$ whose components along the axes x, y, z are n_x, n_y, n_z. The vectors $\boldsymbol{E}$ and $\boldsymbol{H}$ lie in planes normal to $\boldsymbol{n}$ defined by

$$\boldsymbol{n} \cdot \boldsymbol{r} = n_x x + n_y y + n_z z = \text{constant}, \tag{M.1}$$

where $\boldsymbol{r}$ is the radius vector from the origin to a point in the plane considered. Let the plane be located at distance ζ from the origin in direction $\boldsymbol{n}$. This transverse electromagnetic field can be represented by a scalar function ψ given, in its simplest form, by

$$\psi = \exp \mathrm{i}(k\zeta - \omega t), \tag{M.2}$$

where k and ω may be complex and are, respectively, the propagation constant and the angular frequency of oscillation of the field.

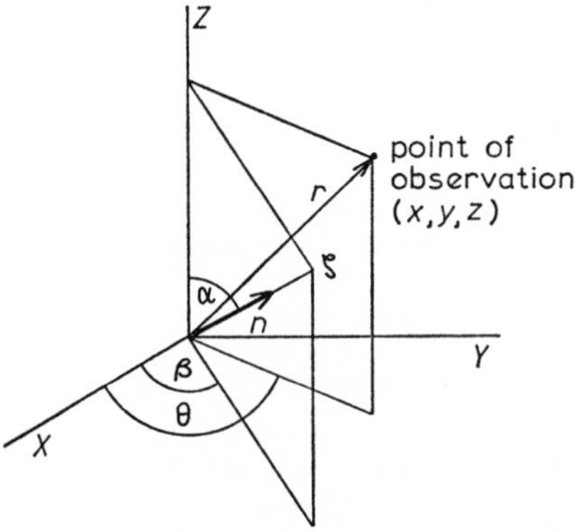

Fig. M.1. The direction of the ζ-axis is that of the unit vector.

The orientation of the unit vector $\boldsymbol{n}$ with respect to the axes is determined by angles α and β, fig. M.1. Hence, we have

$$n_x = \sin\alpha\cos\beta, \qquad n_y = \sin\alpha\sin\beta, \qquad n_z = \cos\alpha. \tag{M.3}$$

At a point along the axis of propagation we have $\boldsymbol{n}\cdot\boldsymbol{r} = \zeta$, so we can combine this (M.1), (M.2) and (M.3) to give

$$\psi = \exp\left[\mathrm{i}(kz\cos\alpha-\omega t)\right]\exp\left[\mathrm{i}k\sin\alpha(x\cos\beta+y\sin\beta)\right]. \tag{M.4}$$

Introducing polar coordinates r, θ and z we have

$$x = r\cos\theta, \qquad y = r\sin\theta, \qquad z = z,$$

and we can write (M.4) in the form

$$\psi = \exp\left[\mathrm{i}(kz\cos\alpha-\omega t)\right]\exp\left[\mathrm{i}kr\sin\alpha\cos(\theta-\beta)\right]. \tag{M.5}$$

At the point $t = z = 0$, we have

$$\psi = \exp\left[\mathrm{i}kr\sin\alpha\cos(\theta-\beta)\right] = f(r, \theta), \tag{M.6}$$

say. We follow STRATTON [1941] in noting that $f(r, \theta)$ is periodic in θ and can be expanded in a Fourier series whose coefficients are functions of r alone; thus we have

$$f(r, \theta) = \sum_{n=-\infty}^{\infty} f_n(r)\mathrm{e}^{\mathrm{i}n\theta}, \qquad f_n(r) = \frac{1}{2\pi}\int_0^{2\pi} f(r, \theta)\mathrm{e}^{-\mathrm{i}n\theta}\mathrm{d}\theta,$$

which give

$$f_n(r) = \frac{1}{2\pi}\int_0^{2\pi}\exp\left[\mathrm{i}kr\sin\alpha\cos(\theta-\beta)-\mathrm{i}n\theta\right]\mathrm{d}\theta. \tag{M.7}$$

From (M.7) and (A.24) we get

$$\exp\left[\mathrm{i}n\left(\frac{\pi}{2}-\beta\right)\right] J_n(kr\sin\alpha) = \frac{1}{2\pi}\int_0^{2\pi}\exp\left[\mathrm{i}kr\sin\alpha\cos(\theta-\beta)-\mathrm{i}n\theta\right]\mathrm{d}\theta. \tag{M.8}$$

This is of the form

$$\mathrm{i}^n\mathrm{e}^{-\mathrm{i}n\beta}J_n(x) = \frac{1}{2\pi}\int_0^{2\pi}\mathrm{e}^{\mathrm{i}x\cos(\theta-\beta)-\mathrm{i}n\theta}\mathrm{d}\theta. \tag{M.9}$$

References on page **398**

THE PLASMA DISPERSION FUNCTION

In our treatment of Lamb theory the integral in the equation

$$Z(\omega_m - \omega_{ab}, \gamma, Ku) = \mathrm{i}Ku \int_0^\infty \mathrm{d}\tau \exp\left[\mathrm{i}(\omega_m - \omega_{ab})\tau - \gamma\tau - \tfrac{1}{4}K^2u^2\tau^2\right] \tag{N.1}$$

arose when we were considering the case of an inhomogeneously broadened laser transition (i.e. when the Lorentzian function describing the homogeneous linewidth (γ) and the Gaussian function describing the inhomogeneous linewidth (Ku) both had to be taken into consideration). This function may be expressed in terms of the following function of a single complex variable, ζ,

$$Z(\zeta) = 2\mathrm{i} \exp(-\zeta^2) \int_{-\infty}^{\mathrm{i}\zeta} \exp(-t^2)\mathrm{d}t, \tag{N.2}$$

where

$$\begin{aligned} \zeta &= \xi + \mathrm{i}\eta, \\ \xi &= (\omega_m - \omega_{ab})/Ku, \\ \eta &= \gamma/Ku. \end{aligned} \tag{N.3}$$

The function (N.2) is referred to as the "plasma dispersion function" since it also arises in treatments of the propagation of electromagnetic radiation in plasmas. The equivalence of (N.2) and (N.1), when the relationships (N.3) hold, may be demonstrated by evaluating the integral in (N.2) over a path in the complex t-plane (cf. appendix A) such that

$$\begin{aligned} \int_{-\infty}^{\mathrm{i}\zeta} \exp(-t^2)\mathrm{d}t &= \mathrm{i}\int_0^{\xi} \exp(-t^2)\mathrm{d}t'' \qquad (\text{at } t' = -\infty), \\ &= +\int_{-\infty}^{-\eta} \exp(-t^2)\mathrm{d}t' \qquad (\text{at } t'' = \xi), \end{aligned} \tag{N.4}$$

where $t = t' + \mathrm{i}t''$.

The first integral on the right hand side of the above is zero, and the second integral on substitution back into (N.2) leads to (N.1) if the various quantities are evaluated according to (N.3).

The plasma dispersion function has been extensively studied and tables of values (FRIED and CONTE [1961]) are available for different values of the variable ζ.

In the case when $\gamma \ll Ku$ (i.e. in the Doppler limit), $\eta \sim 0$, so that $Z(\zeta)$ becomes a function of a real variable (ξ). In this case it may be shown that

$$Z(\xi) = \mathrm{i}\pi^{\frac{1}{2}} \exp(-\xi^2) - 2\xi Y(\xi), \tag{N.5}$$

where

$$Y(\xi) = [\{\exp(-\xi^2)\}/\xi] \int_0^{\xi} \exp(t^2)\mathrm{d}t. \tag{N.6}$$

The imaginary part of $Z(\xi)$ is then given by

$$\begin{aligned} Z_{\mathrm{i}}(\xi) &= \pi^{\frac{1}{2}} \exp(-\xi^2) = \\ &= \pi^{\frac{1}{2}} \exp\left[-\frac{(\omega_n - \omega_{ab})^2}{(Ku)^2}\right]. \end{aligned} \tag{N.7}$$

This is the value for $Z_{\mathrm{i}}(\xi)$ that we have used in deriving (9.127). In this limit, also, it may easily be seen from (N.5) and (N.6) that

$$Z_{\mathrm{r}}(\xi)/Z_{\mathrm{i}}(\xi) = -2\pi^{-\frac{1}{2}} \int_0^{\xi} \exp(t^2)\mathrm{d}t. \tag{N.8}$$

This is the relationship we have used in order to derive (9.132) from (9.131).

References on page 398

THE BIVARIATE GAUSSIAN DISTRIBUTION

In order to derive the bivariate Gaussian distribution (10.103) of two, not necessarily independent, variables x_1, and x_2, we begin by considering the joint probability density of two statistically independent, Gaussian random variables y_1 and y_2, with variances μ_1 and μ_2 respectively. Since the variables are independent, then the joint probability density is just the product of the two probability densities associated with the variables individually, i.e.

$$p(y_1, y_2) = p(y_1)p(y_2) = \frac{1}{2\pi\sqrt{\mu_1 \mu_2}} \exp\left[-\frac{y_1^2}{2\mu_1} - \frac{y_2^2}{2\mu_2}\right]. \tag{O.1}$$

We now define two new random variables x_1 and x_2 by the rotational transformation,

$$\begin{aligned} x_1 &= y_1 \cos\theta - y_2 \sin\theta, \\ x_2 &= y_1 \sin\theta + y_2 \cos\theta. \end{aligned} \tag{O.2}$$

Since the means of y_1 and y_2 are each zero, the means of x_1 and x_2 are also zero. The variances, σ_1^2 and σ_2^2 of x_1 and x_2 respectively, may be derived as follows using (O.2),

$$\sigma^2 = \langle x_1^2 \rangle = \langle y_1^2 \rangle \cos^2\theta + \langle y_2^2 \rangle \sin^2\theta - 2\langle y_1 y_2 \rangle \cos\theta \sin\theta.$$

$$\therefore \sigma_1^2 = \mu_1 \cos^2\theta + \mu_2 \sin^2\theta, \tag{O.3}$$

and similarly

$$\sigma_2^2 = \mu_1 \sin^2\theta + \mu_2 \cos^2\theta. \tag{O.4}$$

A measure of the correlation between the two functions x_1 and x_2 can be defined by the correlation coefficient given by

$$\rho = \langle x_1 x_2 \rangle / (\sigma_1 \sigma_2). \tag{O.5}$$

From this definition, it follows that $-1 \leqq \rho \leqq 1$.

From (O.2) and (O.5) we get

$$\begin{aligned}\rho &= \langle (y_1 \cos\theta - y_2 \sin\theta)(y_1 \sin\theta + y_2 \cos\theta)\rangle/(\sigma_1\sigma_2) = \\ &= \langle y_1^2 \cos\theta \sin\theta - y_2^2 \sin\theta \cos\theta\rangle/(\sigma_1\sigma_2) = \\ &= (\mu_1 - \mu_2)\cos\theta \sin\theta/(\sigma_1\sigma_2). \end{aligned} \tag{O.6}$$

If we solve (O.2) for y_1 and y_2 we obtain

$$\begin{aligned} y_1 &= x_1 \cos\theta + x_2 \sin\theta, \\ y_2 &= -x_1 \sin\theta + x_2 \cos\theta. \end{aligned} \tag{O.7}$$

The Jacobian of the above transformation is given by

$$|J| = \begin{vmatrix} \dfrac{\partial y_1}{\partial x_1} & \dfrac{\partial y_2}{\partial x_1} \\ \dfrac{\partial y_1}{\partial x_2} & \dfrac{\partial y_2}{\partial x_2} \end{vmatrix} = 1, \tag{O.8}$$

and so on substituting from (O.7) for y_1 and y_2 and replacing μ_1 and μ_2 and $\cos\theta$ and $\sin\theta$ using (O.3), (O.4) and (O.6) we obtain from (O.1) the following,

$$p(x_1 x_2) = \frac{1}{2\pi\sigma_1\sigma_2(1-\rho^2)} \exp\left[\frac{-1}{1-\rho^2}\left\{\frac{x_1^2}{2\sigma_1^2} + \frac{x_2^2}{2\sigma_2^2} - \frac{\rho x_1 x_2}{\sigma_1\sigma_2}\right\}\right] \tag{O.9}$$

as the normalized joint probability density of the two dependent random variables x_1 and x_2.

References on page **398**

References

ABELÈS, F., 1950, Ann. phys. (12) **5**, 706.

BORN, M. and E. WOLF, 1964, Principles of Optics, 2nd Edition (Pergamon Press, Oxford) pp. 378–380.

BOWMAN, F., 1962, Introduction to Determinants and Matrices (English University Press, London).

BREENE, R. G., 1957, Rev. Mod. Phys. **29**, 94.

COCHRAN, J. A., 1965, Bell System Tech. J. **44**, 77.

COURANT, R. and D. HILBERT, 1953, Methods of Mathematical Physics, Vol. 1 (Interscience Publishers, New York) pp. 65–68.

FOX, A. G. and T. LI, 1961, Bell System Tech. J. **40**, 453.

FRIED, B. D. and S. D. CONTE, 1961, The Plasma Dispersion Function (Hilbert Transform of the Gaussian) (Academic Press, Inc., New York).

GANTMACHER, F. R., 1959, Matrix Theory, Vol. I (Chelsea Publishing Co., New York).

GOLDSTEIN, H., 1950, Classical Mechanics (Addison-Wesley Press, Cambridge, Massachusetts).

HEADING, J., 1958, Matrix Theory for Physicists (Longmans, Green and Co., London) p. 54.

HILDEBRAND, F. B., 1952, Methods of Applied Mathematics (Prentice-Hall, Englewood Cliffs, New Jersey) p. 421.

HOCHSTADT, H., 1966, SIAM Rev. **8**, 62.

LANCZOS, C., 1957, Applied Analysis (Pitman & Sons Ltd., London) p. 366.

MANDEL, L., 1963, Progress in Optics, Vol. 2 (North-Holland Publishing Company, Amsterdam) pp. 183–244.

MIELENZ, K. D., 1959, J. Res. Natl. Bur. Standards (A) **63A**, 297.

NEWMAN, D. J. and S. P. MORGAN, 1964, Bell System Tech. J. **43**, 113.

RICE, J. R., 1964, The Approximation of Functions (Addison-Wesley Publishing Co., Reading, Massachusetts) p. 34–40.

SANSONE, G., 1959, Orthogonal Functions (Interscience Publishers, New York) pp. 2–3.

SCHWARTZ, J. T., 1961, Introduction to Matrices and Vectors (McGraw-Hill Book Co., New York).

STONE, J. M., 1963, Radiation and Optics (McGraw-Hill, New York) pp. 159–166.

STRATTON, J. A., 1941, Electromagnetic Theory (McGraw-Hill Book Co., New York) pp. 371–372.

STREIFER, W. and H. GAMO, 1964, On the Schmidt Expansion for Optical Resonator Modes. In: Proceedings of the Symposium on Quasi-Optics, Polytechnic Institute of Brooklyn, Brooklyn, New York, June, 1964, pp. 351–365.

AUTHOR INDEX

SUBJECT INDEX